T0138454

MATERIALS OF THE MIND

MATERIALS OF THE MIND

Phrenology, Race, and the Global History of Science, 1815–1920

JAMES POSKETT

THE UNIVERSITY OF CHICAGO PRESS
CHICAGO AND LONDON

The University of Chicago Press, Chicago 60637
The University of Chicago Press, Ltd., London

Published 2019
Printed in the United States of America

28 27 26 25 24 23 22 21 20 19   1 2 3 4 5

ISBN-13: 978-0-226-62675-8 (cloth)
ISBN-13: 978-0-226-62689-5 (e-book)
DOI: https://doi.org/10.7208/chicago/9780226626895.001.0001

Library of Congress Cataloging-in-Publication Data

Names: Poskett, James, author.
Title: Materials of the mind : phrenology, race, and the global history of science, 1815–1920 / James Poskett.
Description: Chicago : The University of Chicago Press, 2019. | Includes bibliographical references and index.
Identifiers: LCCN 2018045366 | ISBN 9780226626758 (cloth : alk. paper) | ISBN 9780226626895 (e-book)
Subjects: LCSH: Phrenology—History.
Classification: LCC BF868 .P67 2019 | DDC 139—dc23
LC record available at https://lccn.loc.gov/2018045366

∞ This paper meets the requirements of ANSI/NISO Z39.48-1992 (Permanence of Paper).

FOR ALICE AND NANCY

CONTENTS

	Introduction	*1*
1.	Skulls	*19*
2.	Casts	*51*
3.	Books	*78*
4.	Letters	*115*
5.	Periodicals	*150*
6.	Photographs	*193*
	Epilogue	*241*
	List of Figures	*265*
	Acknowledgments	*269*
	Notes	*273*
	Bibliography	*331*
	Index	*357*

INTRODUCTION

The phrenologist George Combe was enthralled by the prospect of a shrinking world. Arriving in New York City in 1838, he had just traveled from Britain to the United States in fifteen days aboard the SS *Great Western*, only the second commercial steamship to complete the transatlantic crossing. Looking back across the harbor, Combe reflected on the fact that he had "in this brief period sailed over an extent of the earth's surface, equal to that accomplished in four hours by the globe turning on its axis." This thought marked the beginning of a two-year lecture tour, taking Combe from New York to Virginia. The effect was visceral. Combe admitted that he "never before had so strong an impression of the diminutive size of the globe which we inhabit."[1] He was not alone. Whether in Cape Town or Sydney, phrenologists imagined themselves as part of an international movement, connected by the rapidly advancing communication technologies of the nineteenth century. Skulls were collected in China and Africa, societies cross-circulated journals between Edinburgh and Calcutta, and translations of French phrenological works were imported into Boston and Rio de Janeiro.

Materials of the Mind reconstructs the global history of phrenology—the most popular mental science of the Victorian age—through a close study of its material culture. This novel mental science, which maintained that the brain was the organ of the mind, started life in late eighteenth-century Vienna. Throughout the Napoleonic Wars, the German physician Franz Joseph Gall traveled across continental Europe, presenting his craniological principles to audiences in Berlin, Copenhagen, and Amsterdam. Curious listeners learned how the brain was divided into three major regions devoted to intellectual, moral, and animal faculties. Within each region, phrenologists identified discrete organs, ranging from "amativeness" at the back of the head to "benevolence" on the top.

In 1828, Gall died in Paris—his "doctrine of the skull" had not traveled far.[2] But by the middle of the nineteenth century, phrenologists were self-consciously promoting their work as part of an international scientific movement. Phrenological books were the global best sellers of the day. In 1828 Combe published an unassuming work entitled *The Constitution of Man*. As it was bound in boards, without any illustrations in the first edition, few would have anticipated the excitement it provoked among nineteenth-century readers. By 1900, this book had sold far more copies than Charles Darwin's *On the Origin of Species*.[3] Combe's works were translated into at least six languages, including Bengali and Japanese, and found readers in cities ranging from New York to Shanghai.[4] In an age in which character dominated public discourse, phrenology emerged as a powerful political language.[5] Addressing topics ranging from slavery to prison reform, it was promoted as a universal science of character, just as applicable on the plantations of South Carolina as in the prisons of New South Wales. And, contrary to what a number of historians have suggested, phrenology did not simply fade away as the decades wore on.[6] Rather, it generated waves of support, shifting emphases according to the configuration of world politics and communication networks. In the 1860s, Lorenzo and Orson Fowler made use of the steam press to massively increase the scale of phrenological publishing in America, converting a new generation of readers in the second half of the nineteenth century.[7] In India, phrenology gained increasing support throughout the 1870s among those seeking to forge a national Hindu identity.[8] And in China and Japan, phrenology ultimately enjoyed its heyday in the wake of the industrial wars of the early twentieth century.

Despite this wealth of interaction, the history of phrenology—like the history of science more generally—is usually treated within neat national contexts. The pioneering work of Steven Shapin and Roger Cooter set the tone for years to come. Together, they situated phrenology within the social landscape of early nineteenth-century Britain.[9] This was followed by a range of studies that located phrenology within conventional national frameworks, from the July Monarchy in France to Catholicism in Ireland.[10] Even those histories that ventured outside of Europe, examining phrenology in colonial India and the Cape Colony, tended to treat empires as relatively fixed structures, ignoring connections across traditional regional and imperial spaces.[11] In contrast, this book reconsiders the history of phrenology as part of a global history of science. I chose phrenology for a number of reasons. But chief among them was the prominent role it has played within the historiography of the sciences. The combined work of Shapin and Cooter proved foundational in establishing a social history of science. At a time when there is much debate on the merits or otherwise of writing global histories of science, returning to phrenology seems particularly appropriate.[12]

Materials of the Mind therefore provides an answer to a very basic but important question facing historians of science today: What exactly is a global history of science? In this book, I argue that historians of science need to study the relationship between the global as an analytic category and the global as an actors' category. On the one hand, phrenology traveled across national, regional, and imperial borders in material form. Skulls, plaster casts, books, and letters were all exchanged over incredible distances, connecting diverse communities of phrenological thinkers across the globe. That's the analytic element—how the transit of phrenology, as well as failures of transit—affected the content and political uses of science. Yet, on the other hand, phrenology was itself a science of the relationship between "the external world" and "the constitution of man."[13] That's the actors' element. Phrenologists wanted to explain how and why human nature varied across the planet, mapping the different races of man as they went. My claim is that these two things are connected. How the sciences conceptualize the world is closely related to how the sciences are communicated.[14]

THE MAKING OF A GLOBAL SCIENCE

Whatever Combe's enthusiasm on arriving in the United States, the world in which he lived was not connected in the same way as the world today.[15] Even during the course of Combe's own lifetime, the politics and technologies of communication changed significantly. The railroad, steamship, and telegraph all changed everyday perceptions of time and distance, while the nation-state emerged in the wake of repeated revolutions across the Atlantic world. Over the same period, European powers radically transformed the day-to-day workings of empires in Africa, Asia, and the Pacific.[16] The sciences too underwent significant changes. Geologists, astronomers, and evolutionary thinkers all vied for attention and credibility. Printed on a steam press in 1844, Robert Chambers's *Vestiges of the Natural History of Creation* proved an international sensation, distributed by railroad and steamship to readers in Liverpool, Philadelphia, and Bombay.[17] Similarly, in the 1830s, the Society for the Diffusion of Useful Knowledge published the *Penny Magazine,* cultivating new audiences for the sciences among a weekly audience of over 200,000. With outposts in Canton and New South Wales, the Society for the Diffusion of Useful Knowledge, like many phrenological societies, recognized the political and epistemological significance of operating internationally.[18] At the same time, surgeons, collectors, and surveyors across the colonial world took advantage of cheap postage, generating prestige and income by exchanging specimens with wealthy gentlemen back in London, Paris, and Philadelphia. These emerging audiences and practitioners were reflected in the organization of the sciences, both in Europe and beyond. The 1820s saw the foundation of new scientific societies catering to different social groups and disciplines: the Astronomical Society in London, the South African Institution in Cape Town, and the Philosophical Society of Australasia in Sydney.[19] Phrenology sat at the intersection of a range of emerging disciplines, many of which proved controversial in their own right. It drew on the comparative anatomy of George Cuvier in Paris and the anthropology of Samuel George Morton in Philadelphia. Phrenology also borrowed from the latest geological theories concerning the antiquity of the earth alongside cartographic projects to map Asia and the Americas. Phrenology, then, was a quintessential nineteenth-century science. It combined materialist philosophy, international collaboration, and practical investigation. And it tied all this to a reformist vision of world politics, opening up the study of the mind to new audiences, whether working-class artisans in Britain or freed slaves in the West Indies.

Johann Spurzheim, the German physician who inspired Combe to take up phrenology, was born just twelve years before his Scottish counterpart. But his experience of a scientific and connected world was markedly different from Combe's. Spurzheim arrived in London in 1814, the same year that *The Times* newspaper was first printed on a steam press. But beyond Fleet Street, there were relatively few signs that the sciences were about to be transformed by an "industrial revolution in communication."[20] When Spurzheim traveled from London to Liverpool in 1816, he did so in the back of a mail coach, trundling along dirt tracks at about five miles per hour.[21] It would be another thirty years before the London and North Western Railway connected the two cities, reducing journey times from days to hours.[22] Without the penny post or the *Penny Magazine,* the sciences were largely the preserve of a political and intellectual aristocracy. The Royal Society, founded in 1660, dominated England's scientific scene. Publication in the *Philosophical Transactions* was reserved for Fellows of the Royal Society, although new periodicals such as *Nicholson's Journal* represented the beginnings of a challenge to this model of scientific authorship.[23] Unlike Combe therefore, Spurzheim could not take advantage of a mass audience for science. His books were printed by hand in London and Paris before being sold in small numbers locally. Following his tour of Britain, Spurzheim also made his way to the United States in the hope of conducting a lecture tour. With the SS *Great Western* yet to begin construction, he traveled by sail, taking over thirty days to complete the transatlantic voyage. The crowded conditions and uncomfortable weeks spent at sea proved fatal, and Spurzheim died from typhoid in Boston shortly after his arrival in 1832.[24]

The development of new theories, practices, audiences, and institutions for science in the nineteenth century was closely connected to this changing world. And the fact that phrenology emerged precisely during this period is no coincidence. Writing a global history means providing an analysis of this transformation rather than simply a descriptive account of the spread of science.[25] Understanding global history as an analytic tool is key for addressing the concerns of scholars such as Sarah Hodges and Frederick Cooper. Hodges is justly critical of "scholarship that reproduces rather than critiques globalisation." She also suggests that "the embrace of the 'global' authorises a turning away from analyses of power."[26] Similarly, Cooper argues that historians of the global "risk being trapped in the very discursive structures they wish to analyse."[27] Hodges and Cooper are right to be critical. However, it simply does not follow that global history is committed to the kind of politics they ascribe to it. Global history can and should direct our attention to the development of uneven power relations, particularly once we understand communication as a practice grounded in a material world of capital.[28] This book, therefore, does not merely chart the spread of phrenology. Rather, it presents a critical analysis of the relationship between the politics of phrenology and the circulation of particular objects.[29]

Despite addressing very different geographies, this book actually shares much in common with the social history of phrenology developed by Shapin and Cooter. Rather than a departure, the global should be seen as an opportunity to reengage with this historiography. Many of the arguments put forward by Shapin and Cooter are now broadly accepted. Most importantly, historians now understand that social and political interests play a key role in determining the content and uses of science. In Edinburgh, phrenology was taken up by middle-class merchants, doctors, and lawyers, beneficiaries of the 1832 Reform Act and its Scottish counterpart. Phrenology also excited interest among working-class members of mechanics institutes up and down the country, from Sheffield to Walthamstow. This political world penetrated the very anatomy of the skull. Spurzheim's *Physiognomical System* presented bone and brain in perfect harmony (fig. 1). This implied that anyone, no matter what their social status, had the power to investigate human nature. Mental philosophy was no longer the preserve of the professors at the University of Edinburgh.[30]

Histories of phrenology that followed extended this approach to a variety of other national contexts. Today, we have histories of phrenology in Britain, France, Germany, Australia, India, and the United States.[31] However, in the adoption of this closed approach to context, other important aspects of the early social history of science have been lost. Shapin and Cooter in fact developed a much more complex picture of the relationship between science and society than many of their successors. Two concerns are worth recovering in further detail. First, Shapin and Cooter were critical of histories that treated the social and political as fixed structures waiting to interact with science. Cooter in particular argued that science should not be seen as "epiphenomenal to society."[32] Instead, historians were encouraged to interrogate how science produced social and political relations. In the case of phrenology, disputes over the anatomy of the skull were much more than expressions of existing social and political interests. An anatomical lecture was actually one of the many ways through which social and political differences were constituted in the first place.[33] Second, both Shapin and Cooter recognized that society and politics were actors' categories just as much as historians'.[34] In the nineteenth century, phrenologists and their opponents argued over the boundaries between science and society, just as intellectual and social historians did in the 1970s.[35] For men like Combe, the social import of phrenology was understood as a marker of truth. It "is either the most practically useful of sciences, or it is not true," declared the *Phrenological Journal* in 1829. However, for its nineteenth-century critics, the social interests underpinning phrenology were presented as evidence of deceit.[36] In sum, Shapin and Cooter's history of science was social in two complementary ways. On the one hand, it was a history of how social and political interests influence the theories and practices of science. On the other, it was also an account of how historical actors deployed science to conceptualize society and politics themselves.

In this book, the global takes on the same analytic role as the social did for earlier historians of science. First, I am critical of histories that take certain spaces—whether nations, areas, or empires—as fixed structures through which sciences interact. Instead, historians need to

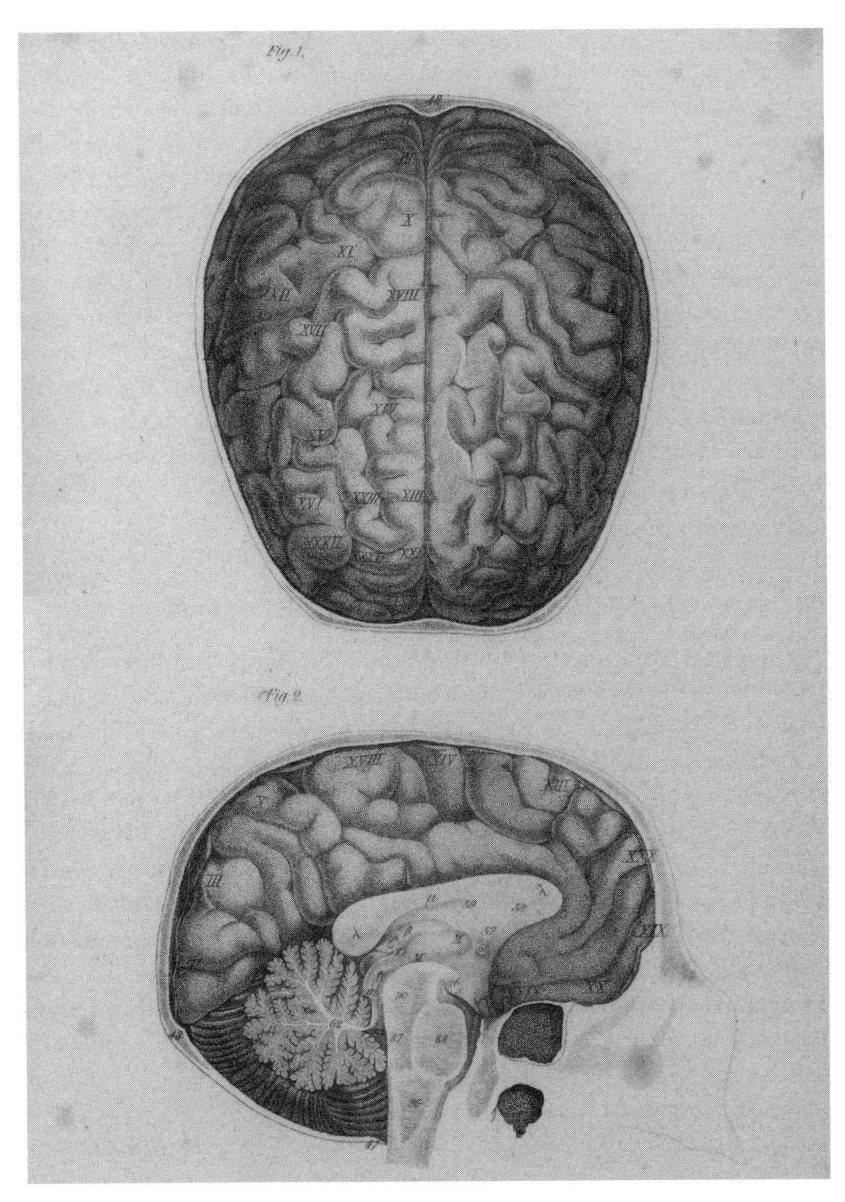

Fig. 1. Johann Spurzheim, *The Physiognomical System of Drs. Gall and Spurzheim* (London: Baldwin, Cradock and Joy, 1815), pl. 2. Reproduced by kind permission of the Syndics of Cambridge University Library.

investigate how scientific exchange produces different social and political geographies.[37] Second, as with the social, the global is also an actors' category. Phrenologists, like historians today, reflected on the political and epistemological implications of working at different scales and over different spaces. What's more, the very meaning of the term *global* underwent a considerable shift in this period: it was a notion in the making, rather than a fixed concept. In the early nineteenth century, *global* might simply mean spherical—as in "globular"—whereas by the late nineteenth century, *global* started to take on the meaning we associate with it today: something "planetary" or "universal."[38] *Materials of the Mind* is a history of how material exchange regulated the uses of phrenology. But it is also a history of how phrenologists conceptualized that same material world, and man's place in it. It is the connection between those two ideas that sits at the heart of my understanding of what it means to write a global history of science.

MATERIAL CULTURES OF THE MIND

The material mind was grounded in a material world. For phrenologists, books and letters were just as much a part of the material culture of science as skulls and plaster casts. However, not all objects were equal. Expensive atlases like Joseph Vimont's *Traité de Phrénologie* did not move along the same paths, nor at the same speeds, as the *Journal de la Société Phrénologique de Paris*.[39] One phrenologist in Florida was dismayed to learn that Vimont's folio was not readily available in the United States. He instead settled for a series of lectures reprinted from the French journal.[40] A more detailed analysis of the geography of scientific practice is possible only once we consider these kinds of distinctions. Phrenologists in major colonial cities made prolific use of the printing press: *Graham's Town Journal* featured advertisements for a series of phrenological lectures, illustrated by "Kaffir Skulls," while the Calcutta Phrenological Society employed an engraver to reproduce copies of Combe's plates.[41] However, in the bazaars of a colonial metropolis it was difficult to find a phrenological bust. There were in fact very few artisans outside of Edinburgh and Paris with access to both the raw materials and the expertise needed to make one. Even in the United States, phrenologists continued to import plaster busts from Europe well into the 1830s. Once we consider these different material dimensions, the emergence of phrenology as a global science appears much more lumpy and incomplete. Certain regions such as the Arctic are part of this story only insofar as they acted as sources for specimens. Others regions, such as the Cape Colony and India, are characterized by a rich print culture, yet little interest in practices requiring more specialized equipment such as busts or calipers.

When we think in terms of material culture, it also becomes clear that exchange was far from straightforward. Sending a book from Britain to the United States might result in a favorable notice in a prestigious periodical like the *North American Review*. Yet at the same time, it wouldn't take long for unscrupulous publishers to reprint a pirate copy. Combe learned this to his cost when the Boston firm Carter and Hendee published the first American edition of *The Constitution of Man* in 1829 without his permission.[42] The material realities of exchange were never far from phrenologists' minds. In Kentucky, the slaveholder Charles Caldwell worried that his fragile collection of European plaster casts wouldn't survive the journey from Paris.[43] In St. Petersburg, a copy of Morton's phrenological folio *Crania Americana* was lost in the post.[44] And in Calcutta, a medical student waited for over three years before finally receiving a pair of phrenological calipers sent from Scotland.[45] The ship carrying the package had been delayed when a group of disgruntled lascars attempted to set the vessel alight.[46] The global history of science must therefore also be a history of limits and failures.[47] Phrenology did not travel everywhere, nor all at once.[48] The material realities of communication presented obstacles as well as opportunities for those looking to advance phrenology on the world stage.[49] As the *Phrenological Journal* in Edinburgh recognized, "the progress of Phrenology" had been "rapid and soundly based in some places" but "slow and imperfectly promoted in others."[50]

Once understood in this way, material culture provides a lens through which to study the uneven circulation of science across national, regional, and imperial boundaries. It also allows us to take seriously the spaces occupied and imagined by historical actors. For too long regions such as South Asia and the Pacific have been separated out from Europe and the United States under the banner of imperial history and area studies. These distinctions have more to do with postwar politics in the twentieth century than the world occupied by phrenologists in the nineteenth. For practitioners in China and India, colonialism did not completely demarcate either their material existence or their imagination. At the same time, phrenologists in the United States did not operate in isolation from European empires. The American physician Samuel George Morton relied on colonial explorers to provide skulls for his craniological cabinet. His collection included a "Chinese" skull sent by the Dutch physician Elisa Doornik from Java and a "Bengalee" skull courtesy of the East India

Company officer Henry Piddington in Calcutta.[51] American phrenological books also found their way into colonial outposts, with the New York City publishers Lorenzo and Orson Fowler shipping copies of their *Self-Instructor in Phrenology* to the Sydney Mechanics' School.[52] In light of this, empires should not be taken as given units of historical analysis. They certainly should not be thought of in fixed structural terms, with metropoles and colonies in dialectical relationships. It would equally be a mistake to replace one structuralist language with another, whether it be the "moving metropolis" or a taxonomy of "locals," "mobals," and "globals."[53] Instead, historians of science and empire need to understand how material exchange both constituted and cut across imperial spaces.

It is only by putting it this way that we can follow individuals like Alexander Pearce, a man who meandered across areas and empires in both life and death. Born in Ireland at the end of the eighteenth century, Pearce was transported to Van Diemen's Land in 1819 following a conviction for stealing six pairs of shoes. Once there he escaped with a group of fellow convicts before apparently committing cannibalism. On his capture, Pearce confessed and was tried and hanged at Hobart Gaol on the morning of 19 July 1824. But this wasn't the end of his journey. Following his execution, Pearce's skull was removed and sent to William Cobb Hussy, a phrenological enthusiast in Calcutta. Hussy later forwarded the skull, along with an account of Pearce's life, to Morton in Philadelphia. Once there, it entered his burgeoning craniological collection where it sits today (fig. 2).[54] If we are to account for such episodes, we cannot treat Europe as separate from the Pacific, the Pacific as separate from South Asia, and the United States as separate from the world.[55]

A GLOBAL IMAGINATION

Just as phrenology found supporters in almost every major city, it also had its fair share of opponents. An article in the *Edinburgh Review*, which described phrenology as "a piece of *thorough quackery* from beginning to end," was just the most famous example of the kinds of disputes that reverberated around the globe.[56] The editor of the *South African Commercial Advertiser* in Cape Town characterized phrenology as "Materialism and Infidelity," a science doomed to "annihilation."[57] In Calcutta, the schoolmaster David Drummond believed phrenology to be "a gross deception," while in Paris the physiologist François Magendie coined the very term "pseudo-science" to describe the doctrine.[58] Given this criticism, phrenologists obsessed over the uptake of their science.[59] Hewett Watson,

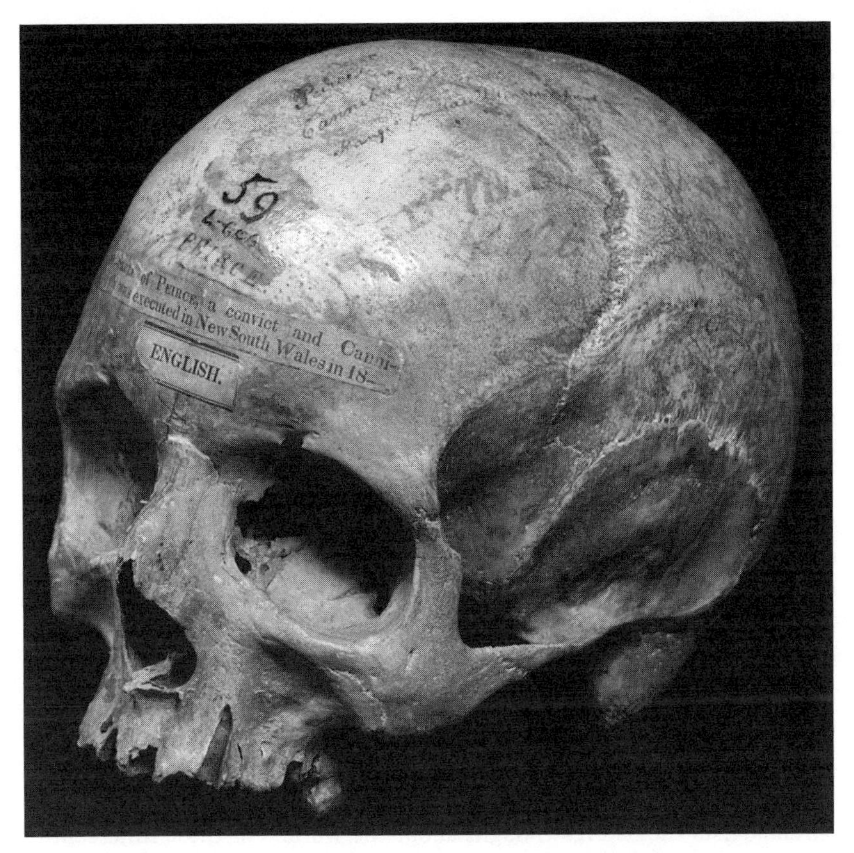

Fig. 2. The skull of Alexander Pearce, University of Pennsylvania Museum of Archaeology and Anthropology. Courtesy of Penn Museum, object number 97-606-59.

Combe's successor as editor of the *Phrenological Journal* in Edinburgh, rejoiced in the fact that phrenology was "spreading over the world." In his *Statistics of Phrenology*, Watson reminded readers that "additional societies for its cultivation are springing up every year" and that "the books of the leading phrenologists are selling by thousands."[60] Practitioners in India and the United States also imagined themselves as part of an international movement. The president of the Calcutta Phrenological Society described a community stretching "from the banks of the Ganges to that of the Mississippi," while his counterpart in Philadelphia praised "the untiring efforts of well-instructed phrenologists, in both hemispheres."[61]

Whether published in New York, Paris, or Calcutta, every phrenological journal contained a section reporting on overseas societies and publications. The first number of the *American Phrenological Journal* featured an account of the Paris Phrenological Society. François-Joseph-Victor Broussais had just finished delivering a "course of lectures on phrenology" that had "excited much attention and interest among the Parisian literati." According to the foreign correspondent, the "collections of phrenological specimens" in Paris were especially "rich and numerous." This was thanks in part to the recent return of Pierre Dumoutier, a phrenologist who traveled aboard the *Astrolabe* on a three-year voyage across the Pacific, collecting skulls and taking plaster casts as he went.[62] The *American Phrenological Journal* concluded that "the French may be considered in advance of any other people in the pursuit of phrenological observation"—an example certainly worthy of emulation.[63] Similar reports permeated phrenological journals the world over, with societies appointing corresponding members and actively arranging to exchange publications.

For those working within earlier traditions of moral philosophy, the spread of mental science held no particular epistemological significance. Dugald Stewart, professor of moral philosophy at the University of Edinburgh, could simply assume that all minds were governed by the same laws. But for phrenologists, things were not so easy. If the mind was material, then what guaranteed that brains in India and America were subject to the same laws of nature? It was this question that led phrenologists to fixate on the material connections that bound their world together. Consequently, there is an important relationship between the global history of material culture presented in this study and the ways in which the phrenologists themselves thought about the world. For these nineteenth-century materialists, the international popularity of phrenology was a marker of truth. Combe made this explicit when he challenged Thomas Stone, one of the foremost critics of phrenology in Edinburgh, to explain how a false science could have so quickly "spread over Europe, and taken root in Asia and America." According to Combe, "nothing but the force of truth" could account for the emergence of phrenology as a global science.[64] This viewpoint was not unique to Britain. The East India Company surgeon George Murray Paterson also explained how he had come to see the correctness of phrenology only after examining over 3,000 heads during his travels across the British Empire. Paterson's "very numerous manipulations in Europe, Southern Africa & the vast continent of Asia" apparently proved that phrenology worked irrespective of place.[65] After striking up a correspondence with Combe from Kentucky, the physician Caldwell also reflected on the significance of phrenology as an international movement. He told Combe it was "gratifying to find that the observations and reflections of two individuals in such distant parts of the world have led them to the same conclusions." This, according to Caldwell, was "a strong presumptive . . . that both are on the right path."[66] In the nineteenth century, as today, global talk proved a powerful ideology.

RACE ON THE WORLD STAGE

Phrenology was both a mental science and a racial science. In New York City, Combe held skulls aloft, inviting his audience to "compare the heads of the Negroes with those of the North American Indians." It was phrenology, Combe told his audience, that best explained the history of the different "races of man." Holding up a Native American skull, Combe pointed out that "the Indian has more Destructiveness, less Cautiousness, less Benevolence." This explained why Native Americans could not be enslaved. According to Combe, "he has retained his freedom by being the proud, indomitable, and destructive Savage which such a combination indicates." In contrast, the "Negro" was "gentler in nature" and so more easily subdued. Combe concluded by comparing the Native American and African skulls, suggesting that "had the Negroes possessed a similar organization, to make useful slaves of them would have been impossible."[67] Phrenology, therefore, both explained and reinforced racial divisions.

Throughout this book, race acts as an exemplar of my broader argument concerning the workings of global history. One of the most striking aspects of racial science is how such an idea became so powerful in so many different places during the nineteenth century. Combe's combination of material culture and racial politics was not unique. Skull collectors could be found in the deserts of Egypt and the jungles of East Timor, while notions of "savagery" were simultaneously employed to characterize Aboriginal Australians and South Asian hill tribes.[68] It is often simply taken for granted that similar kinds of racial science operated across this range of geographies. Phrenologists used the same theories, practices, and objects to understand African slaves in the Americas as they did Pacific Islanders. And by the early twentieth century, Chinese and Indian intellectuals had started adopting racial theories, including phrenology, in the hope of forging new national identities.[69] What made this possible? To date, we don't have an account of how racial science became a global science.[70] This is reinforced by the fact that most histories of racial science are focused on particular areas. We have detailed studies of the relationship between science and race in the United States, South Asia, Africa, and the Pacific.[71] But we don't have a history of how these developments were connected, both intellectually and materially. In this book, I show

how the material process of transit shaped the development of phrenology as a science of race. I also show that the history of racial science extends well beyond the kinds of material culture we might expect, such as skulls and plaster busts. Even the act of writing a letter, or editing a periodical, could be subtly racialized. When we write the global history of science we therefore need to integrate race at a fundamental level.

Charting this history represents the analytic element of my argument. But race was also an actors' category. Just as phrenologists mapped the mind, they hoped to map the races of man. In some cases, phrenologists adopted a comparative approach, as with Combe's contrast between Native American and African skulls. However, phrenologists also understood race as part of a connected history. The Maori, according to one phrenologist, had been "deprived of communication with neighbouring people." This had held back "the organs of intelligence." The same author made a similar argument concerning China. He looked to the day "when the Chinese nation, having reformed its language and thrown aside its prejudice, shall throw open its cities to free communication with the rest of mankind." Only this, the phrenologist argued, would lead to "the progress of the nation."[72] In other instances, phrenologists debated the relationship between the mind and different physical environments. Had the frozen Arctic diminished the organ associated with color perception in the "Esquimaux"? And had the tropical heat dulled the "Hindoo" intellect?[73] While phrenologists tended to shy away from environmental determinism, they nonetheless developed complex accounts of how the history of world—whether it be in terms of war, trade, communication, or environment—had shaped the development of the mind. By following race from the perspective of our historical actors, we therefore get a sense of just how variable global thought was in the nineteenth century.

Historians have long emphasized the need to consider race as an actors' category. While this can seem troubling from the perspective of the present, it allows us to better understand how ideas that may seem abhorrent today proved so powerful in the past.[74] By writing a global history of science, we can push this historiography a step further. Much of the existing history of racial science is written from the perspective of Europeans alone.[75] We certainly know more about what George Combe thought of Africans than what they thought of him.[76] This book is part of an attempt to redress the balance. Put simply, if we're interested in actors' categories, then we also need to ask: who are the historical actors? Across the different chapters I take seriously the perspective of people who have usually been considered simply as objects of phrenological study. Perhaps unsurprisingly, there were many who lambasted phrenology for its support of slavery and imperialism. The African American physician James McCune Smith described the "fallacy of phrenology" in a lecture to a black audience in New York City in 1837.[77] Smith's lecture was timely. Only a few months later, Combe arrived in the city, lecturing on the history of slavery with the aid of his collection of "Negro" skulls. Similar attitudes could be found on the other side of the world. In February 1857, a few months before the outbreak of the Indian Rebellion, the *Bombay Times* printed a letter signed "A Hindu." The author presented a scathing critique of phrenology. This was a doctrine "invented by malice and propagated through jealousy." Phrenology, according to this critic, was simply another means to "degrade the character of the Natives of India in the eyes of Europe."[78]

Yet despite these examples, it would be a mistake to assume that all those who were not white responded in the same way. The reception of phrenology varied, ranging from damning criticism to enthusiastic acceptance. In Barbados, W. D. Maxwell, described by a local newspaper as a "coloured interpreter of the science," gave a series of phrenological lectures. He later traveled to London where he made a name for himself examining heads.[79] Back in India, not everyone was as critical as the correspondent in the *Bombay Times*. In the 1840s, a group of Bengali medical students even set up their own phrenological society, complete with a museum and periodical. In his lectures, Combe used phrenology to justify British imperialism. But in Calcutta, these Bengali phrenologists turned this rhetoric on its head, arguing that it was in fact the East India Company that had stifled mental improvement. Once the East India Company's "monopolies" and "fixed revenue system" were abolished, phrenology would reveal Bengalis to be "one of the most intelligent peoples on the face of the globe."[80]

These cases, and others uncovered in this book, suggest some of the ways in which global history can help us rethink nineteenth-century racial science. Historians traditionally treated race as part of an intellectual history. Race was a "concept" or an "idea," something European men thought about and discussed in books. This intellectual history tended to make a relatively straightforward link between racial thought and a particular politics. It was easy for intellectual historians to understand how racial sciences like phrenology might support slavery and colonialism, but much less easy to understand why African Americans and Indians might take up racial science as well.[81] More recently, historians have emphasized the need to examine race in relation to material culture and performance.[82] Race wasn't just an idea. It was something you did. In Britain, curious audiences gawked at displays of living people from Africa,

America, and Asia. In some cases, human displays were even accompanied by a phrenological examination. Most famously, in 1810, the "Hottentot Venus"—a Khoikhoi woman from Southern Africa—arrived in London and was marketed on posters as "the greatest phenomenon ever seen." It was through popular performances like this that understandings of African racial character were shaped.[83] By concentrating on material culture and performance, historians have been able to complicate the older intellectual history of race. In particular, studies of material culture allow us to better understand the relationship between racial thought and political action. I take the same approach in this book, but push this history beyond individual national or imperial contexts. At various points in the history of phrenology, the politics of oppression hinged on an object, not simply an idea. Skulls were collected from executed colonial rebels. Plaster busts were used in trials of African Americans. And phrenological photographs were reproduced in imperial handbooks.

With all this in mind, we are better placed to make sense of the role played by people outside of Europe in the history of phrenology. There may have been West Indian and Bengali phrenologists, but this did not curb the pervasive racism of the nineteenth century. Indeed, black phrenologists were often ridiculed in the European press. At the same time as W. D. Maxwell was lecturing in Barbados, John Follitt published a series of broadsides in London entitled "Black Lectures." Poking fun at black interest in science, the series even featured "A Black Lecture on Phrenology" (fig. 3). The broadside depicted a West Indian man, racialized as a buffoonish character, with his hands on European and African phrenological busts. Follitt then ventriloquized the black phrenologist, writing in an imagined patois, "I tink it rite to obserb, dat all great an clebber men of every nation are BLACK."[84]

This was the kind of racial discrimination Maxwell would have encountered when he traveled to London to deliver a series of phrenological lectures in the 1860s. On top of this, a West Indian man like Maxwell could not have hoped to command the same kind of material resources as someone like George Combe. Whereas Combe had the capital to finance printing thousands of copies of *The Constitution of Man*, Maxwell's lectures appear as only a few lines in a colonial newspaper. Whereas Combe had access to a vast collection of skulls and plaster casts in Edinburgh, Maxwell made do with homemade props and phrenological charts torn out of old books. And whereas Combe crossed the Atlantic by steamship, Maxwell's forebears endured the Middle Passage. What matters then is not just the intellectual content of phrenology, but who was able to put those

Fig. 3. "A Black Lecture on Phrenology," *Follitt's Black Lectures, No. 1* (1846), Science and Society Picture Library, Science Museum, London.

ideas to political use and how this was achieved in practice. Global histories of science need to recover a variety of voices. But at the same time, we need to recognize that those voices were often marginalized through the very process of global exchange. Circulation could both empower and disempower. Nowhere was this more true than in the sciences of the mind.

This book is divided into six chapters, each focusing on a different type of object. In turn these are skulls, casts, books, letters, periodicals, and photographs. This structure serves a number of overlapping ends. In the first instance, it reflects my belief that material culture should be treated as a continuum. By beginning with museum objects—skulls and plaster busts—and then moving on to less traditional material culture—letters and photographs—I suggest how similar techniques can be used to study a diverse range of scientific material. This structure also helps to identify the different roles played by particular objects in scientific exchange. As the chapters progress, I reveal the contrasting uses and difficulties that certain objects presented. Skulls, for example, were important in establishing phrenological theories of race, but as unique specimens they could not form the basis of a popular movement. Illustrated books could be reproduced and circulated much more widely, but even then, the distributed nature of publication made reception hard to control. Additionally, each chapter addresses a major theme in the history of science, ranging from the emergence of scientific disciplines to the relationship between science and politics. This allows me to draw wider conclusions concerning nineteenth-century science beyond phrenology. Beginning with the collection of skulls in the 1810s and ending with the reproduction of photographs in the 1880s, the six chapters also represent a chronology. By writing a global history, we can therefore challenge the simplistic notion of the rise and decline of certain sciences. Phrenology continued to find new audiences and new political uses well into the twentieth century. This chronology also reflects the changing technologies of communication across the nineteenth century. I discuss the significance of this in further detail in the epilogue, pushing the history of phrenology into 1920s China. But for now, it is important to acknowledge that this is not a history in which the world becomes increasingly flat, and communication is rendered effortless. The steam-printed books and mechanically reproduced photographs of the later nineteenth century in fact introduced new challenges, and did not always solve existing problems. Finally, while each chapter takes in a range of geographies, it is not my intention to fill in a world map. Rather, the aim is to produce a detailed analysis of the relationship between material exchange and the ways in which phrenologists thought about the world—that is, the global as an analytic category and the global as an actors' category. For this reason, the chapters are built around the close study of particular objects: Inuit skulls collected in the Arctic, a plaster bust made in Paris, and phrenological photographs taken in India. Drawing these histories together, *Materials of the Mind* ultimately argues that what it meant to be a universal science of the mind was something that emerged in the context of global material exchange.

CHAPTER 1

Skulls

In the clear water between HMS *Blonde* and O'ahu, two bodies bobbed toward the shore. A few minutes later, Hawaiian chiefs helped British sailors unload a pair of mahogany caskets. The larger coffin bore an inscription. Side by side, in both Hawaiian and English, it read:

Tamehameha II. Elii
no nahina o Awaii
make i Pelikani 28
Makaiki Kaik i ke mahoe
neua o Kemakaihi 1824.
Moa ino no Komakou Elii Iolani.

Tamehameha II., king
of the Sandwich Islands,
died, 14th July 1824, in London,
in the 28th year of his age.
May we ever remember our beloved king Iolani.

Ten months earlier in London, Kamehameha II of the Kingdom of Hawaii had died from measles. His wife soon succumbed to the same fate. Back in the Pacific, local leaders lifted the coffins onto two wooden carts, before covering them in black *kapa,* a Hawaiian cloth woven from plant fibres. A procession soon formed as the deceased king and queen edged closer to their resting place at Pohukaina. Toward the head were twelve Hawaiian warriors cloaked in brightly colored feathers. Behind them were the marines, the chaplain, and the surgeon of HMS *Blonde,* followed by another forty chiefs. George Byron, commander of the British mission, accompanied the grieving princess Naheinaheina while 100 sailors, each wearing a black handkerchief, completed the funeral gathering.[1]

On the morning of 11 May 1825, the British returned two bodies to the Kingdom of Hawaii, but they did not leave empty-handed. When HMS *Blonde* left O'hau on 7 June, it still carried human remains. An exchange, of sorts, had occurred. A few days after the burial ceremony, one of the naval officers accosted an islander. This go-between reluctantly climbed a nearby hill, Luhahi, where the locals buried their dead. He returned with a single human skull.[2] This skull, and three others, left the Sandwich Islands, continued south to Malden Island, before traveling another 15,000 miles, around Cape Horn, to the English Channel. They were then sent on a mail coach from Spithead to George Combe in Scotland, where they entered the burgeoning collection of the Edinburgh Phrenological Society. There these "Sandwich Islanders" were joined by "Thugs" from India, "Esquimaux" from the Arctic, and "Hottentots" from the Cape Colony.[3]

This chapter argues that encounters with death and burial acted as a formative site for phrenological understandings of racial character. Across the colonial world, rituals surrounding death represented a moment in which different cosmologies interacted with one another and often conflicted.[4] Burial was an opportunity both to collect skulls and to reflect on the constitution of the dead.[5] By examining three very different imperial contexts—Ceylon, Egypt, and the Arctic—this chapter illustrates how the practice of collecting was bound up with assessments of racial character. It also emphasizes the need to be attentive to the different ways in which imperialism operated. Encounters with death and burial in the Arctic, as the phrenologists themselves lamented, were not straightforwardly comparable to those in Asia. In fact, the acquisition of human skulls was mediated by contrasting physical environments, political ideologies, and forms of colonial violence.[6] Significantly, collecting practices were also shaped and in many cases contested by colonized people.[7] Following the work of Marshall Sahlins, this chapter draws on historical anthropology to reconstruct both sides of the encounter.[8] What to do with the dead was not obvious, and debate raged over appropriate burial practices. Widow burning was outlawed in India, while new garden cemeteries reflected reformist campaigns back in Europe.[9] The Edinburgh Phrenological Society's collection, much of which survives to this day, was ultimately a product of these conflicting attitudes to death, burial, and human remains.

For the phrenologists, the museum embodied what it meant to take part in a global science. In the late 1820s, the Edinburgh Phrenological Society purchased a "Map of the World" for display alongside its collections on Clyde Street. The intention was to use the new map "to mark the places where skulls are from."[10] Some specimens were even assigned an exact latitude and longitude.[11] Visitors were then invited to connect the physical organization of the museum in Edinburgh with the global geography of mankind beyond. There were different cabinets, each with a glass door, for "National Skulls" and "European Skulls." The layout of the specimens had been carefully arranged, and members were kindly reminded to "replace them carefully" after examination.[12] Later phrenological atlases also often featured maps. Samuel George Morton's *Crania Americana*, discussed further in chapter 3, included a hand-colored chart titled "The World Shewing the Geographical Distribution of the Human Species." Five different races were represented, from the "Caucasian" occupying Europe, North Africa, and India to the "Mongolian" covering Asia and the Arctic. Museum collections also encouraged phrenologists to reflect on the material reach of their science. On receiving an "Ashantee skull" in the 1820s, the *Phrenological Journal* commented that "it is gratifying to know, from the above donation, that we have active friends, who, personally unknown to us, are exerting themselves in all quarters of the globe."[13] From the very beginning, phrenology was therefore imagined as a science that would chart the history and geography of mankind. An early article in the *Phrenological Journal* remarked that "when we regard the different quarters of the globe, we are struck with the extreme dissimilarity in the attainments of the varieties of men who inhabit them." The goal of phrenology was to map the "cerebral development of nations."[14]

Despite this rhetoric, acquiring human remains and making sense of them was no easy task. For a start, skulls arrived in an incredible variety of physical conditions. In the Arctic, the freezing conditions ensured that Inuit skulls remained well preserved. Some were even found with remnants of skin and hair still attached. Sand also tended to preserve human remains. When the surgeon Alexander Moffat seized the skull of a New Zealand chief from the Bay of Islands, he reported that "the Integuments were preserved in a dried state upon the bones."[15] By contrast, skulls buried in acidic soil often showed signs of erosion and were almost always devoid of hair and skin. Other skulls, typically those left unburied in the sun or taken straight from the gallows, were smooth and bleached. The phrenologists even possessed mummified remains, the flesh blackened after thousands of years.[16] These material differences mattered. For the phrenologists, the coarseness of a skull was a marker of mental activity. According to the *Phrenological Journal* the "texture of the Ceylonese and Hindoo skulls is much more delicate and refined than that of the skulls of the natives of New Holland." This apparently explained why "we find the Ceylonese distinguished by refinement, and the New Hollanders by rudeness and harshness of manners."[17]

Ownership proved another headache. In 1828 a number of Yupik skulls were collected from St. Lawrence Island by the surgeon of HMS *Blossom*. Combe had understood that they would be donated to the Edinburgh Phrenological Society, but, on arrival in Britain, the skulls were claimed by the Admiralty. The *Phrenological Journal* issued a complaint, stating, "We understand that they were taken possession of by Government, along with all other specimens of natural history collected during the voyage." The article went on to explain that "Mr Collie intended them as a donation for the Phrenological Society."[18] Such an announcement certainly cemented the status of skulls as commodities that could be owned and exchanged.[19] But, in an age in which the public and private status of museum collections remained fluid, the phrenologists found themselves on the wrong side of ownership.[20] In an effort to recover the skulls, they attempted to assert their intellectual authority over the specimens, writing that "they must be comparatively useless to all but phrenologists."[21] In making this move, the phrenologists brought together their claim of intellectual ownership with a desire for material possession. This partly explains phrenologists' tendency to write directly on skulls. Inscriptions were certainly used to classify collections, but they also allowed phrenologists to assert ownership by tracing particular skulls to particular collectors (fig. 4).

Even if a skull arrived safely in a museum, its authenticity was by no means assured.[22] Many entries in the manuscript catalogue of the Edinburgh Phrenological Society's collection are annotated with the words "authority unknown."[23] This in part stemmed from the fact that phrenologists relied on such a variety of collectors. This was an age before the professionalization of anthropology, with limited attempts to standardize collecting practices.[24] From naval explorers traversing the Northwest Passage to poets living in the Cape Colony, collectors varied in their interests and expertise. More often than not, human remains were procured as part of a broader colonial project, whether it was Napoleon's invasion of Egypt or British attempts to secure the interior of Ceylon.[25] Many collectors were aware of phrenology thanks to the popularity of Combe's books but, as the *Phrenological Journal* complained, this did not guarantee that they were "skilful observers and describers of the manifestations of the human mind." Nonetheless, Robert Cox, the conservator at the Edinburgh Phrenological Society's museum, did his best to collate information arriving from across the world. In this instance, the global was both the problem and the solution. Cox suggested that "by comparing . . . the details given

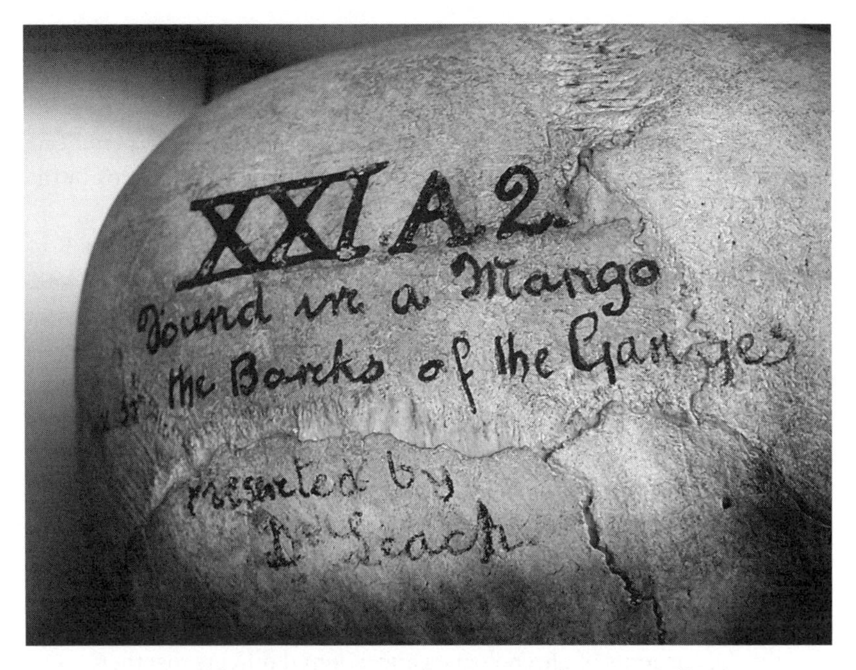

Fig. 4. Detail from XXI.A.2, "Bengali," Anatomical Museum collections, University of Edinburgh.

by different observers, it is possible to discover, with tolerable certainty, the more prominent mental characteristics of the great body of a nation."[26] This strategy is reflected in the pages of the Edinburgh Phrenological Society's catalogue itself. Compiled by hand, the catalogue features hundreds of additions and amendments—crossings out, footnotes, pasted sheets, and cross-references. Cox kept putting off producing a printed copy of the catalogue, which he estimated would extend to "about 200 octavo pages."[27] The phrenologists ultimately relied on a system of collecting that ensured understandings of race were always subject to revision.

EXECUTION IN THE TROPICS

Keppetipola's skull was a war trophy.[28] In November 1818 the British in Ceylon held a court-martial in which the former Kandyan chief was convicted of "levying war, with a view of subverting his Majesty's government, lawfully established."[29] Keppetipola had joined the 1817–18 rebellion in the Kandyan provinces, fighting the British over the course of the monsoons before being captured, tried, and decapitated.[30] After Keppetipola's death, Henry Marshall, senior medical officer for the Kandyan provinces and a graduate of the University of Glasgow, seized the skull and forwarded it to the Edinburgh Phrenological Society.[31] On arrival, the museum conservator etched the number "19" onto Keppetipola's forehead, before placing the skull alongside the other "Ceylonese" specimens in the collection: "Vedahs," "Tom-Tom boys," and "Cingalese."[32]

For colonial officials, Keppetipola's execution was a symbol of British power in Ceylon, particularly over the Kandyan provinces in the interior of the island. While the coastal regions had proved relatively easy for the British to seize from the Dutch in 1796, the Kingdom of Kandy was a different matter. It was only after a number of failed attempts that the British managed to depose King Sri Vickrama Rajasimha and secure the Kandyan Convention of 1815. This treaty effectively brought the region under Crown control while maintaining customary rights for Kandyan chiefs. Even then, British power in the interior was uncertain, and the peace didn't last long. In late 1817 a rebellion broke out that took over a year to put down.[33] In the violence that followed, the British read both the battlefield and the execution block for signs of Kandyan character. Death brought the concerns of phrenologists and colonial officers together.

Marshall, who had served in the Cape Colony and South America, found the rebellion to be unnaturally violent.[34] The Kandyan rebels "showed no mercy, and gave no quarter."[35] One eyewitness gave an account of the massacre of British prisoners near Watapologa, describing how "the executioners, with their large swords, chopped their victims down. . . . When this butchery was complete, they began to strip the dead." Marshall, it seems, wasn't the only headhunter on the island. However, he read a very different meaning into the practice when it was conducted by Kandyan rebels. There was "a reward of ten rupees . . . for the head of every European, and five for that of every other class of soldiers in the English service." Marshall even remembered seeing the heads of European troops impaled on spikes near British outposts.[36] The tropical environment also contributed to the intensity with which Marshall confronted the conflict.[37] Ceylon was described as possessing "an unwholesome climate, producing disease."[38] The fighting itself took place over the course of the monsoons, Marshall recalling the "low swampy ground" and the "heavy rains in the Kandyan provinces."[39] When the rain stopped, the scene was no less unsettling. Marshall remembered seeing "bones . . . whitening in the sun." It was a "sight of horror" with "unburied skulls and thigh bones mixed together."[40] Marshall's attitude toward the conflict was not incidental. For a military man, conduct on the battlefield was ultimately a marker of character. Marshall complained that "the Kandyans were never practically acquainted with the laws of civilized warfare." He pointed to the barbaric practices of former Kandyan kings in which executions consisted of "being killed by elephants, the bodies being exposed, or hung in chains."[41] These assessments fed back into the phrenological accounts. According to an article in the *Phrenological Journal*, Kandyans were characterized by "cowardice and military ignorance." They "lived in a state of the most abject submission to their king," and, consequently, "fear of punishment was the strongest principle which secured their allegiance."[42]

Whatever Marshall's evaluation of Kandyan military practices, he was also well aware that violence worked both ways. The conflict had been "a partisan warfare, which from its very nature and circumstances, was severe and irregular."[43] The governor of Ceylon, Robert Brownrigg, authorized the troops "to inflict a severe punishment on the inhabitants . . . for the purpose of thereby checking the insurrection." Marshall wrote that "the work of devastation commenced; the houses of the inhabitants were forthwith set on fire and burnt to the ground, and all the cattle, grain, &c, belonging to the people, were either carried off by the troops or destroyed."[44] The summary execution of those suspected of helping the rebels also aided in the collection of skulls. John Davy, brother of the chemist Humphry, spent many years in Ceylon as a medical officer and hospital inspector. He was also an avid reader of Johann Spurzheim's phrenological works. Following the fighting, Davy recalled seeing "a human skull that lay by the road-side, under a tree, to which the fatal rope was still attached." A lithograph of the skull of "a Singalese Chief of a secluded part of the Interior" was later featured in Davy's *An Account of the Interior of Ceylon* (fig. 5).[45] Like Kandyan head-hunting, violence also had a political purpose. Correspondence between British colonial officials emphasized the importance of overt displays of military power. The goal of the campaign was to secure the "general and unlimited submission" of the Kandyan people.[46] According to a House of Commons report, "The only way by which any impression could be made on such an enemy was by burning their villages and laying waste their paddy fields."[47]

Marshall admitted that the strategy seemed to work, explaining that "the inhabitants appeared to be horror-struck at the devastation. . . . They ceased to shout at the troops, or to fire upon them."[48] But he was also worried about the relationship between European and Kandyan character. For Marshall, the tropical climate and barbaric warfare pointed to the problem of racial degeneration. The rebellion had been one in which "white and black races, the invaded and the invaders, Christian and Pagan, vied with each other in promoting horrors and barbarities of mutual destruction."

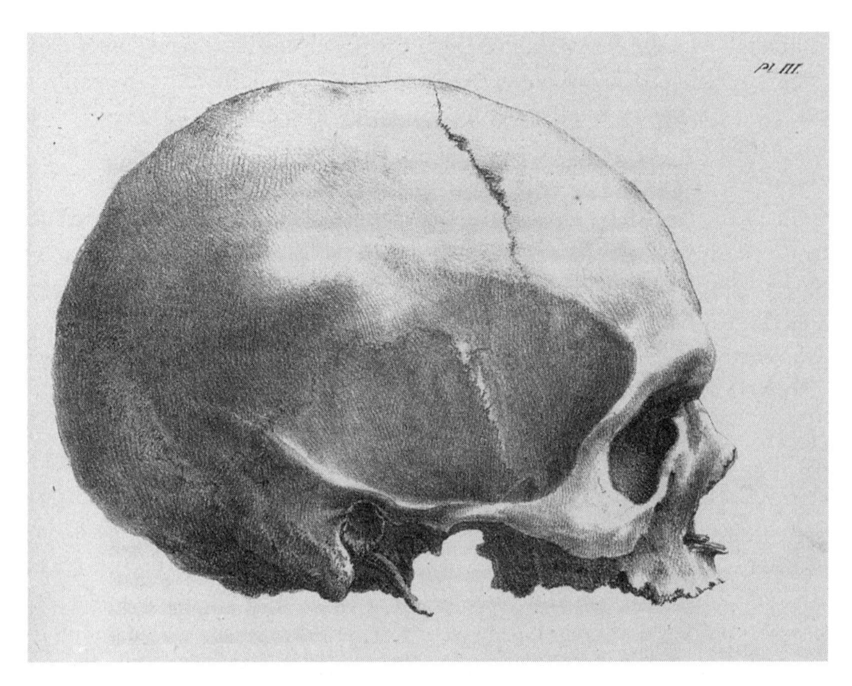

Fig. 5. "Singalese Chief of a secluded part of the Interior," John Davy, *An Account of the Interior of Ceylon, and of its Inhabitants* (London: Longman, Hurst, Rees, Orme, and Brown, 1821), pl. 3. Reproduced by kind permission of the Syndics of Cambridge University Library.

Marshall concluded his discussion, asking, "Are there no means by which civilized nations can carry on war with barbarians, but by retrograding into barbarity themselves?"[49] The Kandyans asked a similar question, albeit the other way around. The following Pali song, dating from the rebellion, recalled the violence of the British campaign: "The English commander . . . pursued and hanged the rebels on trees, thereby stunning them with terror and dismay."[50] For both sides, conduct in war was a measure of civilization.

It was amid these conflicting understandings of violence and warfare that Keppetipola joined the rebellion. Born into an aristocratic Kandyan family, he originally swore allegiance to the British as one of the signatories of the 1815 Convention. Shortly afterward, he was appointed *dissawe* of Uva, one of the precolonial administrative appointments maintained by the treaty. Keppetipola, at least before the outbreak of the rebellion, was thought to be a particularly enlightened chief. He was "distinguished for decorum and propriety of behaviour," "the Kandyan Chesterfield." Marshall approvingly recounted that Keppetipola had allowed himself and his family to be vaccinated by British medical officers in Kandy—"the only chief of consequence" to submit to European medical treatment.[51] In fact, Keppetipola was so well thought of that when the rebellion broke out in Uva in October 1817, the British sent him to put down the insurgents. But not everything was as it seemed. Shortly after arriving in Uva with his men, and despite being armed by the British with muskets, Keppetipola was apparently captured by the rebels. The British suspected he had in fact defected. Before long, it was confirmed that Keppetipola had sided with a pretender to the Kandyan throne: a man going by the name Wilbawe who claimed to be a descendent of the deposed king, Sri Vickrama Rajasimha. Keppetipola had also accepted the position of first *adigar*, the most senior appointment within the Kandyan political hierarchy. With the Kandyan provinces in open revolt, and Keppetipola leading the charge, the British declared martial law in February 1818, and a price of 2,000 rix-dollars was placed on the chief's head.[52]

Keppetipola's defection reinforced a tension at the heart of European understandings of Kandyan character. The phrenologists, like the British in Ceylon, were unsure whether to consider the Kingdom of Kandy as an example of advanced civilization or of feudal barbarism. The *Phrenological Journal* recognized that Kandyans showed all the hallmarks of civilization. After all, the phrenologists noted, "reading and writing are far from uncommon acquirements" in Kandy. The phrenologists were also impressed by the sophisticated natural, geographical, and historical knowledge contained in Kandyan palm-leaf texts, explaining that "their books are all manuscript, and are formed of leaves of trees, confined by boards."[53] Davy also believed that Kandyans showed evidence of civilization, writing that "the form of their head is generally very good, perhaps longer than the European."[54] However, as Keppetipola's treachery and the violence of the rebellion demonstrated, Kandyan character could be deceptive. The *Phrenological Journal* described Kandyans as "ill-natured, false, unkind, though outwardly fair and seemingly courteous."[55] This was matched by "full" organs of "Secretiveness" and "Combativeness" on inspecting Keppetipola's skull. The phrenologists in Edinburgh concluded that the chief was certainly "enterprising" and "ambitious" but ultimately an "unprincipled man . . . actively engaged in endeavouring to subvert the British power in Ceylon."[56] Marshall also felt that, whatever their merits, Kandyans could not be trusted. They were easily "roused by real or imaginary wrongs."[57] Davy even provided a phrenological account of the cause of the insurrection, one that neatly absolved the British of responsibility. According to him, "The rebellion was not the effect of oppression or misrule, but of the innate propensities of the people, concealed till they burst out and showed themselves in acts of violence."[58] And so, from the outset, the phrenologists interpreted Keppetipola's actions in terms of the ambiguous status of Kandyan character: civilized on the one hand, but treacherous and barbaric on the other.

After months of fighting in weather "extremely unfavourable for the march of troops," the British finally captured Keppetipola on 28 October 1818.[59] His treatment was recorded in minute detail by Marshall. The phrenologists in Edinburgh later quoted from this account extensively, in both the *Phrenological Journal* and the manuscript catalogue of the museum.[60] At every stage, from his imprisonment to his execution, Marshall and the phrenologists read Keppetipola's actions for signs of Kandyan character. Like all executions, this was a carefully staged performance in which both the British and Keppetipola himself sought to come to terms with the aftermath of the rebellion.[61] Most significantly, Keppetipola was marched to Kandy. It was here, in the shadow of the Temple of the Sacred Tooth Relic, that Keppetipola would be imprisoned, tried, and executed. By bringing Keppetipola to Kandy, and executing him there, the British confirmed him as an authentic Kandyan, even in death. This was part of a broader strategy in which the British actively distinguished the highland interior from the coastal regions of Ceylon, as well as the rest of South Asia.[62] The phrenologists reproduced this distinction, differentiating between "the aboriginal race" of the "interior" and "naturalized foreigners" found in the "maritime provinces."[63] In light of this, Keppetipola's fate should be contrasted with that of the deposed king, Sri Vickrama Rajasimha, who was transported to Vellore in the Madras Presidency following the 1815 Convention. The British did not believe Sri Vickrama Rajasimha to be an authentic Kandyan. Instead, as a member of the Nayakkar royal line, Sri Vickrama Rajasimha was thought to be a descendant of a class of South Indians who had come to settle in Ceylon in the seventeenth century. Deporting him to Madras was wrapped up in British attempts to rid Kandy of "Malabar" influence from South India.[64] A similar approach followed the aftermath of the 1817–18 rebellion. Less than half of the sixty-three rebels court-martialed were tried in Kandy. Others were taken to Ratnapura, Balangoda, and Colombo, while sixteen were banished, primarily to Mauritius.[65] Bringing Keppetipola to Kandy therefore ensured he would be recognized as one of the "pure Singalese of the Interior." The phrenologists concurred, explaining that "the Kandians . . . are confined to the center of the island." In contrast, "the Malabars" were said to be "originally emigrants from the Indian peninsula."[66]

Marshall visited Keppetipola a number of times during the court-martial. The former chief was apparently keen "to converse on the subject of the insurrection." While Marshall's account certainly reflects his own imperial biases, a close reading points toward a Kandyan perspective on the rebellion. It also suggests how Keppetipola sought to portray himself in the run-up to his death and the seizure of his skull. Significantly, Marshall said that Keppetipola tried to "explain away, or weaken at least, the force of any inference which tended to his inculpation." He was "unwilling to admit that his unhappy condition was an obvious consequence of the policy he had adopted, and the ill success which attended it."[67] This fitted well with the later phrenological account of Keppetipola, as well as the more widespread ambiguity in interpreting Kandyan character. The *Phrenological Journal* explained that, despite their seemingly advanced state of civilization, "the Candians" were characterized by "cowardice and military ignorance."[68] Additionally, Keppetipola's skull was said to show a "very moderate" organ of Conscientiousness, explaining his willingness to defect as well as his inability to take responsibility for his actions.[69] Marshall and the phrenologists also pointed to Keppetipola's religious identity. According to Marshall, "Being a zealous Boodhist, he considered his present misfortune was the result of delinquencies committed during a former state of existence—a belief which repudiates responsibility for offences committed in this life."[70] For the British, Buddhism was another means by which authentic Kandyans could be distinguished from the Hindu "Malabars" of South India.[71] Marshall's account of Keppetipola's Buddhism, therefore, not only served to explain his refusal to take responsibility for the rebellion, but also reinforced his status as a "Candian" back in Edinburgh.

Why did Keppetipola act as he did? And why might he have rejected the idea that he was at fault? Of course, there is the obvious fact that, prior to the court-martial, Keppetipola might simply have been trying to defend himself. After all, the British had, in a calculated display of imperial benevolence, pardoned eight other chiefs convicted of treason.[72] Marshall certainly believed that Keppetipola was trying to secure similar leniency.[73] However, Keppetipola's words and actions also point to subtleties in Kandyan Buddhism that are missing from the phrenologists' interpretation. It seems likely that Keppetipola was in fact trying to ensure that he died a good death. For Christians like Marshall, this meant something very different—namely, the confession of sin. But for Kandyans, like most Buddhists across South and Southeast Asia, a good death meant focusing on the positive aspects of one's life. Crucially, Kandyan Buddhists believed that the quality of a person's last conscious thought played a pivotal role in the form and place of their subsequent rebirth. Recalling past meritorious deeds was therefore an important part of the death ritual.[74] Once we recognize this, we can better account for the phrenologists' misinterpretation of Keppetipola's final moments. This becomes even more clear in Marshall's description of the day of the execution itself.

On the morning of 25 November 1818, Keppetipola was removed from his jail in Kandy. From there he was taken, at his "own request," to the Temple of the Sacred Tooth Relic in the center of the city.[75] This was a site of enormous political and religious significance. Both the British and the Kandyans understood that control of the relic of Buddha signified the power to govern the Kingdom of Kandy. In fact, the relic had even been seized by the rebels during the 1817–18 rebellion, and Keppetipola himself had used it during the "coronation" of the pretender king. The British recognized that "the sovereignty of the island" was invested in this object, and consequently allocated considerable military resources toward its recovery, something that was only accomplished toward the end of the rebellion.[76] Taking Keppetipola to the Temple of the Sacred Tooth therefore served overlapping interests, even if these operated according to very different cosmologies. For the British, this ritual provided another means to affirm Keppetipola's status as an authentic "Boodhist." Marshall recalled how the former chief had been "allowed a short time to perform the ceremonies of his religion."[77] Again, this was part of a British strategy of governance that depended upon the construction of a pure Kandyan people of the interior, despite the fact that "Moors" and "Malabars" had been active in Kandyan Buddhist culture for centuries.

For Keppetipola, the Temple of the Sacred Tooth simply allowed him to proceed with the ritual he had begun in jail. On arrival Keppetipola met a "priest" at the steps of the temple, before proceeding inside. Kneeling, he then "detailed the principal meritorious actions of his life—such as the benefits he had conferred on priests, together with the gifts he had bestowed on temples, and other acts of piety." Significantly, immediately after this, Keppetipola "pronounced the Proptannawah, or last wish; namely, that, at his next birth, he might be born on the mountains of Himmalaya, and finally obtain Neerwannah, a state of partial annihilation."[78] Once again, Keppetipola's words and actions should be read, not in terms of the pure Kandyan identity advanced by both Marshall and the phrenologists, but instead as a manifestation of the local culture surrounding death. Here, Keppetipola explicitly links his *prarthana*—his wish to secure favorable rebirth—with an intense focus on the positive contributions of his life.[79]

With the ceremony complete, Marshall accompanied Keppetipola to Bogambara, about a mile from the Temple of the Sacred Tooth. The former chief was allowed a short time to wash his hands and face. Keppetipola then "tied up his hair in a knot on the top of his head, and sat down on the ground, beside a small bush, grasping it at the same time with his toes."[80] He was focused on attaining a state of calm prior to death. Kandyan Buddhists understood well that dying in a state of anger, agitation, or confusion would interfere with the process of rebirth.[81] Keppetipola then took out a small "prayer-book" and begun "reciting some prayers or verses" before handing the tiny volume to a "native official." He continued to repeat the Pali verses aloud, and "while he was so employed, the executioner struck him on the back of the neck with a sharp sword." According to Marshall, "at that moment he breathed out the word *Arahaan*, one of the names of Boodhoo."[82] Whatever Marshall's colonial motivations, recalling the words of Buddha during one's final moments was an established part of Kandyan rituals surrounding death on the island. In doing so, Keppetipola hoped to ensure his final conscious thoughts would be pure and pious. He might then be reborn in his favored location—the Himalayas—a tantalizing pointer to a broader understanding of Buddhist and Kandyan identity, one that stretched beyond the hills and beaches of Ceylon.[83]

The execution of Keppetipola was ultimately a moment of redemption. In staging the decapitation of the Kandyan chief, the British hoped to cement their power in Ceylon while simultaneously portraying themselves as benevolent masters. This proved a difficult balancing act given the policy of overt displays of violence during the rebellion itself. It is precisely for this reason that Marshall described what had in fact been a gruesome event—in which the executioner required two attempts to sever Keppetipola's head—as "most solemn and impressive."[84] The British understood decapitation by sword, both at home and in Kandy, as a mode of execution befitting a higher class of man.[85] Marshall described how "men of rank were decapitated with a sword, while they sat on the ground; losing the head, as with us, considered the most honorable mode of receiving capital punishment." In contrast, "individuals of the lower ranks were hanged, and the whole body attached to a stake."[86] In opting to decapitate Keppetipola, the British once again confirmed his status as a high-ranking Kandyan chief. They also hoped to emphasize their role in upholding the more "civilised" aspects of Kandyan culture. After all, the Convention of 1815 had attacked "the cruelties and oppressions of the Malabar ruler" along with the "arbitrary and unjust infliction of bodily tortures and the pains of death without trial."[87] The performance of the court-martial and the execution, both of which proved central to phrenologists' interpretation of Keppetipola's character, were therefore part of a broader attempt to reconcile the violence of the rebellion with the propriety of British rule.

This image of benevolence looks less stable once we consider what happened immediately after the execution. While Keppetipola might have agreed that his status befitted execution by decapitation, he would have been unlikely to see the merits in having his skull seized and transported to Edinburgh. In fact, higher-caste Kandyans were often embalmed and cremated following death, an honor that the British apparently denied Keppetipola. Certain bones, particularly the skull, were also often salvaged from the funeral pyre to be treasured as relics.[88] Even if Marshall did not fully understand Kandyan burial customs, he still would have understood the significance of taking Keppetipola's skull within British culture. Prior to the 1832 Anatomy Act, the seizure of human remains was associated with criminal bodies and the unsavory practices of the "body snatchers."[89] As a consequence, Marshall is uncharacteristically reserved about what he did with the skull. Following the execution, Keppetipola's head was apparently "placed on his breast," which Marshall believed to be "according to Kandyan custom."[90] Marshall must have taken the decapitated head, complete with its integuments, shortly afterward. But before it could be sent to Edinburgh, the skull needed to be prepared. The hair, skin, and flesh all needed to be removed in order to reveal the bone itself. It would then need cleaning and polishing. In northern India, collectors simply left the decapitated heads of "thugee" criminal tribesmen to be eaten by insects.[91] In the tropical climate of Ceylon, Marshall might have tried a similar approach. However, as a trained surgeon, it seems more likely he would have used the techniques he learned in Glasgow, dissecting away the flesh before dissolving what was left using soda or corrosive sublimate.[92] These violent details are left out of Marshall's account precisely because they conflicted with the imperial ambitions underpinning the carefully staged court-martial and execution. These were moments in which the British aspired to present themselves as civilized Christians in contrast to the barbaric "Malabar" interlopers in Kandy. Burning the flesh from the head of a decapitated Kandyan chief did not fit with this image.

THE SANDS OF TIME

Crawling on his hands and knees, Orlando Felix hauled himself into the abandoned tomb.[93] Surrounded by gold plates and bronze statues, he finally spotted the treasure he had been looking for: the mummified remains of two ancient Egyptians, at least 3,000 years old.[94] Seemingly unconcerned about the cultural damage he might inflict, Felix promptly wrenched the head from one of the mummies, stashing it with the other antiquities he had collected during his year-long trip down the Nile in the late 1820s. Like many other human remains in this period, the head soon found its way to Edinburgh, passing through the hands of numerous collectors. On his return to Cairo, Felix initially gave the ancient Egyptian artifact to his brother, a Royal Navy captain stationed in the Mediterranean following Napoleon Bonaparte's defeat at the Battle of the Nile. But Felix's brother, perhaps understandably, didn't know what to make of this odd gift. So he packaged it up, and had it sent to the Episcopalian minister Edward Craig of St. James's Chapel in Edinburgh. Felix's brother no doubt thought that Craig, trained in the classics at Oxford and attuned to the biblical significance of Egyptian history, would appreciate the artifact more than he. Active in Scottish debates surrounding educational and religious reform, Craig was also familiar with phrenology and soon presented the head, along with a letter from Orlando Felix, to the Edinburgh Phrenological Society.[95]

Like Keppetipola's skull, mummified remains of ancient Egyptians acted as tokens of British imperialism. However, the mechanisms and relations of power in Egypt operated under a very different regime from that of Ceylon. From the late eighteenth century onward, Ottoman Egypt acted as a focal point for British and French imperial antagonism, culminating in Napoleon's invasion in the blistering summer of 1798. The young French general hoped that by securing Egypt he would undermine British trade between the Mediterranean and India, the first step in securing a global French empire from the Atlantic through the Indian Ocean and beyond to the Pacific. But Napoleon's invasion was much more than a military and economic exercise. It was also deeply imbedded in a specific form of cultural imperialism. Over the course of the nineteenth century, Europeans repeatedly reworked their relationship with the Egyptian past and present.[96] For the French, this culminated in the publication of the *Description de l'Égypte* between 1809 and 1829, a multivolume series of folios charting the history, geography, and natural resources of the land. The *Description* worked in tandem with French military power, depicting Europe, and more specifically Napoleon himself, as the legitimate heir to ancient Egyptian civilization. The preface, written by the French mathematician Joseph Fourier, declared that "this country, which has transmitted its knowledge to so many nations, is today plunged into barbarism." Whereas in Ceylon, imperialism relied on a geographic separation between the highlands and the lowlands, in Egypt, colonial power projected a temporal separation between "ancient" and "modern." This division was manifest in the physical organization of the *Description* itself, with separate volumes for pre- and post-Islamic history.[97]

When the combined British and Ottoman forces finally defeated the French in September 1801, they followed Napoleon's lead. The British in particular ensured that all antiquities collected by the French, including the Rosetta Stone, were transferred to London. Together these formed the founding collection of the Egyptian Gallery, which opened at the British Museum in 1808. At the center of the new gallery sat the sarcophagi of Nectanebo II and Hapmen. These ancient cultural artifacts were explicitly portrayed as symbols of a British connection to Egypt through the Bible and classical civilization.[98] Throughout the first half of the nineteenth century, allusions to Egypt popped up time and time again as European scholars worked through ideas of progress. Ancient Egypt even figured prominently in Charles Lyell's *Principles of Geology* as well as in the phrenologically inspired *Vestiges of the Natural History of Creation.* The British hoped to secure control of the region by, in Thomas Carlyle's words, rescuing Egypt from "the wrecks of time."[99]

Felix was at the center of this world. As a military officer, he had fought Napoleon's troops at the Battle of Waterloo in 1815. Following this, he spent a few years in Ireland where, during quieter days, he took the time to study ancient Egyptian history as well as the latest French work on deciphering hieroglyphics. In 1826, Felix arrived in Cairo as part of a British military delegation charged with securing the patronage of the new governor and de facto ruler of Ottoman Egypt, Mohammad Ali.[100] Following his rise to power in the wake of the French surrender, Mohammad Ali had instigated a program of self-conscious modernization, building on the Ottoman "new order" reforms.[101] But Felix was not impressed. On attending a cavalry review he complained that "all the movements were very slowly and very loosely done." He also found Cairo to be a terrible specter of the "antique land" he had read about in historical works. Living in the city was "detestable." The streets were "narrow" and "crowded with camels, donkeys, dogs and horses." Felix did see some evidence of progress in the silk and cotton factories, as well as the new schools established by Mohammad Ali. However, he ultimately concluded that the degenerate state of modern Egypt, in contrast to its glorious ancient past, was a result of the "dislike of the Musselman to improvement." Cairo, like the Orient more generally, was a city of "splendour and rags."[102]

This distinction—between ancient civilization and modern barbarism—shaped both Felix and the phrenologists' attitudes toward human remains in Egypt.[103] On receiving the mummified head, the *Phrenological Journal* explained that the "ancient Egyptians" were "a people remarkable for intelligence, taste, enterprise, and all the elements of civilization."[104] There was a "fair balance of the animal, moral, and intellectual brain."[105] Crucially, cerebral organization provided a novel way to link ancient Egypt to Europe. The phrenologists in Edinburgh argued that "the mummies confirm our uniform experiences, that the Egyptian head belonged to the Caucasian variety of Blumenbach, to which the European also belongs."[106] Additionally, the Edinburgh Phrenological Society compared their new specimen to other Egyptian remains in their collection, explaining that "the same general character is found in all mummy heads which we have yet met with, indicating a European or Caucasian race of the human species."[107] These comparisons relied on access to transimperial networks. The Edinburgh Phrenological Society's museum was ultimately a site in which British, French, and Ottoman colonial violence intersected. Spurzheim had acquired three mummified heads in Paris from French collectors returning from Naploeon's invasion, casts of which he later donated to the Edinburgh Phrenological Society.[108] The museum also held skulls of "modern Egyptians." One of these had been acquired by William Wilde, an Irish surgeon touring the Levant.[109] It had been collected in the aftermath of the siege of Acre in 1832, part of Mohammad Ali's broader campaign against the Ottoman Empire in Syria.[110] On inspecting this skull the phrenologists charted the differences between "modern" and "ancient" Egyptians. In contrast to the mummified heads, "modern Egyptians" were "characterized by extreme narrowness of breadth as compared with the height of their foreheads." In light of this configuration, the contemporary inhabitants of Egypt were said to be a "mixed variety" of "Arab and Negro," something the phrenologists attributed to the licentiousness of the slave trade in North Africa.[111]

Felix chose not to visit the slave market, but he did engage in a series of sexual liaisons. He recalled that, while in Cairo, his "love" had been a "deep copper Abyssinian who spits fire like a cat."[112] However, he quickly grew tired of the city, and when the opportunity presented itself, Felix set off down the Nile, hoping to turn himself into an explorer. In 1826, Lord Prudhoe, the younger son of the Duke of Northumberland, arrived in the city, looking for a reliable companion.[113] By this time Egypt had developed into an important stop on the Grand Tour, with travel books like Robert Richardson's *Travels along the Mediterranean* promoting its "ancient grandeur" among European gentlemen.[114] Prudhoe had attended the University of Cambridge and would have been familiar with Greek and

Roman accounts of Egypt. Collecting antiquities was also an established means to secure favor and promote oneself as a patron back in Britain, with a new generation of collectors acquiring artifacts for the British Museum.[115] Hoping to make a name for themselves, the two men set out from Cairo on 11 December 1828.[116] And it was during this trip that Felix collected the mummified head that later reached the Edinburgh Phrenological Society.

Felix's attention to the material traces of the past paralleled that of the phrenologists. When he arrived at a tomb, Felix would make a sketch of it along with any hieroglyphics. He also made extensive notes, commenting on the relative sophistication of the architecture and the inscriptions. At Jebel Barkal he found the "sculpture and hieroglyphics . . . vile." This, he suggested, indicated that the monuments dated from "Greek, or Roman time" rather than the glorious period of Egyptian "early antiquity."[117] All these details served the phrenologists well. The letter accompanying the mummified head even included a copy of the hieroglyphic tracings Felix had made there. He explained that the character "is the name of Ramses, in the Egyptian hieroglyphic" and that the tomb "had been erected by Ramses II." The letter went on, describing the experience of entering the tomb in detail. Incredibly, Felix suggested that the mummy might be that of the Ramses II himself:

> Under one of the chambers was a small vault, containing two mummies, a man and a woman, richly and completely gilt. . . . In the chamber where the mummies were, the king was dedicator, and no other name appeared. It is always the person to whom the tomb belongs that dedicates it. I therefore thought these might be the mummies of Ramses II . . . and his wife.[118]

Material details like these helped to bolster the phrenological analysis. While Felix initially believed the mummies might be that of Ramses II and his consort, he later admitted that "it is more probable to have been those of some Greek priest and his wife who usurped the chamber." The phrenologists, however, rejected this. The specimen was certainly an "ancient Egyptian" and most likely Ramses II himself. In making this argument, the phrenologists invoked the particular material condition of the head, which was very different from those collected in the frozen Arctic or from an execution block in Ceylon. The *Phrenological Journal* explained that "the mummies excited great curiosity. They are not the bare skulls, but the *heads*—integuments, hair and all, even the eye-lashes visible, and several teeth entire." What's more, the gilding reported in Felix's letter was actually there for the phrenologists to see in Edinburgh. The *Phrenological Journal* described the "rich gilding all over the skin" that "adheres to the dark leathern visages." It was this material evidence that the phrenologists used to counter the idea that the mummies might be Greek or Roman. The phrenologists argued that the head must be Egyptian, as "the Greeks and Romans never embalmed and preserved the dead, and much less gilded them from head to foot."[119]

This attention to the materiality of mummified remains also served to reinforce the phrenological distinction between ancient and modern Egyptians. The *Phrenological Journal* suggested that the head might have "shrunk" over time and that this should be taken into account when conducting an analysis of the development. In fact, "allowing for some diminution of size in 3000 years," the phrenologists argued, the head "must have been . . . above the average European size." A similar interest in the material condition of the mummy played to the racial separation of ancient and modern Egyptians. The mummification process had in fact preserved the hair. Inspecting this closely, the phrenologists noted that it was "of a reddish-brown; curling as it does in the European head, but having nothing of the woolly curl and quality of the hair of the Negro."[120] Throughout this period, debate raged over whether Egyptians were Africans or Europeans.[121] The phrenologists had already suggested that this problem could be solved by separating ancient from modern. Now the phrenologists also invoked a material reading of the mummified remains, drawing on contemporary aesthetic understandings of European and African bodies. The *Phrenological Journal* went on, tying this material argument to a geographic and environmental one:

> The geographical position of Egypt brings it, of the African continent, nearest to that part of Asia which may be called the Caucasian region; between which and Egypt there is a much easier communication, for a migrating nation, than by the deserts which separate Egypt from other habitable parts of Africa.

The article concluded by suggesting that "Africa is large enough to have held other races of men besides Negroes."[122]

The separation of ancient and modern people was also manifest in the very practice of travel and collecting in Egypt. Although Prudhoe and Felix made little reference to their assistants, it is important to remember that their journey depended upon a large Egyptian entourage. The two men actually traveled on camelback, following the course of the Nile, while their supplies and antiquities were transported by boat.[123] They also relied on the military power of Mohammad Ali, whose conquest of Nubia in 1820 had opened up a range of ancient sites beyond the Second Cataract to European explorers.[124] Felix even sometimes traveled with an Ottoman military bodyguard—a janissary.[125] And although Felix rarely mentioned it, most monuments were in fact inhabited during this period by seminomadic peasants and traders.[126] When he did make note of the local population, it served to promote an image of primitiveness. On arriving in Korosko in Nubia, Felix was astounded by the people he met. While he usually made sketches of hieroglyphics or temples, this time Felix recorded the "dance of death of the Upper Nubians" (fig. 6). This image neatly divides ancient and modern, civilized and barbaric. In the foreground, the viewer is presented with wiry figures, in a trance-like state, dancing to the sound of a drum. In contrast, the background depicts an ancient temple alongside an oasis in an otherwise empty landscape. For Felix, the contemporary inhabitants of Egyptian lands were little more than wildlife. In a revealing letter, suggestive of the violence that went

Fig. 6. "The death dance of the Upper Nubians," Orlando Felix, ca. 1829, SD.380, Case 85, Shelf SC, Box 13, Victoria and Albert Museum.

hand in hand with collecting, Felix recalled "how we shot birds and beasts and human beings" and "how we wandered through Thebes and Philae." Felix had in fact been badly injured when a "Nubian" at Korosko, presumably from the same community as those that he had sketched, attacked him with an axe. His life had been saved only when Prudhoe scared off the assailants by firing a rifle.[127]

Felix and Prudhoe's relative inattention to the life of the people they met and worked with served a very particular purpose.[128] The acquisition of antiquities depended upon depicting Egyptians as both uninterested in and unworthy of their own cultural heritage.[129] When the British consul in Egypt, Henry Salt, acquired the colossal bust of Ramses II from Thebes, he did so under the pretense that it had been neglected by both the Ottomans and Mohammad Ali.[130] Prudhoe also complained that the ancient temple at Contra Latopolis, "which was in good preservation last year," had been "utterly destroyed" when he visited again in 1829.[131] The argument was the same with human remains. When Prudhoe and Felix arrived at the immense mortuary temple of Ramses II—the Ramesseum—they were dismayed by the looting. The site was "inhabited by plunderers . . . who are seen emerging from pits and bearing coffins and bodies on their shoulders." Similarly, at Asyut, mummies were "pulled out by dozens in their winding sheets." Of course, Felix himself was engaged in his own act of grave robbery. What mattered, however, was not that the bodies were taken, but rather why. Egyptians traders raided tombs and bodies for "some idols, papyri or trinkets" to sell to "travellers as well as students." It was a "system of spoliation."[132] In contrast, Felix portrayed himself as a learned Christian gentleman. He wasn't in it for the money; rather, the acquisition of human remains was part of a project to link ancient Egypt to Europe through the Bible. Felix had spent the last few years developing a chronological list of the pharaohs, proudly claiming that he had "found the daughter of the pharaoh who saved Moses and the king under whom Exodus took place." Prudhoe concurred, arguing that the study of antiquities would help "prove the correctness of bible history on the subject of Egypt."[133] These attitudes were also reflected in phrenological accounts. The mummies were presented as a material link to the Bible. According to the phrenologists, the Egyptian head in the museum dated from "Joseph's lifetime, 1722 years before Christ."[134]

Irrespective of this pious rhetoric, the study of ancient human remains was not always deemed in good taste, even in Europe. With the rise of "Egyptomania" in Britain, public unrollings and even dissections of mummies became a regular feature of the entertainment circuit.[135] One show, run by the Charing Cross surgeon Thomas Pettigrew in the 1830s, came in for savage criticism. Periodicals described Pettigrew as "the principal unroller on this filthy occasion," while the dissection of the mummy was deemed a "disgusting amusement."[136] Despite the negative association, phrenologists often relied on these kind of shows.[137] In 1835, two members of the local phrenological society attended the unrolling of "Kabooti, an Egyptian Mummy," at the Belfast Natural History Society. While there they took measurements of the "dimensions of the head" of this ancient Egyptian woman, later printed in the *Phrenological Journal*.[138] Wary of bad press, the phrenologists were therefore careful to stress that their interest in human remains stemmed, not from mere "curiosity," but rather an attempt to develop a "Phrenological Standard of Civilization."[139]

Still, it wasn't just British periodicals that phrenologists and tomb raiders like Felix needed to worry about. Egyptians too were increasingly challenging the narratives advanced by European collectors. While Mohammad Ali had initially been relaxed about granting permits to excavate ancient sites, he soon realized that antiquities could be used to bolster his own vision of the modern Egyptian nation-state.[140] In August 1835 Mohammad Ali issued a decree, denouncing European collectors for "destroying ancient edifices, extracting stones and other worked objects and exporting them to foreign countries." If these "foreigners" continued, "soon no more ancient monuments will remain in Egypt." He therefore "judged it appropriate to forbid the export abroad of antiquities found in the ancient edifices of Egypt."[141] The decree also suggested that antiquities should be sent to Rifa'a al-Tahtawi, an Egyptian political and legal scholar, who would organize their display in Cairo. Al-Tahtawi himself had recently argued that Egyptians needed to take ownership of their ancient past in order to emerge as a modern nation, writing in 1834 that "since Egypt has undertaken to adopt civilization and instruction on the model of European countries, it would be better to preserve the ornaments and works which their ancestors have left them." Terms like "foreigners" and "ancestors" helped to challenge European accounts that separated modern Egypt from the ancient world. Clearly, Egyptians were just as capable of using antiquities to define the racial bounds of both the nation and civilization.[142]

Comparing the histories of skull collecting in Egypt and Ceylon reveals a shared phrenological fascination with death. However, the differences between these two imperial contexts are also stark. In Egypt, colonial expansion relied on a separation between "ancient" and "modern" peoples. Phrenologists adopted the same distinction when examining human remains, hence the excitement surrounding ancient Egyptian heads. In contrast, British colonialism in Ceylon relied on a geographic separation, between the highlands and the lowlands. Phrenologists believed that the rebels in mountainous Kandy represented a distinct race. The contrast between Ceylon and Egypt was also environmental. For a science obsessed with materiality, the physical condition of human remains carried tremendous weight. The fact that ancient Egyptian heads were so well preserved allowed phrenologists to make further claims based on fragments of skin and hair. Finally, there were also important differences between the ways in which Egyptian and Kandyan people resisted and responded to phrenology. In Egypt, Mohammad Ali began to use antiquities, including human remains, to forge a new sense of national identity. In Ceylon, however, Kandyan agency was limited by British military power. Keppetipola certainly had his own conception of what death meant, but this did not stop his skull from ending up in the hands of the phrenologists. Nonetheless, by recovering both sides of the encounter we can better understand the ways in which phrenologists read and misread other cultures. In Edinburgh, Keppetipola's skull and ancient Egyptian remains sat alongside specimens from across the colonial world. Among these were six Inuit skulls. By following phrenology to the Arctic, we can once again get a sense of how racial science stretched across imperial spaces, as well as how it varied.[143]

THE FROZEN NORTH

Royal Navy explorers both collected Inuit skulls and buried them. In July 1821, Captain William Parry arrived on the Upper Savage Islands, opting to anchor HMS *Fury* in light of unfavorable conditions. He ordered his crew to make "the usual observations for longitude and variation," before setting off to climb a nearby hill. Accompanied by Captain George Lyon of HMS *Hecla*, Parry surveyed the Arctic landscape. In the distance, the two naval officers spotted "the remains of some Esquimaux habitations." These seemed worth a closer look. On descending the hill, and approaching the abandoned Inuit winter settlement, Parry came across "small rude circles of rough stones" amid which he found "one human skull."[144] Without hesitation, he picked it up and took it back to HMS *Fury*, where it joined a growing collection of natural history specimens—narwhal tusks, polar bear skins, and samples of limestone.[145] Parry and Lyon spent the next two years in the Arctic, trying to find a way through the elusive Northwest Passage. When the ice set in and they could make no further progress, the two explorers set up camp and lived among the Inuit, hoping to push on through the following summer. During the winter, Parry and Lyon became closely acquainted with the Inuit, undertaking detailed observations that were later published in John Murray's official account of the voyage. Death and burial in particular prompted the two naval officers to reflect on the differences between European and "Esquimaux" character. On at least two occasions, Parry and Lyon directly assisted with the burial of a deceased Inuk. It was also during these long periods of overwintering that they collected additional skulls. On his return to Britain in 1823, Parry donated these specimens to the Edinburgh Phrenological Society. He probably learned about phrenology from Lyon, who corresponded with Combe, or from John Ross, a fellow Arctic explorer who attended meetings of the Edinburgh society.[146] Eager to record the provenance of the new specimens, the museum conservator, Cox, once again took out his pen. Across the crown, Cox inked the words "Esquimaux, Found in the Snow, by Captain Parry."[147]

Although it only occasionally features in surveys of imperial history, the Arctic was an important site for early nineteenth-century colonial interests.[148] Davis Strait was a major commercial whale fishery, providing significant revenue in the form of whale bones and blubber.[149] Greenland was also under Danish colonial rule.[150] Most significantly, the Northwest Passage represented a potential means for Britain to project itself more effectively into the Pacific, something that seemed increasingly pertinent in light of the loss of the American colonies. The project was deemed so important that the Longitude Act was amended to provide a staggered reward for navigating the Northwest Passage: £5,000 for reaching 110° W, £10,000 for 130° W, and £20,000 for a complete voyage.[151] Between 1819 and 1825, Parry made three separate attempts to navigate the Northwest Passage. But despite the potential reward, this was no easy task. Parry understood better than most that success would depend on working closely with the Inuit themselves. He relied on them for shelter, for food, and, crucially, for geographic information.[152] It is only by recognizing this particular relationship of interdependence, that we can properly account for Parry's collection of Inuit skulls and the phrenologists' assessment of their character.

In January 1823, nearly two years after collecting that first skull, Parry and Lyon were overwintering with the Inuit at Igloolik. There was no prospect of making any progress through the ice, with the sea frozen and temperatures dropping to as low as -23°C. During this time, Lyon met an Inuit woman, Pootooalook, who had fallen seriously ill. She was found in her snow shelter in an "extremely debilitated state," unable to suckle her young child. The child itself, "about three years of age," was unwell and "almost starved." In keeping with the relationship of gift exchange that had developed in the Arctic, Lyon suggested to Pootooalook's husband that he come to HMS *Hecla* to pick up some medicines. But things took a turn for the worse. Pootooalook's husband arrived the next day, reporting that his wife's condition had deteriorated. Lyon then insisted that Pootooalook herself be brought to HMS *Hecla*. According to Parry, Pootooalook's husband "joyfully assented" and was supplied with "a sledge and dogs" to assist him. Keen again to emphasize British benevolence, Parry described how the mother and child were "comfortably lodged in Captain Lyon's cabin, and attended with all the care that their situation required." However, despite some signs of improvement, Pootooalook passed away aboard HMS *Hecla* on 24 January 1823. Her baby daughter died two days later.[153]

The cause of death itself generated much discussion. Parry and Lyon, in keeping with European attitudes more generally, considered the Inuit to be unclean. Parry recalled the "indescribable filth of the passages" he had crawled through to enter Inuit snow shelters. In an article entitled "On the Character and Cerebral Development of the Esquimaux," the *Phrenological Journal* also explained that "the habits of Greenlanders and other Esquimaux tribes are described on all hands as most filthy and disgusting." This, the phrenologists argued, was tied to a "deficient" organ of Order.[154] As was the case in Ceylon, the environment itself was understood as a potential source of disease. Parry explicitly blamed the "coldness and moisture" of Pootooalook's dwelling for her illness.[155] When Pootooalook was brought to HMS *Hecla*, she was attended to by the ship's surgeon. Lyon then described how he introduced "a system of cleanliness." This was designed to combat the "most filthy state" that Pootooalook and her child had been found in. Lyon carefully washed the Inuk woman's face, which was "covered with so thick a coating of dirt," and provided "warm broth" and "dry bedding." These actions were undoubtedly done out of personal affection, as well as a broader sense of Christian charity. They also reflect the very close relationship that developed between the Inuit at Igloolik and British explorers. But whatever his motivations, Lyon's interaction with Pootooalook, both before and after her death, invited a contrast between Inuit and European character. Parry himself remarked that "the circumstances attending the death and burial of this poor woman and her child" offered "an insight into some of the customs of the Esquimaux."[156]

Both the phrenologists and naval explorers believed the Inuit to be exceptionally coldhearted. This was a theme that was repeated time and time again both in the Arctic and back in Britain. It also later served to justify the acquisition of Inuit skulls. Parry and Lyon, invoking Christian ideals of mourning and burial, were shocked by the Inuit response to Pootooalook's death. Parry recalled how Pootooalook's sister, Ootooguak, had failed to conduct herself in a manner consistent with his expectation of decorum. She was found "laughing on deck, . . . not caring to hurry herself to come to the house of mourning." Worse, Ootooguak seemed "in high spirits, laughing and capering on deck as if nothing had happened." Lyon generated a similar impression when a group of Inuit boarded HMS *Hecla* the day after Pootooalook's child had died. The body of the deceased, which Lyon had carefully wrapped in a blanket, "excited neither disgust nor any other feeling amongst them more than a block of wood could have done."[157] These attitudes all fed back into phrenological assessments of Inuit character. The *Phrenological Journal* even quoted directly from Lyon's account of Pootooalook's death, explaining that the Inuit possessed a "peculiarly unfeeling nature." This apparently matched a "feebly manifested" organ of Benevolence found on the skulls collected by Parry.[158]

If Parry and Lyon had been unsettled by Inuit attitudes to mourning, they were in for another shock when it came to burial. Arctic graves were deemed woefully inadequate—another example of Inuit callousness. Because of the difficulty of digging in the ice, the Inuit typically buried their dead in very shallow graves covered with a mound of stones and snow.[159] This certainly did not fit the moral and aesthetic values promoted back in Britain, with the growth of landscape cemeteries and magnificent tombs modeled on Père Lachaise in Paris.[160] The two Christian gentlemen watched in horror as Pootooalook's child was buried in a grave "not above a foot deep" with just "a few loose slabs of snow" heaped over the body. Lyon noted disparagingly that "a fox could have dug through it in half a minute."[161] He even questioned whether the term "grave" should really be applied to "so insecure and rude a covering for the dead." Lyon was so disturbed by the burial that he later had his men disinter the child, "without any one being the wiser." The tiny body was fitted with weights and buried at sea. Lyon explained that "I deemed this requisite, lest the general thaw, which was soon expected, should leave the poor little creature a prey to wolves and dogs." Once again, the environment mediated attitudes to burial. On another occasion, this time following the death of a man named Pekooya and a woman named Kaimookhiak, Lyon noted that "the snow was so shallow, that one day's strong thaw would leave them lying bare on the ground." He later recalled grimly that "the bodies were nearly picked clean of the flesh as soon as discovered by the dogs."[162] According to Parry, "None but the immediate relatives of the deceased cared a jot about the matter."[163]

The phrenologists, like Parry and Lyon, read Inuit death rituals as evidence for a lack of compassion. But they also tied burial to an assessment of religious sentiment. The *Phrenological Journal* argued that the "Esquimaux" lacked properly developed "religious feelings." According to the phrenologists, the lack of care for the dead indicated that the Inuit "have no religion or superstition whatever, and not the most distant idea of a future state." Only a race that did not believe in heaven could treat a body so poorly. Most worryingly, the Inuit were said to have "no idea of a beneficent Supreme Being." The phrenologists then linked this assessment of Inuit religious life directly to the contours of the skull. The organ of "Causality," the *Phrenological Journal* explained, "leads mankind to infer a Presiding First Cause from the marks of wisdom which every where present themselves in the material universe." In short, phrenology explained why only some races were capable of undertaking natural theology. Tellingly, the organ of Causality showed only a "moderate" development in the Inuit skulls in the Edinburgh Phrenological Society's collection.[164] This, especially when combined with a deficient organ of Benevolence, explained to the phrenologists why the Inuit were not inclined to properly care for the dead, at least in a manner consistent with Christian expectations.[165]

These attitudes all reinforced the very system of collecting that the phrenologists relied upon. In a move reminiscent of the grave robbers in Egypt, Parry and Lyon justified their acquisition of skulls on the basis that the Inuit did not care for the dead and did not believe in an afterlife. The Inuit at Iglooklik even apparently helped Parry collect skulls. Once again emphasizing their lack of compassion, Parry recalled how

> our new friends, who not only treated the matter with the utmost indifference, but on observing that we were inclined to add some [skulls] to our collections, went eagerly about to look for them, and tumbled, perhaps, the craniums of some of their own relations into our bag without delicacy or remorse.

With only a hint of irony, Parry concluded by remarking on "how little pains these people take to place their dead out of the reach of hungry bears or anatomical collectors."[166] On other occasions, the collection of skulls followed from the practice of exploration. The Inuit built stone cairns—or *inuksuit*—to serve multiple functions. These landmarks identified places, paths, and animal migration patterns. *Inuksuit* also acted as grave markers.[167] Pootooalook's body, for instance, was covered with "a quantity of heavy stones."[168] Parry and Lyon frequently relied on these *inuksuit*, and the accompanying Inuit geographic knowledge, to guide them through the Arctic environment. Shortly after discovering a skull on the Calthorpe Islands, Parry described "a curious path made by the natives, two feet in width, and formed by removing the stones in places where they were naturally abundant."[169] Lyon even made a sketch, later printed under the title "An Esquimaux Grave," featuring the very *inuksuk* he had used to find his way (fig. 7). The image also depicted two wolves cradling a human skull. By following Inuit paths, Arctic collectors were also following Inuit bodies. This reliance on local geographic knowledge also points to a broader need to incorporate Inuit perspectives in writing histories of collecting in the Arctic. Contrary to the assertions of the phrenologists, the Inuit did believe in an afterlife, and their rituals surrounding death certainly did not indicate an unfeeling nature. The Inuit also reflected on British at-

Fig. 7. "An Esquimaux Grave," in George Lyon, *A Brief Narrative of an Unsuccessful Attempt to Reach Repulse Bay* (London: John Murray, 1825), 68. Reproduced by kind permission of the Syndics of Cambridge University Library.

titudes to burial, which they too found unsettling. Understanding these conflicting cosmologies surrounding human remains helps to explain exactly why naval explorers and phrenologists came to the conclusions that they did. It also illustrates the existence of a counter narrative that was carefully written out of accounts produced in the museum.

While Parry and Lyon were shocked when they discovered bodies mauled by wolves, the Inuit understood both wild and domestic animals as an integral part of burial culture. Dogs in particular occupied a special place in Inuit life. They provided companionship, acted as a means of transport, and assisted with hunts. During the winter, if supplies ran low, dogs were even sometimes eaten. Unlike other animals, dogs were also given individual names. Tellingly, they were often given the name of a recently deceased member of a tribe. Inuit burial rituals were borne out of this close relationship with animal life, as well as the Arctic environment itself. It simply was not possible to bury the dead deep in the rock and ice. Dogs and wolves provided a practical and spiritually coherent means to dispose of human remains. This practice was even reflected in mourning rituals, in which dogs would be left deliberately unfed for three days following the death of a relative. Far from implying a lack of sensitivity, the Inuit ensured that the bodies of their loved ones were properly integrated into the wider community, a community that included animals.[170] Building *insuksuit* was also understood as a means to honor the dead. A large stone cairn could simultaneously act as a grave and a waypoint, one that generations might follow to and from a camp long after the death of an ancestor.[171] Again, Inuit burial practices were premised on reconnecting the deceased with the world of the living, whether through animals or geographic knowledge.[172]

What then did the Inuit think of European rituals surrounding death? In June 1822 Parry organized the burial of two British sailors who had died the previous month. James Pringle of HMS *Hecla* had fallen from the mast and cracked his head, while John Reid had passed away following a painful inflammation of the lungs. Parry was determined that the two men should have a Christian burial. He recalled how, in contrast to the Inuit, the crew of HMS *Hecla* had constructed "a handsome tomb of stone and mortar." In order to ensure that no animal could access the bodies, the mound was carefully plastered over. It seems Parry might even have been thinking of the garden cemeteries of Europe. He ordered his men to cover the top of the graves "with tufts of purple saxifrage," a colorful perennial. Each man was also provided with a headstone inscribed with his name and age. Parry recalled how the ceremony had been conducted "with every solemnity that so mournful an occasion demanded," culminating with a volley of gunfire and a service aboard HMS *Fury*.[173] Fittingly, the Inuit were not impressed with any of this performance. Headstones looked like very poor *inuksuit*. They were small in comparison to a typical Inuit cairn, and, most confusingly, they seemed to encode no geographic information. Without this, British graves failed to properly integrate the dead back into the daily practices of the living. In later ethnological accounts the Inuit described British tombs as "skinny." The Inuit could tell these "inskshuks" had been "built by white men," as they were "pointing westwards," seemingly to nowhere. It is also precisely because *inuksuit* acted as repositories of geographic knowledge that the Inuit were so outraged when the British tampered with their graves. No wonder Parry never made it through the Northwest Passage. After all, disturbing an *inuksuk* was believed to bring bad luck.[174]

Understanding these cosmologies also helps make sense of Parry's account of the Inuit collecting skulls for him. This was not simply an example of colonial mythmaking, although Parry misread the motivation. Oral histories suggest that the Inuit at Igloolik really did offer up some skulls. (It is, however, important to recognize that many other skulls were acquired without consent, particularly those taken from *inuksuit*.) Nonetheless, Parry's interest in human remains made sense to the Inuit in terms of their own creation story. Animals again figure prominently. The Inuit believed that "there once was a girl called Uinigumasuittuq who was married to her dog. . . . She gave birth to six babies—two were Inuit, two were intimidating half-Indian half-dogs, and two were half-white half-dogs." Variants of this story are found across the Arctic, in which European and Inuit people are separated from one another in time and space, although sharing a common origin. Clearly, the Inuit were developing their own understanding of human genealogy at the same time as the phrenologists. "The two half-white half-dog babies were put into the sole of a kamik with two stems of grass and let go into the ocean. . . . There were only the two babies in the boot sole—a girl and a boy—but that's how the white people multiplied." Far from being passive recipients of European explorers, the Inuit saw themselves as possessing incredible agency. They had "created the white people." It was within the context of this creation story that the Inuit came to terms with Parry's obsession with human skulls. In direct contrast to the phrenological account, the Inuit actually saw their offering as an act grounded in compassion. "Paarii and his people came around here for the skull of their mother." The Inuit offered a select number of human skulls, which they understood as genealogically linked to Europeans, in order that Parry might reconnect with his Inuit ancestors.[175] The people of Igloolik, it seems, understood the significance of repatriation well before the British.[176]

Encounters with death and burial provided an extraordinary opportunity to both acquire human remains and make cultural assessments about the races of man. By examining three contrasting contexts, this chapter has demonstrated how different physical environments, imperial politics, and forms of colonial violence played into phrenological understandings of race. Recovering these collecting practices also meant recognizing both sides of the encounter. Contrary to the assertions of scholars such as Susan Pearce, there was no "European tradition" of collecting.[177] Even as their bodies were being seized, colonized people exerted their own agency, again, in a variety of ways. Many rejected Christian understandings of death and mourning, as illustrated in examples from Kandy and Igloolik. Furthermore, the idea of a "European tradition" of collecting was itself a nineteenth-century construction. Men like Orlando Felix wrote colonized people out of their accounts precisely to justify the acquisition of skulls. Once again, these narratives did not go unchallenged. Nineteenth-century Egyptians increasingly incorporated human remains and ancient artifacts into their own national identity, while twentieth-century Sri Lankan nationalists successfully campaigned for the return of Keppetipola's skull.[178] More broadly, this suggests that historians of science in general, and historians of collecting in particular, need to apply the lessons of historical anthropology if we are to make sense of the production of colonial knowledge.[179]

Encounters with death and burial proved a particularly powerful moment in the social life of these objects. But the acquisition of a skull was in fact just the start of a very long journey. In the Arctic, Inuit skulls were stored for years aboard HMS *Fury* as Parry tried to make his way through the ice. They spent the time sitting alongside a vast array of navigational equipment, partly explaining why the phrenologists were able to note down the exact longitude and latitude of the islands from which they had been taken. Six artificial horizons, seventeen chronometers, three theodolites, and one "forty-inch, triple-object glass, achromatic Telescope" accompanied Parry to the Arctic.[180] Other skulls moved through the hands of multiple collectors before finally arriving in the museum. The "twenty-one skulls . . . of different Asiatic tribes" presented by George Mackenzie to the Edinburgh Phrenological Society had in fact been acquired by his son, William, an East India Company officer in Madras. William Mackenzie had not collected the skulls himself though. Instead, he had paid "a native on whom he believed he could rely" to recover these human remains "with great care from the burying-places of the respective castes." The *Phrenological Journal* admitted that, with the specimens having changed hands between at least two collectors already, their "authenticity" might be in question. Nonetheless, the skulls were deemed "unquestionably Asiatic."[181] Even when an object entered the museum, this was not the end of its journey.[182] Skulls were polished, written on, inspected, and often mounted on a pedestal. Some skulls were even sawed in half, and then held together with a hook, allowing curious phrenologists to inspect the inside.[183] A number of skulls also went in and out of the museum, accompanying Combe on his international lecture tours to the United States and continental Europe. The Inuit skulls collected by Parry in the Arctic were later held aloft to audiences in New York City.[184] Combe even lent them to the phrenological enthusiast Samuel George Morton while in Philadelphia. Over a few weeks Morton measured the skulls and had an artist produce engravings that later featured in *Crania Americana*.[185] Recognizing their value, Combe reminded Morton to "box & send them" back to him before he returned to Britain.[186] Still, as they were unique specimens, there was a limit to how useful skulls really were for communicating phrenological knowledge. A skull certainly made a good impression during a lecture, and the rhetorical force of a museum collection was also important for establishing phrenology as a science of man. But, by itself, a skull couldn't reach a global audience. As the following chapter explores, phrenological theories of race might have been forged in the context of colonial skull collecting, but they were reproduced and recirculated as plaster casts and prints.

CHAPTER 2

Casts

Eighteen Aboriginal warriors locked in deadly combat—it was a sensational sight. At ten o'clock each morning, Philemon Sohier opened the doors of his Phrenological Museum in Melbourne. For many, the highlight was "the Native Black's Revenge," a haunting diorama depicting intertribal warfare in the Australian outback. Every day crowds of people paid one shilling each to see the incredible display of plaster and wax casts that Sohier had been manufacturing since the early 1850s. Few could doubt their authenticity. These models had been "made from nature, from casts taken by Professor Sohier, before many witnesses." The Aboriginal Australians stood in the same room as "Patara, the New Zealand Chief." Sohier's audience listened attentively as they learned of Patara's apparently violent past. He had been "the principal in the murder of the Rev. Volkner," a German missionary. Curious visitors willing to venture upstairs were then greeted with another phrenological account of savagery. "Chong-Sigh" and "Hing-Tzan," two Chinese immigrants, had been found guilty of murdering a European woman in Melbourne. The catalogue recounted how "these two monsters murdered this unfortunate girl while she lay in a state of senseless intoxication." Their victim had apparently been found with her head severed from her body. It wasn't long before the popularity of Sohier's Phrenological Museum extended beyond Melbourne, with copies of the casts sent to New South Wales for an exhibition in Sydney.[1] Similar phrenological museums opened in Paris, London, and Boston, many of them exchanging plaster casts in the hope of presenting a complete history of the different races of man.[2]

This chapter argues that phrenological theories of racial difference circulated in a material world of manufacture and exchange. Whereas the previous chapter concentrated on the making of racial knowledge in relation to colonial collecting practices, this chapter uncovers how nineteenth-century racial theories were reproduced and circulated. Plaster casts manufactured in Edinburgh and Paris were packaged up and loaded aboard mail coaches, trains, and steamers bound for India, the United States, and the Pacific. Not all of them arrived in one piece. Race was not just a "concept" or an "idea."[3] Rather, understandings of race and nation were forged and contested as they circulated in material form.

For the phrenologists, plaster casts represented perhaps the most tangible embodiment of their global movement. London shop windows soon featured phrenological paraphernalia, with the *Phrenological Journal* declaring that "no one can walk along the streets of the metropolis and not be struck with the number of situations in which *phrenological busts and casts* are exposed for sale."[4] By 1836 Hewett Watson estimated that over 15,000 casts had been produced in Britain alone. "The numbers both of the originals and the duplicates is rapidly on the increase," Watson reported in his *Statistics of Phrenology*.[5] Phrenological busts were even on display at the Great Exhibition of 1851 in Hyde Park.[6] Before long, similar displays could be found across the world. During his tour of the United States, George Combe recalled coming across "shops in which Phrenological casts are extensively sold" in Philadelphia, Boston, and New York City.[7] In Melbourne, over 100,000 visitors a year came to see Sohier's Phrenological Museum in the 1860s.[8] Back in Britain, any budding phrenologist with a few shillings to spare could purchase a copy of a cast from Luke O'Neil and Son. Operating out of 125 Canongate in the Edinburgh Old Town, Luke and Anthony O'Neil ran an international business. Trading on their access to the vast collections of the Edinburgh Phrenological Society's museum, they cultivated a reputation for "authentic" specimens, with customers writing to them from Denmark, the United States, and India.[9] An entire series of thirty-nine casts comprising the "skulls of different nations" would set you back £2 15s. Alternatively, individual busts could be purchased for between 1s. 6d. and 2s. 6d.[10]

Reproducing phrenological knowledge in plaster was by no means straightforward. When the American physician Charles Caldwell visited Britain in the summer of 1821 he promptly purchased over thirty busts from O'Neil and Son. Worried that they might not survive the journey back to Kentucky, Caldwell warned that "it is essential that the cask containing them be very strong, and that they be packed in the most skilful and secure manner."[11] The busts did arrive intact, but other phrenologists were not so fortunate. When the editor of *Zeitschrift für Phrenologie* in Heidelberg ordered a series of casts in 1843 they arrived "smashed to pieces."[12] Combe himself experienced similar problems during his lecture tour of the United States. Shortly after arriving in Connecticut he was dismayed to find that "very many of my casts are broken, or dreadfully worn through rubbing."[13] And in Philadelphia, where he allowed the Fowler brothers to make copies from his collection, Combe complained of "the great damage which you have done to my casts in taking duplicates." As a result of "sheer carelessness," the Fowlers had "broken off . . . Napoleon's nose."[14]

Not only were casts liable to break, but they were also difficult to manufacture in the first place. Producing casts required the skills of specially trained artisans along with materials that were often hard to access, except in newly industrializing cities. If you wanted to be immortalized in plaster, Bristol was a good place to die. The quarries on the banks of the Severn provided an abundant source of calcium sulphate. Ground in local mills, and sold in brown paper bags by the half peck, this white powder was the most common form of raw plaster used in Britain at the time.[15] Even then, it needed careful preparation in order to prevent the development of air pockets.[16] But again, Bristol housed a burgeoning community of skilled practitioners including Cosimo Cosimi and Charles Lucassi.[17] Like Anthony and Luke O'Neil, they were Italian migrants from northern Lucca, traveling to Britain in search of work following the end of the Napoleonic Wars.[18] It is precisely for this reason that a number of casts, chief among them the bust of the Indian social reformer Rammohun Roy, were manufactured in Bristol.[19] Edinburgh was another leading center of production, as was Paris, with its gypsum mines surrounding Montmartre.[20]

Without access to the materials or skills needed to manufacture plaster busts, phrenologists outside of Britain and France were dependent upon European imports. Prior to the rise of the Fowlers' phrenological business empire in America, practitioners in the United States routinely purchased casts from O'Neil and Son.[21] In the 1830s, an article entitled "Method of Moulding and Casting Heads" was even printed in *Annals of Phrenology* in Boston with the hope of developing local expertise. Most colonial phrenological societies also continued to rely on European imports. After waiting for over three years, the Calcutta Phrenological Society finally received "two busts with organs marked" from Edinburgh. The postage alone had cost over £1.[22] Phrenologists in Melbourne were equally dependent, at least until the 1850s. However, following the discovery of gold at Bendigo in 1851, a new wave of European immigrants arrived looking to make their fortune.[23] When the gold rush ended, some of them turned to phrenology. Maximilian Kreitmayer and Sohier, the two artisans responsible for the majority of busts manufactured in Melbourne, had both trained in Europe. Sohier was the brother of a French clergyman, while Kreitmayer was German, although he had mastered his art at St. Bartholomew's Hospital in London.[24] Understanding these divergent geographies of manufacture is crucial for appreciating the material limits of phrenology as a global science.

Despite these difficulties, phrenologists from New York to Melbourne manufactured thousands of plaster casts of skulls, heads, and brains over the course of the nineteenth century. A considerable number were presented as material evidence of racial difference. These included plaster copies of skulls such as those described in chapter 1. This process of reproduction was essential for turning these troublesome specimens into legitimate objects of inquiry. Plaster allowed phrenologists to present what they described as an "average specimen." When the Edinburgh Phrenological Society published a "catalogue of casts of skulls of different nations," they explained that each skull had been "selected from a number of the same tribe or nation, so as to present, as nearly as possible, a type of the whole."[25] The authority of the cast was further reinforced through its reproducibility.[26] Most fledgling phrenological societies, such as those in Glasgow and Warrington, purchased a collection from O'Neil and Son, rather than procuring skulls themselves.[27] Adopting the language of print, Combe described these as "published" specimens.[28] In the end, the material and rhetorical production of an "average specimen" allowed the phrenologists to mitigate the questionable authority of their original collections. While individual accounts might be untrustworthy, a cast purchased from O'Neil and Son was sure to illustrate "the general qualities of that variety of the human species."[29]

This kind of thinking, in which an individual is presented as representative of some homogeneous group, will be familiar to historians of race. But it does not tell the full story. When Rammohun Roy passed away in Bristol, his physician, the phrenological enthusiast John Estlin, arranged for a cast to be taken of his head. Significantly, the bust of Rammohun was considered noteworthy precisely because he was believed to be unique. The *Phrenological Journal* in Edinburgh explained that the bust represented "the character of that distinguished Hindoo,—so different from that of his countrymen in general."[30] By preserving Rammohun's likeness on his deathbed, right down to his mustache, the phrenologists marked him out as materially distinct (fig. 8). This was a common strategy, in which phrenologists produced both an "average specimen" and a contrasting case: exceptions that effectively proved the rule. When the Sauk leader

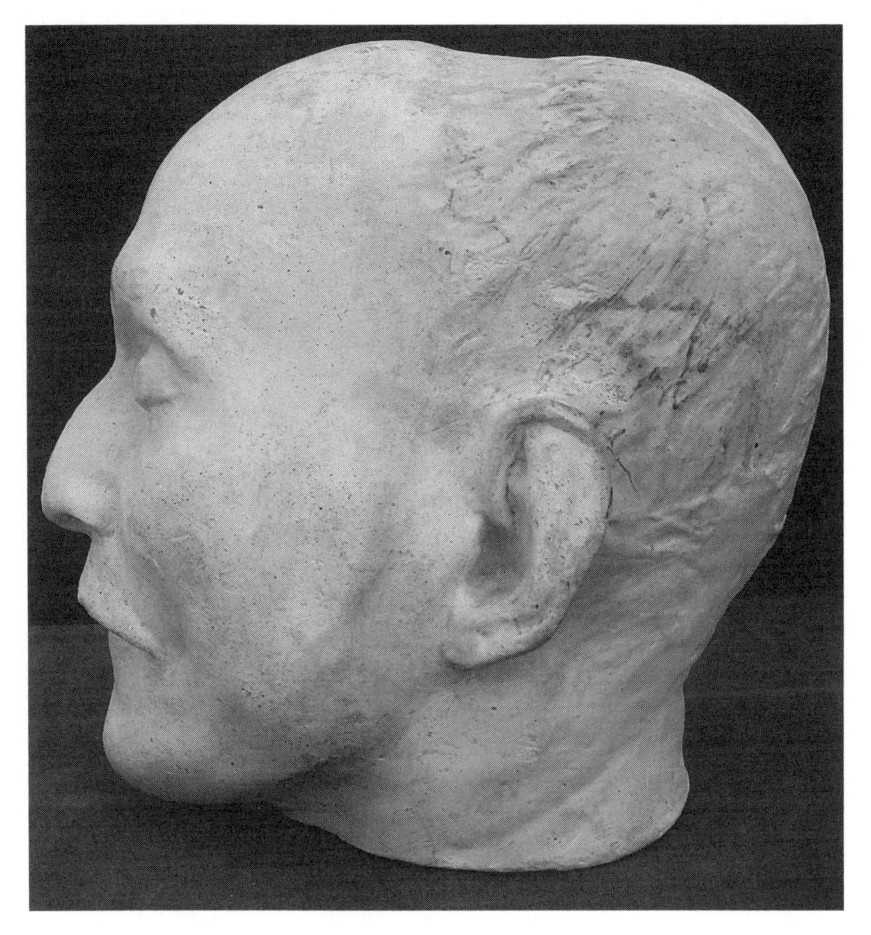

Fig. 8. Plaster cast of Rammohun Roy's head, 1992-34/32. Courtesy of Science Museum, London.

Black Hawk was captured in 1832 following his resistance to the Indian Removal Act, he was subject to one final indignity. The Fowlers in New York City were given permission to take a cast of his head.[31] Like Rammohun, this highly individualized representation was routinely contrasted to the typical "Indian." Black Hawk had an head comparable to "the better class of white men," so much so that "in him you see the Indian character in its best light."[32]

Picking up on this understudied aspect of the history of race, this chapter follows a single phrenological cast manufactured in Paris in the early 1830s. Taken from a Haitian slave named Eustache Belin, this bust provides a rare opportunity to examine how understandings of African character were constructed and contested as they traveled. The phrenologists who examined Eustache's bust were united in identifying him as an exceptional individual, unlike any other slave they had seen. But despite this apparent consensus, attitudes toward African character and its relationship to slavery were anything but uniform.[33] As copies of the bust traveled between Paris, Edinburgh, and Philadelphia, Eustache's character was reworked time and time again to serve different political goals, from the maintenance of West Indian slavery to the campaign for immediate abolition. In the end, it is only by examining this world of material and political exchange, a world that stretched across the British, French, and American empires, that we can understand the development and uses of nineteenth-century racial thinking.

REMEMBERING HAITI

Eustache experienced the violence of both slavery and revolution firsthand.[34] Born a slave in the French colony of Saint-Domingue in 1773, Eustache spent his youth toiling in the sugar mill on the estate of Paul Belin de Villeneuve near Limbé in the northern part of the island.[35] Villeneuve's plantation was characteristic of the increasingly brutal and industrialized system of slavery operating in the West Indies toward the end of the eighteenth century, with Saint-Domingue producing nearly a third of the world's sugar.[36] Each slave on the estate was branded on the right chest with their master's name: VILLENEUVE.[37] And in the sugar mill itself, Villeneuve introduced a new machine for processing the crop. The machine featured a double roller, increasing the yield as the sugarcane made two passes through the mangle. But mechanization did not eliminate the misery associated with sugar farming. In fact, the new system absolutely relied on slave labor, as Eustache or one of his companions needed to manually pull the sugarcane back through the mangle and feed it into the second roller. It was not uncommon to lose a hand.[38] Many of the slaves on Villeneuve's plantation ultimately decided that life would be better on the run, and fled to the hills to the east. Labeled "maroons" by the French, runaways from Villeneuve's estate included "Jean-Pierre," born a "créole" at Gros-Morne on Saint-Domingue, and "Éveillé" who had been transported from the Congo.[39] Together, these fugitive slaves formed an essential part of the resistance that culminated in the Haitian Revolution of 1791.

The years that followed witnessed escalating levels of violence and instability on the island. In March 1792 the gens de couleur of Saint-Domingue were granted full political rights by the National Assembly in Paris in order to persuade them to fight with the white planters against the rebel slaves.[40] But allegiances quickly shifted. In 1793 the white population took up arms against the French troops stationed on the island, arguing that the revolutionary ideals of 1789 had ultimately undermined the stability of the colony. When the commissioner, Léger-Félicité Sonthonax, arrived from Paris, charged with bringing the crisis to a close, he found the greatest threat was no longer the slaves but the white planters themselves. In August 1793 Sonthonax decided that he had no option but to abolish slavery on Saint-Domingue in order to guarantee the allegiance of the black population against the antirevolutionary whites. Years of violence followed as different groups competed for control of the island. The Spanish invaded from the east, and the revolutionary slave leader, Georges Biassou, vied for control of the west. Eventually Biassou's former ally, the general Toussaint Louverture, secured control of the French colony, defeating both Biassou and a British expeditionary force. However, with Napoleon Bonaparte's rise to power in 1799, the relative stability brought about by Louverture's rule soon came under threat. The new emperor of France quickly reestablished slavery in the remaining French West Indian colonies and was determined to secure control of Saint-Domingue once again. But although Louverture was captured in 1802, Napoleon's forces were ultimately defeated, and Jean-Jacques Dessalines declared Saint-Domingue an independent nation, free from slavery, under the aboriginal name Haiti in November 1803.[41] Over the course of just thirteen years, over 300,000 people had been killed.[42]

When Eustache arrived in France in 1812, these events were still fresh in the minds of the political and intellectual elite. Eustache became a curiosity in early nineteenth-century Paris, not because he was black, but because he was a participant in a defining moment in French history, one that continued to shape the meaning of both the French Revolution and the politics of race.[43] Had he taken part in the violence of 1791? Did he fight for or against Louverture or Biassou? Had he witnessed the burning of Cap Français in 1793, in which 10,000 people died? And how had he reached Paris unscathed? Incredibly, stories started to emerge suggesting that Eustache was not like any other slave. Newspapers began to report that Eustache, rather than turning against his ruthless master, had in fact helped Villeneuve escape the violence of 1791, risking his own life in the process. Eustache's apparent rejection of the rebellious conduct of his fellow slaves was often repeated in the Parisian press between his arrival in

1812 and the eventual abolition of slavery in the French Empire in 1848. Often referred to as "le bon nègre," he appeared in didactic moral works such as *Les Bons Exemples* as well as *Leçons de Philosophie Morale.*[44] Apparently convinced by these reports, as well as Eustache's own testimony, the Institut de France awarded the former slave the Prize of Virtue in August 1832. In its report the Institut reinforced Eustache's status as an exception, contrasting both his thoughts and his actions with those of the other Haitian slaves. According to the Institut, "The idea of murder . . . did not associate itself, in the mind of Eustache, with that of liberty."[45]

Whatever Eustache's actions back in the West Indies, his seemingly unique constitution was carefully constructed and contested in order to serve different intellectual and political ends, both in Paris and beyond. The improvability or otherwise of African character was at the center of disputes concerning slavery during the early nineteenth century.[46] And with this came a variety of subtly different racial theories promulgated by leading French naturalists. The physician Julien-Joseph Virey, a firm polygenist, argued that "negroes do not civilize themselves" in the 1824 edition of his *Histoire Naturelle du Genre Humain*. The work was accompanied by engravings comparing the facial profile of Europeans, Africans, and apes.[47] It was also during Eustache's time in Paris that Georges Cuvier examined Sara Baartman, the Khoikhoi woman known then as the "Hottentot Venus," at the Jardin des Plantes. She later featured as the only illustration of a human in his *Histoire Naturelle des Mammifères*.[48] Cuvier himself was skeptical about the possibility for African improvement, rejecting Lamarck's theory of transmutation and insisting that Caucasians, Mongolians, and Ethiopians represented separate varieties. (Cuvier did, however, accept that all the races of man shared a common origin).[49] Other naturalists, often those who opposed slavery in one way or another, argued that Africans were in fact capable of improvement. Victor Schoelcher, a member of the Société Française pour l'Abolition de l'Esclavage, argued that Lavater's facial angle was simply a means to deny Africans "nearly all intelligence." According to Schoelcher, the apparently small size of African crania was due to a lack of intellectual exercise.[50] In its opening number, the *Bulletin de la Société de Géographie* also presented evidence that Africans in Senegal showed signs of improvement, albeit due to increased contact with European traders.[51]

It wasn't long before the phrenologists in Paris, organized around the physician François-Joseph-Victor Broussais, entered the fray. Following the July Revolution of 1830, Broussais and his colleagues developed a close connection with the Orléanist political regime.[52] Eustache had in fact been introduced to them by a government officer working for the minister of the interior. The former Haitian slave represented an exceptional individual, one through which to work out a phrenological account of African racial character. The plaster cast of Eustache's head was first presented at a meeting of the Paris Phrenological Society, held at the Hôtel de Ville in August 1834 (fig. 9). The phrenologists in Paris immediately equated Eustache's extraordinary actions with the shape of the cast, announcing that "one thing must strike you, it is the projection of this part of the front, it is the organ of benevolence." It was this phrenological configuration, distinguished by "courage and devotion," which explained why Eustache had warned his master about the impending revolt. Impressing further the uniqueness of Eustache's character, the Parisian phrenologists claimed that there was "no similar example" in any existing collection.[53]

The cast itself had been manufactured a few years earlier at the workshop of Pierre-Marie-Alexandre Dumoutier at 32 rue de l'École-de-Médecine. Dumoutier might have been a failed medical student, but he was a skilled artisan, opening a phrenological museum on rue de Seine in 1836 housing over 900 casts of skulls, heads, and brains that he had manufactured. He was also a founding member of the Paris Phrenological Society in 1831 and knew Broussais from his time as an anatomy assistant at the Faculty of Medicine.[54] The plaster had been mixed using burnt gypsum, extracted from the labyrinth of mines running beneath Paris and the quarries surrounding Montmartre. One tunnel even ran directly under rue de Seine, just south of the Institut de France in the old eleventh arrondissement, where Dumoutier later lived and worked.[55] Once the plaster was ready, the cast was manufactured in multiple stages. Dumoutier chose to make five separate molds and join them together in order to preserve the contours of each section. Eustache, seemingly a willing participant, first had his face covered with plaster to produce a mold of the front of his head. By reducing the weight of the plaster, this method carefully preserved the facial features, including the lips, nose, cheeks, and closed eyelids.[56] Eustache usually kept his hair short, but Dumoutier shaved off what was left.[57] This made producing the four remaining molds much easier. It also ensured that any phrenological analysis based on the shape of the cast would be less open to contest. Although the joins between the separate molds are visible on surviving copies of the cast, Dumoutier was evidently much more comfortable working with this method than were the artisans in Bristol. When Combe received the bust of Rammohun Roy, he complained that there was "an awkward appearance of scraping in a line passing across the head from immediately before the opening of the

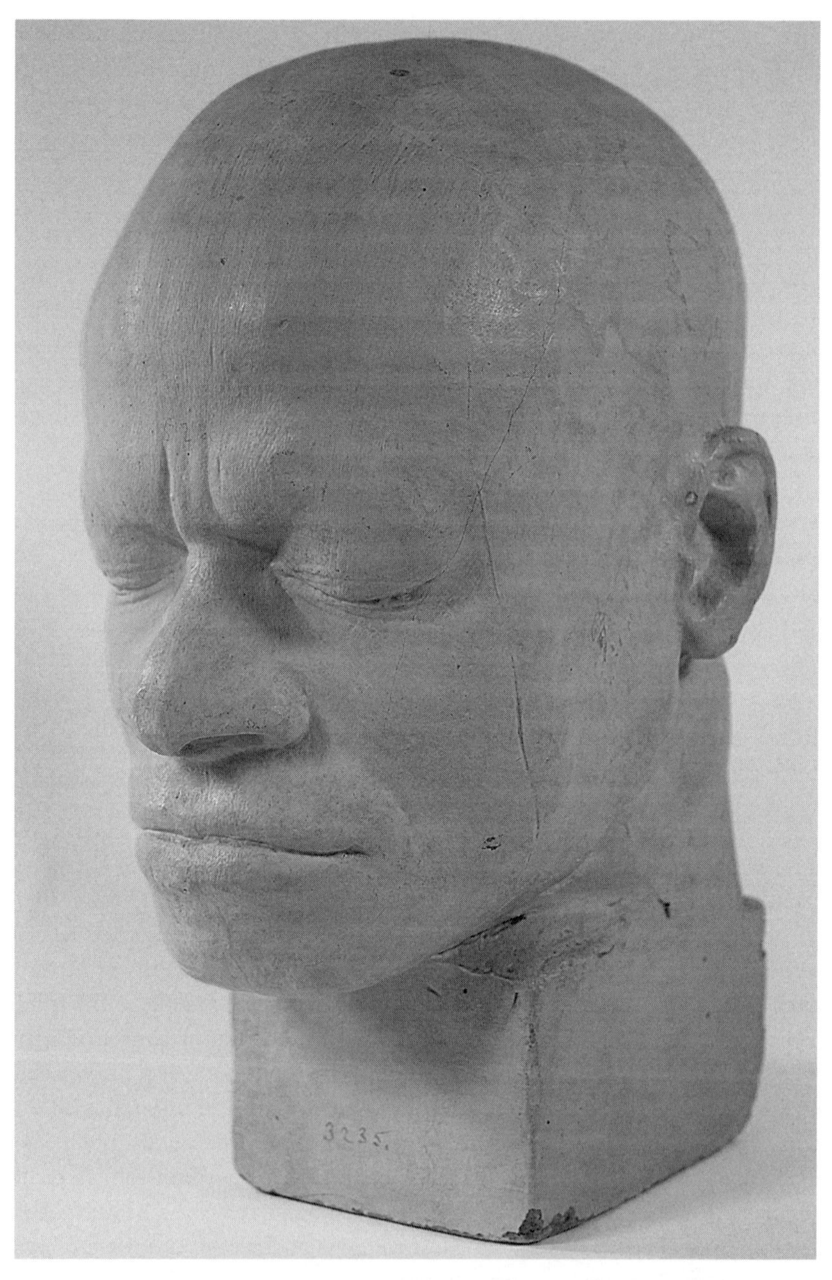

Fig. 9. Plaster cast of the head of Eustache Belin, WAM 03235, Warren Anatomical Museum in the Francis A. Countway Library of Medicine, Harvard University.

ear." These imperfections, which are clearly visible on copies of the cast (fig. 8), directly interfered with attempts to draw certain phrenological inferences.[58] In contrast, Dumoutier's casting process was so precise that it is possible to distinguish the veins on Eustache's head, adding to the sense of authenticity and, crucially, uniqueness. The divergent geographies of skill, even between Bristol and Paris, therefore had a direct effect on the veracity of phrenological evidence.

Eustache's death in 1835, alongside the foundation of the Société Française pour l'Abolition de l'Esclavage in 1834, marked renewed interest in his character.[59] François-Joseph-Victor Broussais this time presented the cast to a much larger audience during his phrenological lectures for the Faculty of Medicine in the summer of 1836. With over 3,000 people in attendance, he soon had to relocate to a much larger rented auditorium on rue du Bac.[60] At one o'clock in the afternoon on Friday 8 July 1836, Broussais began his lecture "The Application of Phrenology to History." He opened with a broad description of "negro" character, presenting a cast of a skull taken from Dumoutier's collection and announcing, "You see how the anterior part of the forehead is depressed in that race." Holding up another cast of a skull taken from "a native of High Guinea," Broussais then concluded that "intelligence, and especially reflection, are often deficient." Crucially, this negative account of African character was marshaled in defense of a particular understanding of French colonialism. Broussais went on to tell his audience that "the black people who do progress in civilization are those placed under the yoke of Europeans."[61]

It is important to understand that, despite his belief in European superiority, Broussais was in fact siding with French naturalists like Schoelcher who believed that Africans were capable of improvement. Additionally, Broussais's insistence that African improvement relied on the continuation of European colonialism also reflected the defensive politics of the Orléanists. Even though King Louis Philippe himself had been a member of an early French antislavery group, the Société de la Moral Chrétienne, the Orléanist regime ultimately adopted a piecemeal approach to abolition. In the wake of the July Revolution, the landed and professional elite were apprehensive of mass political movements like those in Britain and the United States. Both the phrenologists and their Orléanist colleagues also recognized that slavery represented a substantial source of financial revenue at a time when the French economy was struggling.[62] The rhetoric of the improvement of African character bolstered a strategy that suited both the political elite in Paris and the planters in the colonies: a series of laws designed to "ameliorate" the condition of the slaves was passed between 1831 and 1845. These included provisions for education and religious instruction, as well as ordinances on feeding and punishment. This was all wrapped up in an argument, common among proslavery campaigners abroad, that slaves needed adequate "preparation" before they could be trusted with freedom. Unsurprisingly, this had the effect of deferring abolition to some unspecified time in the future.[63]

Eustache, as an exceptional slave, acted as a pivot in Broussais's argument. Immediately after showing his audience the skull from "High Guinea," Broussais lifted up another cast, announcing, "In contrast, you have the head of Eustache." Like Rammohun, Eustache was explicitly presented as a contrasting case. He was "a true negro" (as the audience could confirm for themselves from Dumoutier's carefully manufactured cast) but one in which "the higher sentiments predominated." Relating the story of how Eustache came to warn his master of the impending slave revolt, Broussais informed his audience that this man was "distinguished above his whole race, a kind of anomaly." For Broussais, Eustache's cast was material evidence of the capability of African character for improvement. But he was also evidence of just how much more work needed to be done before the majority of slaves could reach this level of attainment. In fact, immediately after showing his audience the cast of Eustache, Broussais lifted up a cast of Sara Baartman's head, taken by George Cuvier following her death in 1815 at the Jardin des Plantes. To make sure his audience did not draw too liberal an inference from Eustache, Broussais told them that Baartman's cast revealed "like most negroes . . . an extremely limited intelligence."[64] Exhibited between the "native of High Guinea" and the "Hottentot Venus," Eustache's cast ultimately allowed Broussais to present himself as a champion of African improvement while still maintaining the political status quo with regard to slavery.

MOVING AND MEASURING

In Paris, Eustache was so well known that Broussais could rely on his audience's knowledge of both the history of the Haitian Revolution and the widely popular story of "le bon nègre." In fact, when an account of Eustache's character was published in the *Journal de la Société Phrénologique de Paris* in 1833, the author simply wrote, "I will not tell you about his life . . . ; you have all heard or read the story."[65] However, by following Eustache to Britain and the United States, it soon becomes clear that the reproduction and movement of the cast introduced fresh challenges for phrenologists looking to pin down African character.

Shortly after Broussais's lectures in Paris, a copy of the cast of Eustache's head arrived in Britain. It had been purchased from Dumoutier by William Gregory, secretary to the Edinburgh Phrenological Society, on a recent visit to the French capital. Once in Edinburgh, Eustache was incorporated into the existing collections on Clyde Street. The bust was inscribed with the number 159 on the pedestal and recorded in the Edinburgh Phrenological Society's museum catalogue, well away from the anonymous "Negro" allotted number 19. Combe and his colleagues in Britain valued the bust of Eustache for the same reasons that they were so interested in Rammohun's. Unlike the other casts of African skulls in their collection, Eustache's facial features were clearly visible, as was his "prodigious development of the organ of Benevolence." There was "in the collections no specimen which can be in this respect compared with it."[66]

Initially, the arrival of the cast in Edinburgh provided a chance to confirm the claims already made in Paris. If phrenology was to operate as a global science, then plaster casts needed to be made into a reliable means to communicate knowledge about the human body. Gregory and the conservator of the Edinburgh museum, Robert Cox, therefore produced a detailed set of measurements taken from Eustache's bust that were then printed in the *Phrenological Journal*. Approving of Dumoutier's work, Cox and Gregory noted that "there was no hair on the head when the cast was taken." With its circumference of 22 inches, the two men argued that "it will be obvious to every phrenologist, from the foregoing table, that the head of Eustache is of very considerable size." The Parisian phrenologists' analysis of the different organs had also apparently been accurate. Cox and Gregory wrote that "we entirely concur in the statement of Dr Broussais. . . . Benevolence, rising to a great height." Finally, Cox and Gregory reinforced the distinction between European and African character with this carefully constructed exception. According to the phrenologists, Eustache's head had "quite the appearance of a European."[67]

This experiment was then repeated in Aberdeen. The instrument maker James Straton concurred with both the Parisian and the Edinburgh phrenologists, describing Eustache as "one of the finest specimens of human nature ever known." Straton had recently developed a novel method for calculating the internal capacity of the head, which he was keen to try out on Eustache.[68] Like Cox and Gregory, Straton was pleased to find that the head had been shaved, noting that it was "impossible to measure them accurately" if casts "have masses of stucco representing hair." Straton got to work by marking the locations of different phrenological organs directly on the cast with a piece of chalk. Taking out a pair of calipers he had manufactured himself, Straton then measured the distance between each point. Through a complicated process of adding and dividing these measurements, Straton arrived at values for the breadth, length, and height of the head. These were then multiplied to give the final "cubic measure." According to Straton, Eustache's head had an internal capacity of 155 cubic inches, "far above the Negro average" of 123 cubic inches. Consequently, Eustache earned a place next to European men, like Horace Smith and Robert Owen, in Straton's taxonomy of phrenological specimens published as *Contributions to the Mathematics of Phrenology* in 1845.[69]

Despite these attempts to portray a consensus with regard to Eustache's character, this process was in fact much more complex. The arrival of the cast in Britain also revealed that access to information concerning a particular individual had the potential to seriously restrict the usefulness of plaster as a medium of reproducing phrenological knowledge. Eustache certainly possessed an outstanding head, but the phrenologists in Scotland actually knew very little about his life and the events of the Haitian Revolution. They did not have access to the report of the Institut de France, of which Broussais was a member, nor the French periodicals and popular books that recounted the tale of "le bon nègre." The *Phrenological Journal* in Edinburgh ultimately admitted that "we are sorry that so few particulars relative to Eustache are given by Dr Broussais." The Edinburgh Phrenological Society did subscribe to the *Journal de la Société Phrénologique de Paris*, but Broussais had not bothered to give many details about Eustache precisely because his Parisian audience already knew about him. In Edinburgh, Combe and his colleagues ultimately acknowledged that the usefulness of a plaster cast was tied to different geographies of print, writing "distant readers, who do not possess these advantages, are left very much in the dark."[70]

The Aberdeen, Warrington, and Nottingham phrenological societies all soon purchased copies of the bust from O'Neil and Son. But without access to the information available in Paris, Eustache was simply presented as another example of "National Phrenology."[71] It wasn't until much later in the 1840s, after the bust had traveled back and forth across the Atlantic, that abolitionists started to take notice, with William Armistead of the Leeds Anti-Slavery Association publishing an account of Eustache's life in his *A Tribute for the Negro*.[72] By the 1860s, Eustache was being compared to "Uncle Tom" of Harriet Beecher Stowe's antislavery novel in *The Popular Manual of Phrenology*.[73] But this connection between Eustache and antislavery was not obvious in Britain in the 1830s. It required the efforts of phrenologists and abolitionists on both sides of the Atlantic to rework the original narrative developed by Broussais in Paris.

The phrenologists in the United States initially faced similar problems to those in Britain. These were exacerbated by the fact that they didn't even have access to a copy of Eustache's bust. In 1834 the Boston Phrenological Society, organized around the abolitionist Samuel Gridley Howe, published a summary of Broussais's original article from the *Journal de la Société Phrénologique de Paris*. In it the author simply referred to Eustache as "a young negro . . . remarkable for numerous acts of benevolence."[74] There was no mention of what these acts were, nor the Haitian Revolution, nor even the fact that Eustache was a slave. Without access to the Parisian world of print and gossip on which Broussais's original description of Eustache relied, Howe and his colleagues in Boston, even though many of them were interested in the antislavery cause, could say very little. The fact that *Annals of Phrenology* referred to Eustache as "young," when he was in fact over sixty years old, is indicative of their lack of familiarity with the episode.

These shortcomings were partially overcome in October 1838 when a copy of Eustache's bust finally arrived in the United States. In the summer of that year, the new museum conservator, Abram Cox, began gathering together the casts from the Edinburgh Phrenological Society's collection and packaging them up. From there, they were sent to Glasgow on a mail coach before being loaded onto a packet ship moored on the Clyde, soon bound for North America. Combe was waiting at the Custom House in Boston to collect the package containing Eustache's bust. These were accompanied by four other crates rammed with "skulls, casts, and drawings" prepared for Combe's imminent lecture tour.[75] These props were essential if Combe's visit was to be a success. The American public more than anyone else relished visually stimulating lectures.[76] The customs official, George Bancroft, was no exception. As he checked over the goods, Bancroft spotted the casts and engaged Combe in conversation, advancing his own theory that Immanuel Kant's works could be read in phrenological terms. Bancroft, clearly an enthusiastic admirer, soon let Combe on his way, informing him that his packages were "entitled to be landed free of duty, as articles of science."[77] But although they moved freely into the United States, the meanings associated with individual specimens did not travel so easily.

In the case of Eustache, Combe did his best to relate a life history that was, until then, relatively unknown in the United States. Lecturing in

New York City at the Mercantile Library Association, Combe lifted Eustache's bust up for his audience to see.[78] As he did so, Combe described how "during the insurrection of the Blacks at St. Domingo, the disinterested exertions of Eustache on behalf of his master, Mr. Belin, were unbounded." At least, unlike the article in *Annals of Phrenology*, Combe's audience now understood that Eustache was in fact a slave, one who had witnessed the Haitian Revolution firsthand. Combe then explained that, thanks to Eustache's "address, courage, and devotion . . . four hundred other Whites . . . were saved from the general massacre." Impressing on his audience the idea that Eustache should be understood in contrast to the general slave population, Combe concluded that he was "a most benevolent Negro." This could be confirmed by inspection of an "extremely developed" organ of Benevolence.[79]

And so, unlike Broussais in Paris, Combe had to deal with a lecturing environment in which his audience were not already familiar with the life history of the bust on show. The material form of the plaster—particularly as a marker of racial difference—therefore took on an increasingly prominent role. When arranging for a lecture theater in Pennsylvania, Combe was gratified to learn that the room at the Philadelphia Museum would be "lighted with gas." He nonetheless fretted that those seated in the raised gallery at the back would be unable to see his specimens.[80] At the beginning of each lecture, Combe began by lining up all his busts and skulls on a table. As he enumerated the different faculties of the mind, Combe collected individual specimens, holding them up for the audience to see. The gas lighting at the Philadelphia Museum was particularly helpful, with shadows throwing into relief the contours of different casts. The specimens seemed to go down well, with those in attendance in New York City celebrating Combe's use of "appropriate illustrations" whereby "the memory is greatly aided, and the judgement much gratified."[81] Combe was also fond of comparing busts to skulls, as illustrated in the frontispiece to the publication of his American lectures (fig. 10). This method of comparison allowed Combe to further reinforce the material difference between Eustache's individual features and the average African skull. Before concluding his lecture on Eustache's character, Combe held up the bust for a final time, announcing, "I again show you the cast, as it is one of the most beautiful demonstrations of this organ which we possess."[82] These verbal cues were crucial. After all, as Combe himself recognized, a cast was potentially "unintelligible," even "uninteresting," without a phrenologist to guide you.[83]

In June 1840, Combe and his collection of phrenological busts left the

Fig. 10. Comparing a skull to a bust during a lecture in the United States, George Combe, *Lectures on Phrenology* (London: Simpkin, Marshall and Co., 1839), frontispiece. Reproduced by kind permission of the Syndics of Cambridge University Library.

United States aboard the SS *British Queen*, bound for Portsmouth. Less than a month later, Eustache's cast was sitting back on the shelves of the Edinburgh Phrenological Society's museum on Clyde Street.[84] Combe had ultimately found it difficult to say much about African racial character during his tour. He even admitting to being afraid of alienating his audience by presenting too strong a position on slavery in his lectures.[85] As will be detailed in chapter 4, Combe was still working out his own understanding of the relationship between slavery and race in private correspondence during this period. Nonetheless, Combe's tour did spark much wider interest in Eustache's life in the United States. Reports of Combe's lectures were reprinted up and down the country. Even those in slave-owning Virginia learned about Eustache, as the *Southern Literary Messenger* reprinted Combe's lectures verbatim from the *New Yorker*.[86] What's more, the American campaign for immediate abolition greatly increased its activities in the run-up to the end of slavery in the British West Indies in 1838. This period also saw greater cooperation, not just between American and British abolitionists, but also between American and French antislavery campaigners.[87] It was in fact this connection that ultimately led to a radical rereading of Eustache's bust in the United States and beyond.

AN AMERICAN IN PARIS

In 1837 the American abolitionist David Lee Child arrived in Paris, hoping to galvanize the French antislavery cause.[88] The Société Française pour l'Abolition de l'Esclavage was still sitting on the fence, refusing to back the campaign for immediate abolition led by Cyrille Bissette, a former slave from Martinique. Bissette was one of the few French abolitionists to argue that it was the slaves, not the planters, who should receive compensation.[89] Anglophobia was also still rife among the Parisian political elite, further hampering any chance of the Orléanist government emulating the British example.[90] Child therefore hoped an American perspective might help to break the deadlock, addressing the Société Française pour l'Abolition de l'Esclavage during his visit.[91] While in Paris, Child also learned much more about French attitudes toward the Haitian Revolution. It wasn't long before Eustache's name was mentioned, and, sure enough, Child returned to Boston in 1839 with a copy of the bust.

Three years earlier, the students at the Faculty of Medicine in Paris had listened as Broussais held the same cast aloft, simultaneously defending African improvement and French colonialism in the West Indies. But now in the United States, antislavery campaigners in Boston and Philadelphia deployed Eustache as part of a rereading of the Haitian Revolution, one that demanded the immediate abolition of slavery. A new article on the bust was published in the *Pennsylvania Freeman*, a weekly periodical run by the local antislavery society. Eustache was introduced as part of an unambiguous attack on slavery. Crucially, Child had supplied a wealth of additional details concerning Eustache's life, mostly gathered from periodicals in Paris such as the "educational and entertaining" weekly *Le Caméléon*. This allowed the *Pennsylvania Freeman* to claim precedence over existing accounts that exclusively borrowed from Combe's lectures. Eustache, the abolitionist paper claimed, was "a man not hitherto known in the United States, but worthy to be known throughout the world."[92]

The article opened by contesting the meaning of both African improvement and the Haitian Revolution. In Paris, newspapers subsidized by the proslavery lobby continued to warn of the violent consequences of slave emancipation. *Le Journal du Havre* reminded its readers of the "ferocious murderers, who as a first act of emancipation pillaged and massacred." Similarly, publications in the southern United States recalled "the degradation and barbarism of the St. Domingo negro."[93] The account of Eustache's bust in the *Pennsylvania Freeman* provided an opportunity to counter these arguments. Whereas for Broussais, African improvement was simply a means to perpetuate slavery, for Child and his friends in Philadelphia, Eustache's exceptional character was evidence that slavery could never be justified on the grounds of a lack of African talent. Indeed, the article in the *Pennsylvania Freeman* acknowledged that Eustache was an "extraordinary man," one who proved the capability of Africans in the present, rather than some distant point in the future. According to the *Pennsylvania Freeman*, "No individual case could, perhaps, bear with greater force upon the mooted question of 'inferiority.'" Eustache was "a *black man*, who, in the best points of human character . . . beat all France." *The Pennsylvania Freeman* even picked up on the politics of naming, using the very inscription featured on the pedestal of the bust as a means to highlight the moral and intellectual depravity of proslavery rhetoric:

> Like the ancient serfs of Europe, forefathers of white freemen, and of white slaveholders too—like other coloured men of modern slaveholding countries, and like dogs in all countries, he had only one name—EUSTACE, to which was sometimes added another, signifying whose property he was, Eustace Belin.

For the abolitionists in Philadelphia, slavery was akin to a feudal system of labor: the freedom of Eustache Belin, a name that signified an outmoded system of production, was therefore introduced as part of a broader political process of modernization.[94]

While Eustache was presented as remarkable, he was also portrayed as proof that emancipation did not need to end in violence. As the article went on to explain, "Freedom did not change, it only elevated and hallowed his friendship for his late master."[95] This was part of a wider abolitionist strategy to portray the Haitian Revolution as progressive and successful, an event that pointed to the propriety of ending slavery, rather than its risks. Lydia Maria Child, David's wife, made comparable use of Toussaint Louverture, arguing that "the greatness of his character and achievements proved the capability of the Black Man." Similarly, the African American editor of *Freedom's Journal* in New York City argued that Haiti's "subsequent progress in all the arts of civilisation . . . prove the blacks are capable."[96] In fact, in the very same issue as the article on Eustache, the *Pennsylvania Freeman* urged the United States government to recognize "Haytien independence" in order to benefit from a successful West Indian economy founded on free labor. The abolitionists kept up the pressure, arguing that "this governmental obstinacy is the result of nothing but slavery," well aware that the southern states were campaigning hard to secure their own northern export market and shut Haiti off from the world.[97]

This link between the United States and Haiti was reinforced by the *Pennsylvania Freeman*'s access to new information about Eustache's life. American readers learned for the first time that, shortly after the uprisings of 1791, Eustache and his master had left Haiti aboard a ship bound for Maryland. Although they were briefly intercepted and captured by a British cruiser, Eustache and his master eventually made it to Baltimore.[98] There they joined thousands of other refugees—white planters, gens de couleur, and slaves—seeking asylum in the United States. The majority of white planters headed to the southern states, particularly to Maryland as well as Louisiana—then part of the Spanish Empire. Like New Orleans, Baltimore was a popular choice because of its large Catholic population, giving white planters a sense of community as well as relatively easy access to employment. The gens de couleur tended to gravitate toward northern states like Pennsylvania and New York, rather than risk being captured and put to work as runaway slaves. Across the United States, the arrival of a large black and mixed-race population fueled ongoing debates about the relative risks associated with abolition and the maintenance of slavery. In

South Carolina four slaves were tried and hanged for conspiracy to commit arson in 1802. According to a local newspaper the plot had "originated among the French negroes." The possibility of Charleston burning to the ground, just as Cap Français had in 1793, haunted the local white population. Similarly, newspapers in Virginia blamed the Nat Turner rebellion of 1831 on the influence of French slaves, although there is no evidence that this was in fact the case.[99]

In the northern states, the arrival of refugees from Haiti reinforced divisions within the abolitionist movement. Encouraged by the stories he was hearing from those seeking asylum in Boston, David Walker called for continued violent resistance. In his *Appeal to the Colored Citizens of the World*, Walker urged fellow African Americans to "read the history particularly of Hayti, and see how they were butchered by the whites." And although he rejected Walker's calls for armed resistance, William Lloyd Garrison also linked the Haitian Revolution to the potential for widespread unrest, writing, "What created the bloody scenes of St. Domingo? Oppression! What has infuriated the southern slaves? Oppression!"[100] Other abolitionists, including those behind the *Pennsylvania Freeman*, found this inflammatory approach unhelpful, fearing it would simply reinforce white prejudice surrounding the potential danger of ending slavery. Instead, they emphasized the successful integration of Haitian refugees, particularly the former slaves, into American society. Men like Eustache were proof that the United States might one day emerge as a peaceful interracial republic. The *Pennsylvania Freeman* proudly told its readers that, when in Baltimore, "Eustache devoted the resources which his industry and skill could command to the relief of those lives he had saved." He had even apparently set up a store to support his master, selling some of the sugar loaves they had brought with them from Haiti.[101] The Boston abolitionist Hannah Lee published a remarkably similar account of another Haitian slave who had accompanied his mistress to New York, supporting her as a barber before finally gaining his freedom.[102] Here, Eustache was represented within a common abolitionist narrative, emphasizing devotion, hard work, and charity among Haitian slaves seeking asylum in the United States.

However, for the *Pennsylvania Freeman*'s account to work, it needed some careful editing. How, after all, did Eustache and his master escape from their British captors en route to Baltimore? The *Pennsylvania Freeman* simply explained that "Eustache was a superior cook, and soon rendered himself very useful and agreeable to the officers." Having gained their confidence Eustache "rushed into the cabin . . . with a rusty sword in his hand." He then apparently told the British officers that "if they would surrender at once, no harm should be done to any of them." The *Pennsylvania Freeman* concluded: "They did surrender, and the vessel arrived safely with its prisoners and passengers at Baltimore."[103] For the abolitionists in Philadelphia, Eustache was again portrayed as an exemplary slave, one who did not resort to unnecessary violence to secure emancipation. However, an earlier French account of Eustache's actions presented a very different perspective. In his *Traité de Phrénologie* Joseph Vimont reported that one of the British officers had in fact "called for assistance" when challenged. Vimont then informed his readers that "Eustache with a blow cut off his arm, the two others prayed for their lives."[104] This small but significant act of violence was entirely omitted in the account given in the *Pennsylvania Freeman.* It did not fit either the received phrenological account of Eustache's benevolence or the abolitionist strategy to emphasize the good-natured conduct of former slaves. In this case, different political geographies, not just access to print, explain the varying portrayals of African character.

With Congress debating the recognition of Haitian independence, news of Eustache's life in Baltimore proved timely. The *Pennsylvania Freeman* article was reprinted up and down the country. And as Eustache circulated further and faster, his character was reworked to serve different political and intellectual needs. While the *Pennsylvania Freeman* had given little indication of the religious context of American abolitionism, the Quakers behind the *Friends' Intelligencer* in Philadelphia were much more forthright. For them, slavery was a sin. Their reprint of the article was headed by a passage from Corinthians ("He that is called in the Lord, being a servant, is the Lord's free-man") followed by another from Acts ("God is no respecter of persons: but in every nation he that feareth him, and worketh righteousness, is accepted with him").[105] However, the same account of Eustache's character could just as easily serve the aims of the American Colonization Society, which reprinted the *Pennsylvania Freeman* article under the title "The Faithful Slave" in the *African Repository.*[106] As discussed further in chapter 4, leading Quaker abolitionists were fiercely critical of the American Colonization Society, which sought to permanently resettle freed slaves in West Africa. Nonetheless, for supporters of colonization, the Haitian Revolution showed that African improvement was incompatible with a biracial society. Eustache's life in Baltimore and subsequent return to Haiti were pertinent. The Maryland Colonization Society worked closely with Jean-Pierre Boyer, the Haitian president, to arrange for the transportation of slaves to Haiti as an alternative to West Africa.

Boyer agreed to guarantee loans and land for the emigrants once they arrived, so long as the Maryland Colonization Society funded the costs of transportation.[107] Eustache was presented, like other former slaves returning to Haiti, as contributing to the improvement of an immature black republic. The *African Repository* explained that "there was a vast deal of misery, and but one Eustace in the island of St. Domingo." Nonetheless, "if a soldier was without clothing and pay, a family without bread, a cultivator or mechanic without tools, the new riches of Eustace were dispensed for their supply."[108]

FUGITIVE PHRENOLOGY

Before long, the story of Eustache's life saturated the American periodical press. It continued to be reprinted in publications ranging from religious weeklies such as the *Catholic Telegraph* in Ohio to educational monthlies such as *Parley's Magazine* in Boston.[109] And it didn't take long for a copy of the *Pennsylvania Freeman* to reach the phrenologists Lorenzo and Orson Fowler in New York City. The duo were in negotiations to take over the publication of the *American Phrenological Journal* and were looking for new content. They wanted something that would simultaneously boost the circulation of the struggling periodical as well as encourage visitors to their new phrenological cabinet at 135 Nassau Street. Eustache fit the bill.

In January 1840 the *American Phrenological Journal* published an article entitled "Character of Eustache" accompanied by two original woodcuts of the bust (fig. 11). The cast itself had been given to the Fowlers by the New York Anti-Slavery Office, just down the block at 143 Nassau Street. Through this connection, the phrenologists in the United States were finally in a position to claim authority over the existing British and French accounts. The article opened by pointing to the new illustrations, explaining that "the above cuts are designed to present two different views of the head of a negro, by the name of Eustache, who was eminently distinguished for the qualities of virtue and benevolence." This was followed by a long extract from the *Phrenological Journal* in Edinburgh. However, the *American Phrenological Journal* emphasized the limits of the existing narrative. The article went on to explain that "the above account was published three years before the death of Eustache, when only a few facts were known concerning his life and character." But, as the phrenologists themselves reflected, it was through circulation that new evidence came to light:

> Within the past year, the following history of Eustache appeared in several papers in this country. We present it entire, as every phrenologist cannot fail, with the above data, to derive additional interest from the narrative.

Ultimately, the *American Phrenological Journal* combined narratives from the United States, France, and Britain to produce what it claimed was an authoritative account. Only with this "minute and extended history of Eustache" could phrenologists link his character to the "remarkable cerebral developments."[110]

The publication of the article on Eustache also provided an opportunity to promote the Fowlers' new phrenological cabinet. In late 1839, the brothers staged an experiment, one that was made possible only through transatlantic exchange. Carrying the bust of Eustache the short walk from the Anti-Slavery Office, the abolitionist Theodore Dwight Weld arrived at 135 Nassau Street to meet Lorenzo Fowler. The bust was then placed on a table and covered with a sheet. Lorenzo ran his hands over the cast, feeling for the relevant protuberances, before finally giving his opinion. The bust

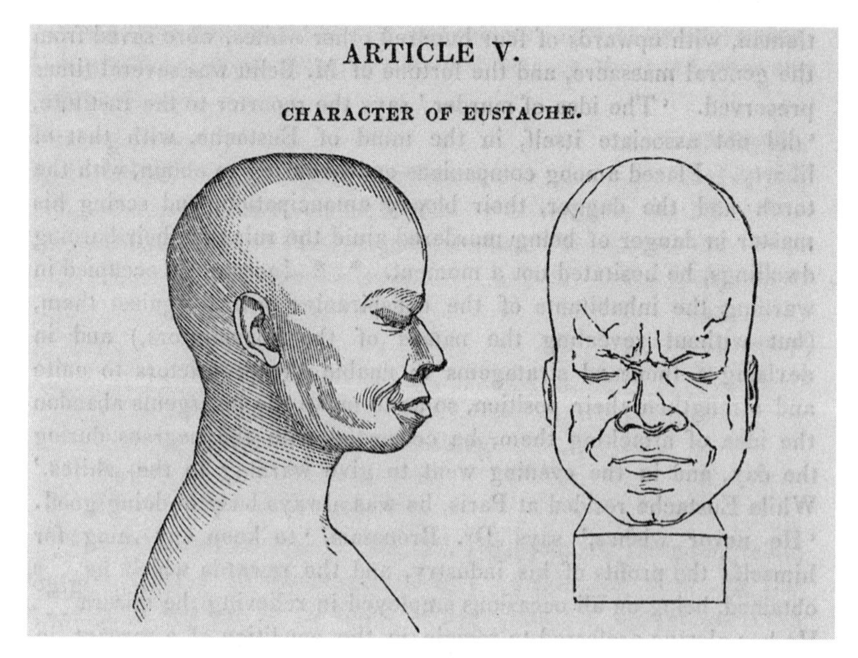

Fig. 11. Woodcuts of the bust of Eustache Belin, "Character of Eustache," *American Phrenological Journal* 2 (1840): 177, Wellcome Collection. CC-BY.

was of an individual with "Benevolence being very large, distinguishing him for good nature and humanity." This was "a man of principle, moral sense, and strong conscientious feelings." Incredibly, when the sheet was removed, Lorenzo claimed to have believed the bust "belonged to a *white* person." The covering of the bust as part of this experiment therefore also reveals the limits of the phrenological emphasis on the organs of the mind. Clearly, both the Fowlers and the abolitionists believed that they would recognize a slave from his facial features as well.[111] Nonetheless, the apparent similarity of Lorenzo's reading to that of the French and British accounts confirmed, according to the *American Phrenological Journal*, the "proof of phrenology." Lorenzo had "attributed to the individual the same character as had been given by the phrenologists in Paris." The analysis was also "in striking accordance with the observations of the Edinburgh phrenologists." And while there were clearly divergent political uses of Eustache's bust as it moved back and forth across the Atlantic, the phrenologists were ultimately united in identifying him as an exceptional man. Eustache's character, the Fowlers concluded, was "the more striking, inasmuch as the individual belonged to a race generally regarded as deficient in those qualities."[112] It was therefore through global exchange that plaster was presented as a reliable medium through which to communicate individual and racial character.

Fowler's experiment on Nassau Street was much more than a party trick. These kinds of phrenological practices also played a major role in the course of slaves' lives. On 25 August 1851, John Bolding, a tailor working in Poughkeepsie, was dragged from his home in front of his wife by Henry Tallmadge, a United States Marshal.[113] Less than a year earlier, Congress had passed the Fugitive Slave Act, allowing southerners to seek the forced return of runaway slaves from the north. A woman traveling from South Carolina had apparently recognized Bolding—who was described as "nearly white"—and reported him as a fugitive slave. Before long, Bolding was bundled into a carriage and forcibly transported by railroad to New York City where he stood before a judge and his alleged master, Robert Anderson. The case rested on proving that Bolding was "of African descent." The defense, organized by the New York abolitionists, called Lorenzo Fowler. Lorenzo repeated the same experiment he had performed on Eustache, this time for the court. He was blindfolded and asked to pronounce on Bolding's "lineage." Lorenzo concluded that he "should not suppose, by passing my hand over his head, that there was any African blood" in Bolding. In a performance akin to Combe's earlier lectures, Lorenzo then held up "the skull of an Indian" and "the skull of an African," comparing them to Bolding before the court. According to the phrenologist, "in the absence of evidence to the contrary" he would "suppose no African blood in him any more than in Mr George Copway, the Indian Chief." Unfortunately, the judge did not take kindly to phrenology and decided that, despite being "less dark in colour than many white men," Bolding was both of African descent and a fugitive slave.[114] Thankfully, the community in Poughkeepsie rallied to his aid. Over $1000 was raised, with which Bolding's freedom was purchased.[115] Still, this episode does go to show how pronouncements on African character were much more than popular spectacles. Nineteenth-century racial theories were in fact embedded in the political and material practices that underpinned Atlantic slavery and its eventual abolition.

By the outbreak of the American Civil War in 1861, copies of the bust of Eustache Belin had reached France, Britain, the United States, and Prussia. This chapter has argued that phrenological understandings of race were grounded in this material world of production and exchange. In the course of an often troublesome process of transit, Eustache had been deployed in debates ranging from Haitian independence to the Fugitive Slave Act. This chapter has also uncovered the role that exceptional individuals played in nineteenth-century theories of race. Although Eustache's character was constantly reinterpreted, phrenologists from Philadelphia to Paris agreed that he was distinct from other slaves. By confirming what made Eustache exceptional, the phrenologists also defined "average" African character.

The very process of exchange ultimately allowed the phrenologists to present plaster as a reliable medium of reproduction. Busts were certainly liable to break. They were also difficult to manufacture. However, both the Fowlers in New York City and Broussais in Paris understood the significance of their comparable readings of Eustache's character, even if they pointed to different conclusions with regard to slavery. Nonetheless, the usefulness of plaster as a material through which to conduct a global science soon reached its limit. A comparison to printed illustrations is instructive, a subject treated in further detail in the following chapter. At the beginning of the nineteenth century, both plaster and printed engravings had a lot in common. They both required the expertise of local artisans, something that was hard to come by in the colonial world. Those artisans also needed access to raw materials, whether it was plaster, a copper plate, or a woodblock. And finally, both plaster and paper presented problems of transit: busts arrived in pieces, and books also degraded over time, particularly in tropical climates.

However, over the course of the nineteenth century, printed illustrations came to dominate as the medium of reproducing visual knowledge on a global scale. Copies of the fourth edition of Combe's *System of Phrenology*, which included a new engraving of Eustache's bust, reached audiences in India and New South Wales, far beyond the original plaster cast.[116] The same was true of Rammohun Roy, whose bust never reached India. Instead, woodcut illustrations of the cast were reprinted in the *Indian Journal of Medical and Physical Science* in Calcutta.[117] They had been copied by a local artisan from an issue of the *Phrenological Journal* imported from Britain. By midcentury, Calcutta housed a thriving community of publishing houses, with access to printing technologies ranging from copperplate engraving to lithography.[118] In contrast, there was no comparable industry dedicated to plasterwork. The East India Company continued to commission statues to be executed in London for display in Madras and Bombay, while both Bengali and European phrenologists based in Calcutta regularly imported casts from O'Neil and Son in Edinburgh.[119] The dominance of print as a medium of transmitting visual knowledge was not inevitable, nor was it obvious in the 1830s. By recognizing these changing geographies and technologies of global communication we can better understand the material forms that both phrenology and racial ideas took. It was only with the industrialization of print and the development of publishing as an international business that plaster lost its edge.[120]

CHAPTER 3

Books

Thomas Pringle was setting out on a long journey and needed something to keep him occupied. A phrenological book seemed a good bet. Tucked away in his luggage was a copy of George Combe's *Essays on Phrenology*, published by Bell and Bradfute in Edinburgh. Set in type and printed by Patrick Neill in 1819, this was Combe's first phrenological book and, as with many of his later works, it reached far beyond its original place of publication. Pringle, a Scottish poet and abolitionist, was boarding a ship bound for Cape Town, where he later became a leading liberal critic of the colonial government.[1] He was looking forward to reading Combe's book during the lengthy voyage. In the dockyard at Deptford, people were already talking about phrenology. Pringle wrote to Combe while waiting for the ship to leave, informing him, "I have met with several people here on whom your book appears to have made a considerable impression." Unfortunately Pringle felt "ill qualified to enter into discussion of the subject—the book you kindly presented to me lying still locked up with my other books at the Wharf." In any case, Pringle had plenty of time ahead of him to come to grips with Combe's principles. The journey to Cape Town was likely to take about three months. During the voyage, Pringle hoped he might be able to test out phrenology on his fellow passengers. He told Combe that he would soon "have both the time to study it and a variety of individuals on whom I may try it."[2]

On arriving in Cape Town, Pringle penned another letter, delivering his verdict. Although "not a decided convert" himself, Pringle offered to help promote phrenology in the Cape Colony. In fact, Pringle donated his copy of Combe's *Essays on Phrenology* to the newly opened South African Public Library.[3] With printing presses hard to come by, imported books like Combe's were particularly sought after.[4] Wealthy settlers paid up to £3 per year to access the South African Public Library's stock of over 4,500 volumes. East India Company officials, many of whom took leave in the Cape, could also use the library at a cost of £2 for a six-month subscription. Readers could find *Essays on Phrenology* in the section dedicated to "Surgery and Medicine" (although Combe would have preferred to be catalogued alongside Dugald Stewart and David Hume under "general treatises of Philosophy").[5]

Beyond Cape Town, phrenological books soon found their way into libraries across the colonial world. The Calcutta Public Library, patronized by the wealthy landowning Tagore family, stocked octavo editions of Combe's *Elements of Phrenology* and *Constitution of Man*.[6] Similar collections of phrenological works were held at the Mercantile Library of San Francisco and the Melbourne Public Library.[7] Combe also personally presented a copy of his *System of Phrenology* to the University of Heidelberg, while, closer to home, he was delighted to learn that students at Queens' College, University of Cambridge, were reading Johann Spurzheim's *Phrenology, or, The Doctrine of the Mind*.[8] For nineteenth-century readers and writers, libraries were spaces in which a range of national and imperial publishing projects overlapped. The Sydney Mechanics' School of Arts held British, French, and American phrenological books.[9] Some readers were even lucky enough to have access to both British and American editions of the same work. The New York Mercantile Library held two copies of Andrew Combe's *Principles of Physiology*, an 1841 edition published in Edinburgh as well as an 1838 edition published in New York City. The same library also stocked an imported English translation of François-Joseph-Victor Broussais's *Cours de Phrénologie*, originally published in Paris, but translated and reprinted by George Routledge in London.[10] With a growing international market for the latest French scientific works, libraries in New South Wales and Calcutta soon purchased copies of Routledge's translation as well.[11]

This chapter argues that the practices of reading, writing, and publishing science in the nineteenth century were not confined within national borders. Rather, publication and reception were activities grounded in a global world of material exchange. For too long, histories of science and the book have been dominated by national case studies.[12] Despite all its strengths, James Secord's work on Robert Chambers's *Vestiges of the Natural History of Creation* is restricted to the British Isles, whereas Chambers was in fact read in locales ranging from Philadelphia to Bombay.[13] Studies of phrenological books, particularly Combe's best-selling *The Constitution of Man*, have followed a similar pattern, treating publication and reception as activities bounded by the nation.[14] However, as the libraries of Calcutta and Cape Town remind us, this is misleading. Right from the start, phrenological books were produced and consumed by authors and readers occupying a much wider world.

Many of the major themes of nineteenth-century scientific publishing look different once we start to think globally. Take format. The popularity of smaller duodecimo and octavo formats is often attributed to the emergence of a British market for portable works to be read in pubs, parlors, and train stations. This is certainly true, but it is also an international story.[15] During the first half of the nineteenth century, the vast majority of books read in both the United States and British colonies were imported from Europe. Duodecimos and octavos were easier and, crucially, cheaper to ship internationally.[16] Travelers also needed something portable to read while stuck at sea. It is not a coincidence that Pringle decided to take Combe's *Essays on Phrenology* with him to Cape Town rather than Franz Joseph Gall's folio atlas. Similarly, both the Melbourne Public Library and the Calcutta Public Library housed collections of phrenological books consisting exclusively of octavos and duodecimos.[17] The New York Mercantile Library even held a tiny octodecimo edition of *The Constitution of Man*.[18] When thinking about the popularity of certain formats, it is therefore important to remember that publishers in Edinburgh and Paris were responding to the demands of both local readers and those in the United States and the colonies.

The story is the same for the development of cheap publishing. New technologies like the steam press enabled publishers to radically increase their output while simultaneously reducing costs. Historians have typically linked this to the growth of working-class readerships in Britain, with mechanics institutes and evangelical publishers promoting science as a means of self-improvement.[19] Phrenology certainly benefited from this. In 1835, William and Robert Chambers in Edinburgh published a "People's Edition" of *The Constitution of Man*. Confident they were on to a good thing, the Chambers brothers used a steam press to print over 5,000 copies of Combe's book. The final product, an octavo in double columns, sold for just 1s. 6d., a quarter of the price of the original edition published in 1828.[20] The steam press unquestionably expanded the readership of phrenology in Britain. However, it also helped phrenologists reach new audiences in the United States and the colonies.[21] The Calcutta Public Library's copy of *The Constitution of Man* was in fact a "People's Edition" printed in Edinburgh, as was the copy in the Melbourne Public Library.[22] By the mid-nineteenth century, the Fowlers in New York City were also printing a new cheap edition of *The Constitution of Man*, further reducing the retail price to just 30 cents.[23] Like William and Robert Chambers in Edinburgh, the Fowlers' use of new printing technologies allowed them to reach audiences far beyond the United States, with the Sydney Mechanics' School of Art purchasing copies of phrenological books printed in New York City.[24]

The emergence of publishing as an international business certainly helped phrenology extend its geographic reach. But the global circulation of printed books also introduced new problems. In some cases novel technologies actually impeded the progress of phrenology. Stereotyping—the manufacture of plaster casts of set type—provides a good example of these tensions. In the 1830s, Combe's official American publisher, Marsh, Capen and Lyon in Boston, commissioned stereotypes of *The Constitution of Man* and *A System of Phrenology*. This practice allowed publishers to quickly and cheaply reprint books that might later become popular. There was no need to reset type or leave it standing in the forme. The stereotype plates could simply be stored and printed from whenever they were needed. Stereotyping also allowed publishers to expand into overseas markets. Marsh, Capen and Lyon even hoped that they might be able to use the new technology to compete with British publishers. Nahum Capen, one of the proprietors, proudly explained to Combe that "we should suppose that we might send you an edition of almost any work cheaper than you would manufacture in Europe." Capen then offered to print 1,000 copies of Combe's *System of Phrenology* and ship them for sale in Britain for just 67 cents each.[25] Stereotypes were also commodities in their own right, with publishers buying and selling plates internationally.[26] In 1836, Capen suggested to Combe that the stereotype plates of *The Constitution of Man* might be sold to a publisher in Quebec. Marsh, Capen and Lyon would charge for the plates, and Combe would benefit from copyright payments.[27] William and Robert Chambers in Edinburgh also sold stereotype plates of their own "People's Edition."[28] After all, a deal with a foreign publisher "would be for the good of science, & for the profit of the parties interested."[29]

Despite these potential benefits, stereotyping made it difficult to keep phrenology up-to-date. This was a problem compounded by the emergence of a global phrenological readership. In the late 1830s, Combe clashed with his American publisher over exactly this. Marsh, Capen and Lyon already owned stereotype plates of *A System of Phrenology* and were refusing to print the latest revised edition sent from Edinburgh. It was simply cheaper to republish from the old plates. This meant that American readers did not have access to the latest British editions of scientific works. Combe worried that they might even be put off reading phrenology altogether. As he explained to Capen, "It is well known that your System is behind the English one, and many persons refrain from buying it accordingly."[30] To make matters worse, Capen suggested to Combe that he hold back on revising his books in the future. Marsh, Capen and Lyon didn't want any more of their stereotype plates to be rendered obsolete.[31] This did not go down well. Combe complained angrily to Capen, summing up the tensions that new technologies of reproduction imposed. In his opinion, stereotyping was "bad practice for works in an advancing science."[32]

For many phrenologists, the spread of printed books was a measure of progress. Combe boasted that his *System of Phrenology* had been "translated into German," *The Constitution of Man* had been "printed at Paris in the French language," and revised editions of his *Elements of Phrenology* were sold in the United States, reaching as far as the "Mississippi Valley."[33] Translation itself was part of an active program designed to extend phrenology across the globe. The Edinburgh Phrenological Society financed new translations, commissioning a cheap German edition of *The Constitution of Man* published in Mannheim in 1845.[34] By the end of the century, phrenology had been translated into Swedish, Spanish, and Bengali.[35] Revised editions also suggested that phrenology was constantly advancing, hence Combe's frustration with the practice of stereotyping. Combe proudly told an American correspondent that he was "sending new editions or new impressions of every one of my books thro' the press." The sale of them was "very great," from which, Combe believed, "the progress of the science may be judged."[36] The international success of works like *The Constitution of Man* apparently proved that phrenology was "wanted by the world."[37]

The global spread of phrenological books was tied to a particular conception of reading. For phrenologists, reading was a physiological activity.[38] The aim of a book like *A System of Phrenology* was to call into action each of the organs of the material mind. The reader would then be presented with the truth of phrenology through their own experience.[39] It was possible to "learn from books" to "distinguish the *form* of each organ." The "vigour of the reflecting faculties," for instance, could be directly "augmented by reading."[40] The organization of phrenological works typically reflected this, with chapters devoted to individual organs, inviting the reader to activate different areas of the brain. Readers were then encouraged to focus on those faculties that needed further development. Combe ultimately presented *The Constitution of Man* as a kind of mental workout, explaining that "the best mode of increasing the strength and energy of any organ and function, is to exercise them regularly and judiciously."[41] This approach relied on a belief in the universality of the reading experience. Hence, for phrenologists, there was a direct connection between the function of a book like *The Constitution of Man* and its global reach. Whether a reader was sitting in the South African Public Library or the New York Mercantile Association, they needed to be able to activate the same organs of the brain by reflecting on the same printed text. This, in turn, would produce converts to phrenology as each reader was made aware of the faculties of their own mind.

While cheap editions of phrenological books were certainly the most widely read, larger and more expensive works still played an important role. Folio atlases such as Gall and Spurzheim's *Anatomie et Physiologie du Système Nerveux* proved central to the formation of phrenology as a discipline.[42] Steel engravings and lithographs helped to sharpen the boundaries between phrenology and other sciences of the mind. It was Gall and Spurzheim's atlas, printed in Paris in 1810, that sparked debate between phrenologists and metaphysicians in Edinburgh.[43] Atlases allowed phrenologists to present the connection between the mind, brain, and skull as part of a visual argument. Joseph Vimont's *Traité de Phrénologie Humaine et Comparée*, for instance, featured a lithographic cross section of the head of a French soldier. This, Vimont claimed, revealed "the true position of the brain in the skull."[44] And although both Gall's and Vimont's works were printed in Paris, they also shaped phrenological understandings of the mind well beyond France. Gall's *Anatomie* found its way into the library of the Pennsylvania Hospital in the United States as well as the Royal College of Surgeons in London.[45] Vimont's *Traité de Phrénologie* was so popular that a Belgian firm pirated the work, including all 120 lithographic plates. With the original French edition in short supply, many Americans had to make do with the inferior Belgian copy.[46] Expensive atlases, therefore, need to be understood as part of an industrializing world of print and exchange just as much as octavos and duodecimos.

Through a close study of a single folio atlas, this chapter recovers how readings of phrenological texts and images were continually reworked and contested as they traveled. Published in Philadelphia in the winter of 1839, Samuel George Morton's *Crania Americana* was a transatlantic sensation (fig. 12). Readers in Boston and Paris marveled at its seventy-eight lithographic plates of Native North and South American skulls. In the accompanying text, printed on the same folio paper, Morton followed Johann Blumenbach in dividing man into five races before linking these races to

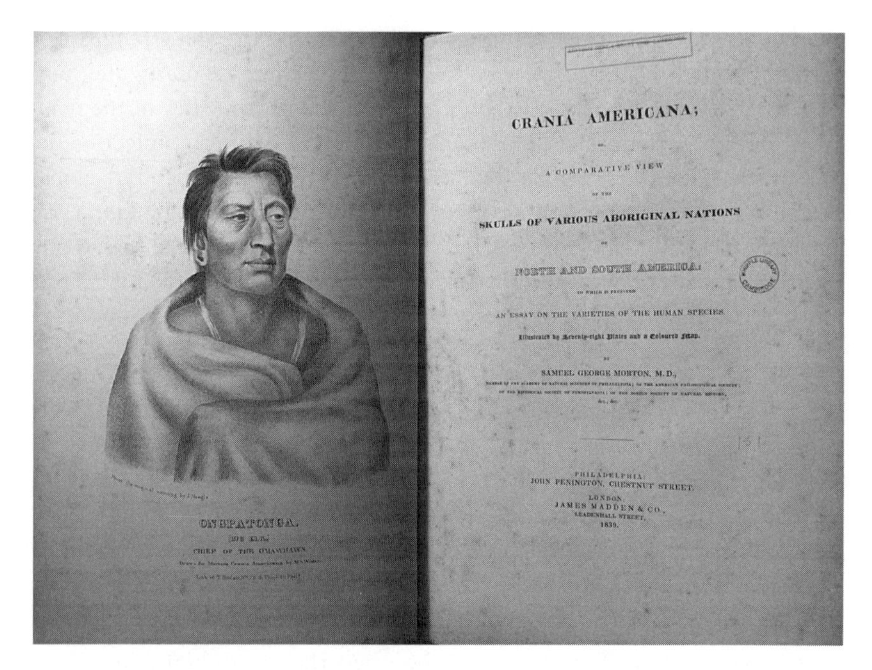
CRANIA AMERICANA;

A COMPARATIVE VIEW

SKULLS OF VARIOUS ABORIGINAL NATIONS

NORTH AND SOUTH AMERICA:

AN ESSAY ON THE VARIETIES OF THE HUMAN SPECIES.

Illustrated by Seventy-eight Plates and a Colourd Map.

SAMUEL GEORGE MORTON, M.D.,

PHILADELPHIA:
JOHN PENINGTON, CHESTNUT STREET.
LONDON,
JAMES MADDEN & CO.,
LEADENHALL STREET,
1839.

Fig. 12. Frontispiece and title page, Samuel George Morton, *Crania Americana* (Philadelphia: J. Dobson, 1839), Whipple Library, University of Cambridge.

skull configuration. By the 1850s, Morton was sufficiently acclaimed to be described as "the Founder of the American School of Ethnology."[47]

Historians today have largely followed this national narrative. Bruce Dain identifies *Crania Americana* as one of the "foundational texts of an American scientific movement."[48] Similarly, Ann Fabian describes Morton's work as "a distinctive American enquiry."[49] In contrast, this chapter resituates *Crania Americana* within the transatlantic world in which it was originally produced and read. There were distinct material, political, and intellectual challenges faced by historical actors when operating transnationally: crates of books were lost at sea amid the swells of an unforgiving Atlantic Ocean, while nationalist attitudes toward American science, particularly following the War of 1812, hampered any chance of a consistent reception across the Old World and the New.[50] Furthermore, while only one edition of *Crania Americana* was ever printed, the work nonetheless circulated in a variety of fragmentary forms. Copies of the prospectus, loose plates, and reviews from periodicals all crisscrossed the Atlantic in the months immediately before and after publication. The character of this two-way flow of traffic challenges conventional national histories of publication and reception. We cannot simply treat *Crania Americana* as a work produced in the United States and received in Europe. A range of actors on both sides of the Atlantic informed practices ranging from printing and publishing to advertising and reviewing.

For readers in the early 1840s, the disciplinary position of Morton's folio was far from obvious. It was only much later in the nineteenth century that *Crania Americana* was consistently referred to as a work of "American ethnology."[51] In contrast, readers and reviewers in the 1840s strived to situate Morton's impressive tome within a range of emerging sciences of man. The British physician James Cowles Prichard was adamant that *Crania Americana* should be read as a contribution to ethnology. Further reviewers positioned Morton's work within disciplines ranging from natural history to geography and medical physiology. Others also believed strongly that *Crania Americana* should be understood as a work of phrenology. It is this connection with phrenology that is most often glossed over by historians today. This is despite the fact that Combe wrote a twenty-three-page appendix for the work at Morton's request. Today, historians often ask whether Morton took phrenology seriously.[52] This is exactly the question nineteenth-century readers also struggled with: Was *Crania Americana* a work of phrenology or ethnology or something else? Beginning with the early arrival of Morton's loose lithographic plates in Birmingham, this chapter demonstrates how debates over the disciplinary position of *Crania Americana* dominated its initial reception on both sides of the Atlantic. The transatlantic world presented both opportunities and challenges for those hoping to claim *Crania Americana* on behalf of a variety of emerging human sciences. And it was precisely through these contested readings that disciplinary categories started to emerge. What it meant for a book to be phrenological was ultimately a product of global exchange.

BANGING HEADS AT THE BRITISH ASSOCIATION

On Tuesday, 7 August 1839, Prichard stood to deliver his paper at the annual British Association for the Advancement of Science meeting, that year held in Birmingham. The mood was charged. Only a month before, the Metropolitan Police had been sent to the city in order to contain Chartist crowds following Parliament's rejection of a major petition.[53] Addressing Section D (Zoology and Botany), less than a mile away from the ruined scenes of the Bull Ring Riots, Prichard delivered his paper "The Extinction of some Varieties of the Human Race."[54] In doing so, he hoped to secure financial and institutional backing for natural historical studies of mankind, a discipline he had only recently begun to identify by the term "ethnology."[55]

Up to that point, the British Association had not been particularly forthcoming. In the early 1830s the organizing committee had been dominated by men from the University of Cambridge, such as William Whewell. They promoted a hierarchical view of knowledge in which the Newtonian sciences occupied the pinnacle. This was reflected in the division of the British Association into different sections, with Section A (Mathematical and Physical Sciences) receiving the vast majority of funding. For fear of provoking religious controversy, the British Association had also excluded most of the emerging sciences of man: metaphysics, human geography, and phrenology were all rejected.[56] Nonetheless, the British Association was far from homogeneous. Under the leadership of the geologist Roderick Murchison in the late 1830s, debate continued on the relative merits of Newtonian and Baconian philosophy alongside the extent to which the British Association should be allied with the government.[57] Attendees at the annual meetings ranged from aristocrats and clergy to mechanics and schoolmasters. And, despite the rejection of phrenology, a medical section was established in 1836 in order to accommodate the large number of provincial physicians who attended each meeting.[58]

Prichard sought to establish a similar section for ethnology, promoting it as a discipline in its own right. What's more, having spoken at the 1832 meeting in Oxford to a lukewarm reception, he knew a simple speech would not be enough.[59] This year, things would be different. On the tables at the side of the lecture hall in Birmingham, Prichard displayed a set of loose lithographic plates that would later feature in *Crania Americana* (fig. 13). They had been sent directly by Morton following Prichard's election as a corresponding member of the Academy of Natural Sciences of Philadelphia earlier that year.[60] On seeing these naturalistic illustrations of Native North and South American skulls, each finely shaded, observers were quick to describe the plates as "splendid" and "beautiful."[61] Thomas Hodgkin, a founding member of the Aborigines' Protection Society, later wrote to Morton himself recalling "the pleasure of seeing a part of thy work displayed at the last meeting of the British Association."[62] Prichard's message was clear. While Britain's industrial towns fell into disrepair, the rest of the world was advancing its understanding of mankind. As the signature on each lithograph announced, these incredible images had been produced in the United States by American naturalists and artists. With Morton's plates there for all to see, Prichard confidently reminded his audience that, for Britain, "it would be a stain on her character, as well as a

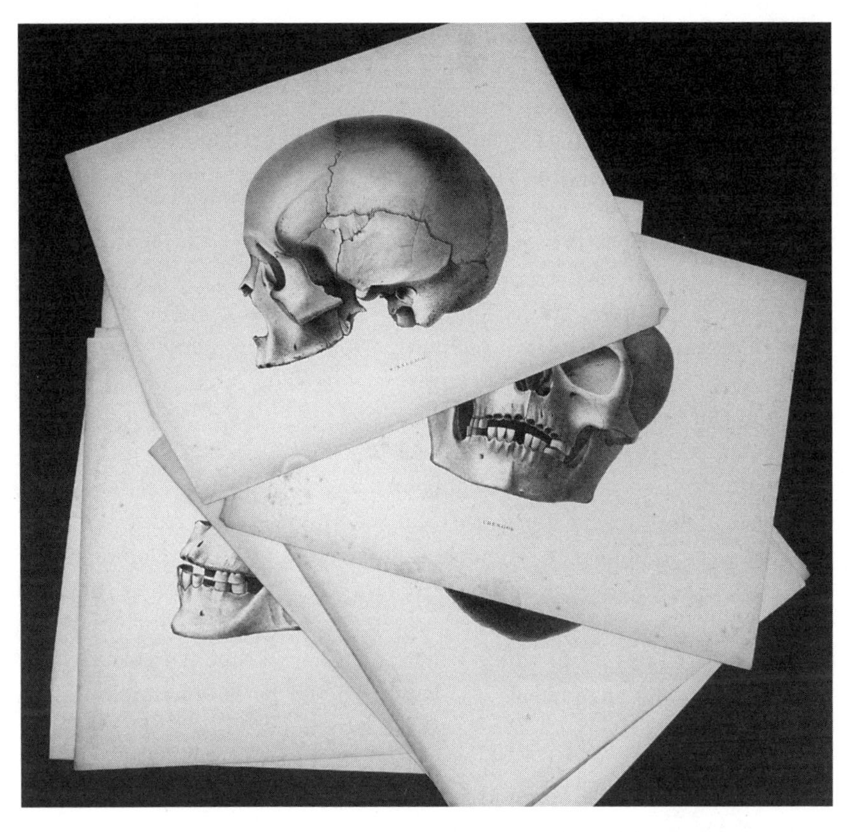

Fig. 13. Loose lithographic plates printed to promote Samuel George Morton, *Crania Americana*, Whipple Library, University of Cambridge.

loss to humanity, were she to allow herself to be left behind by other nations in this enquiry."[63]

The strategy seemed to work, at least in part. By the end of the meeting, Prichard had secured £5 for "Printing and Circulating a Series of Questions and Suggestions for the use of travellers and others, with a view to procure Information respecting the different races of Man." The motion was also supported by members of the British Association organizing committee, including Charles Darwin, secretary to Section C (Geology and Physical Geography).[64] The British Association defined its own reputation on an international stage, and so could not afford to ignore the advancing status of American natural history. Indeed, William Harcourt, in his first presidential address at the York meeting of 1831, had warned against a world in which "colony after colony dissevers itself from the declining empire, and by degrees the commonwealth of science is dissolved."[65] Harcourt chose an imperial metaphor to express his fear that specialization would lead to the disintegration of national science. Prichard played to these concerns. Morton's plates were presented as an opportunity for Britain both to reengage with its lost colony and to maintain a disciplinary connection with ethnology.

The specific choice to present craniological plates, rather than some other ethnographic illustration, also reflected Prichard's hope of carving out a new disciplinary space. The 1830s saw the development of novel visual languages as disciplines including geology, astronomy, and zoology all sought to gain institutional footing.[66] The same was true for ethnology.[67] In the first edition of *Researches into the Physical History of Mankind,* published in 1813, Prichard had followed Enlightenment scholars in assuming the importance of language and skin color for determining the races of man.[68] Not much had changed by the second edition of 1826, in which Prichard cross-referenced linguistic and racial development in a long appendix.[69] However, in 1836 Prichard began publication of the third edition of his *Researches* in five volumes. The first volume included nine lithographic plates, all of which depicted human skulls. An entirely new chapter had also been added, entitled "National Forms of the Skull," in which Prichard declared, "Of all peculiarities in the form of the bony fabric, those of the skull are the most striking and distinguishing. It is in the head that we find the varieties most strongly characteristic of different races."[70] The questionnaire he prepared following the British Association meeting reflected this too, in which he reminded recipients that "the head is so important as distinctive of race."[71] Prichard also demarcated three broad forms of skull including the "symmetrical or oval," the "narrow and elongated," and the "broad and square-faced."[72] With this in mind, we can better appreciate Prichard's choice of Morton's images in 1839. It was precisely during this period that Prichard moved away from a focus on language and toward identifying the human skull as the distinguishing feature of different races. In doing so, he hoped to develop a new visual language, one that would secure ethnology as a distinct discipline.

Not everyone at the Birmingham meeting was convinced. Once Prichard had finished speaking, Hewett Watson, editor of the *Phrenological Journal* in Britain during Combe's absence, stood to respond. His presence was no accident. From 1838 onward, the *Phrenological Journal* had organized its own annual Phrenological Association "independent of the British Association, although holding its meeting at the same times and places." By following the same circuit as the British Association, phrenologists such as Watson hoped to advance their own study among attendees "interested in the sciences relating to organic nature, and to *man*."[73] Prichard's paper was a good opportunity. As the *Phrenological Journal* later reported, Watson "felt it a duty on his own part, to state some phrenological facts bearing on the communication of Dr. Prichard . . . , a department of knowledge, towards which Dr. Prichard was known to be hostile."[74] This claim was not without foundation. In his 1835 *Treatise on Insanity*, published in London, Prichard had previously suggested that "phrenology will not continue to make proselytes . . . , it will be ultimately discarded as an hypothesis without foundation."[75] When Combe responded in private correspondence, Prichard simply dismissed him, explaining there was "want of sufficient evidence on so difficult a question."[76]

Despite Prichard's response, this debate actually had little to do with the quality of evidence. Indeed, both ethnologists and phrenologists alike privileged the human skull as the seat of national difference. Rather, at the British Association meeting in 1839, phrenology and ethnology vied for recognition as legitimate and independent disciplines: for Prichard and Watson, there could be only one natural historical study of mankind. Watson understood well that any funding Prichard might secure, no matter how small a sum, would amount to institutional acceptance from the British scientific establishment, the first step toward securing a separate Ethnological Section at future meetings.[77] In the hope of derailing these plans, Watson reminded Prichard's audience in Birmingham that the Edinburgh Phrenological Society's museum "contains probably the best collection of national crania in existence . . . and that in applying the funds of the Association in seeking further evidences, it would be going for that which was distant and dear, before that which was at home and of easy access." He then addressed the speaker directly: "May I ask Dr. Prichard whether he has examined the evidences contained in the museum I have alluded to?" Prichard responded, "No. I have not had the opportunity of doing so," with Watson firing back, "That is enough. I can say no more to one who asserts the insufficiency of evidences he has not examined."[78] By December, this exchange had made its way back to Philadelphia. Combe complained to Morton that "Dr Prichard asked for funds from the British Association for investigating, but without the aid of Phrenology, the very points which you have accomplished."[79]

Watson's rhetorical move ultimately failed, and Prichard was awarded the £5. But his intervention illustrates how, in the same year in which *Crania Americana* was published, phrenologists and ethnologists each sought to establish distinct studies of mankind. With each discipline privileging the skull, Morton's plates were subject to competing interpretations. Following Watson's critique, another supporter of Prichard, George Thompson, rose to respond. He announced that it was "his conviction that the uncivilized races had heads equally well-formed as were those of their destroyers." Thompson also added, apparently "with a sarcastic laugh," that these races were not so different from "large-headed Englishmen . . . , even Mr. Watson himself." This riled the phrenologist. Watson pointed to Morton's plates, "only a few feet distant." Challenging the ethnological account, he directed the audience toward "the figure of a Pawnee skull," describing it as "villainously low."[80] Here, Watson's language contrasts sharply with Prichard's. While Prichard opted for geometric labels ("symmetrical" or "narrow"), the phrenologist saw Morton's plates in moral terms ("villainous").

This contest, over the correct reading of both *Crania Americana* and its accompanying lithographs, came to dominate its reception in both Europe and the United States. In Britain, Prichard had a head start. The transatlantic publishing context separated these early lithographic plates from Morton's own text as well as Combe's phrenological appendix. With these out of the way, Prichard was free to read Morton's unique illustrations in terms of a new visual language, one he claimed as distinctive of ethnology. From then on, the phrenologists were playing catch-up.

A PHRENOLOGICAL APPENDIX FROM SCOTLAND

Shortly after Combe's arrival in Boston during the winter of 1838, Morton introduced himself. The two were quick to strike up a friendly correspondence. Morton's stories of his time at the University of Edinburgh as a young man certainly went down well with Combe, particularly when Morton informed him, "I occasionally attended the meetings of the Phrenological Society." At this time, prior to his contact with Prichard, Morton still saw *Crania Americana* primarily as a contribution to phrenology. On first mentioning the book to Combe, he wrote, "Although I am imperfectly informed on the subject of Phrenology, I have been for some years engaged in collecting a mass of facts which will bear directly on the science." Before long, Morton had offered Combe the chance to write "a paper on the Phrenological development of the American Race" to accompany the book.[81] Impressed by the quality of the lithographs, Combe quickly took Morton up on his offer and began working on the appendix in early 1839.

In correspondence, Morton urged Combe to use his essay to explain "the principles of Phrenology, & their application to the heads of the American Race."[82] The phrenologist did not disappoint. Combe presented the study of national character as an essentially phrenological question. In his opening paragraph, he complained that the topic had "been investigated by philosophers in general, without any knowledge of, or reference to, the functions of the different parts of the brain." In Combe's eyes, Blumenbach's attention to skin color and Prichard's early interest in philology marked them out as very different, ultimately flawed, intellectual enterprises. What made Morton's work phrenological, according to Combe, was his attention to the materiality of the mind. He followed "a more perfect method of investigation," one characterized by attention to "the relative magnitudes of the whole brain, and the relative proportions of the different parts of the brain, indicated by the national skulls."[83] Combe also wanted to guide readers away from interpreting *Crania Americana* within existing natural historical traditions. This was certainly a legitimate concern, with the *North American Review* giving notice of Morton's prospectus under the heading "natural history" in 1838.[84] For Combe, natural history without natural philosophy was just cataloguing. Only phrenology allowed one to understand the workings of politics, empire, and industry. Earlier in the century, the phrenologists in Edinburgh had clashed with their opponents over exactly this: the extent to which social and political issues should be manifest in mental science.[85] This problem now made its way across the Atlantic. Without phrenology, Combe argued in Philadelphia, "these skulls are mere facts in Natural History, presenting no particular information as to the mental qualities of the people."[86]

For Combe, as for Prichard, the plates were crucial. Although unaware of Prichard's interest in *Crania Americana* at this point, Combe too hoped to claim Morton's illustrations on behalf of a new visual language, but this time characteristic of phrenology. Morton concurred in his own preface, explaining that Combe's essay would enable the reader "to apply Phrenological rules to every skull in the series here figured."[87] But in order for this argument to work, Combe needed to introduce a point of reference. Shortly after completing the draft in April 1839, he convinced Morton to commission an additional lithograph, featured as plate 71 in the final volume (fig. 14). It depicts a "Swiss" skull from the collection of the Edinburgh Phrenological Society, one that Combe carried with him as part of his lecture tour. It was with this illustration that Combe hoped to secure his phrenological reading. In the appendix he explained that "by comparing the dimensions of this Swiss skull as they appear to the eye in the plate, with those of the other skulls delineated in this work,

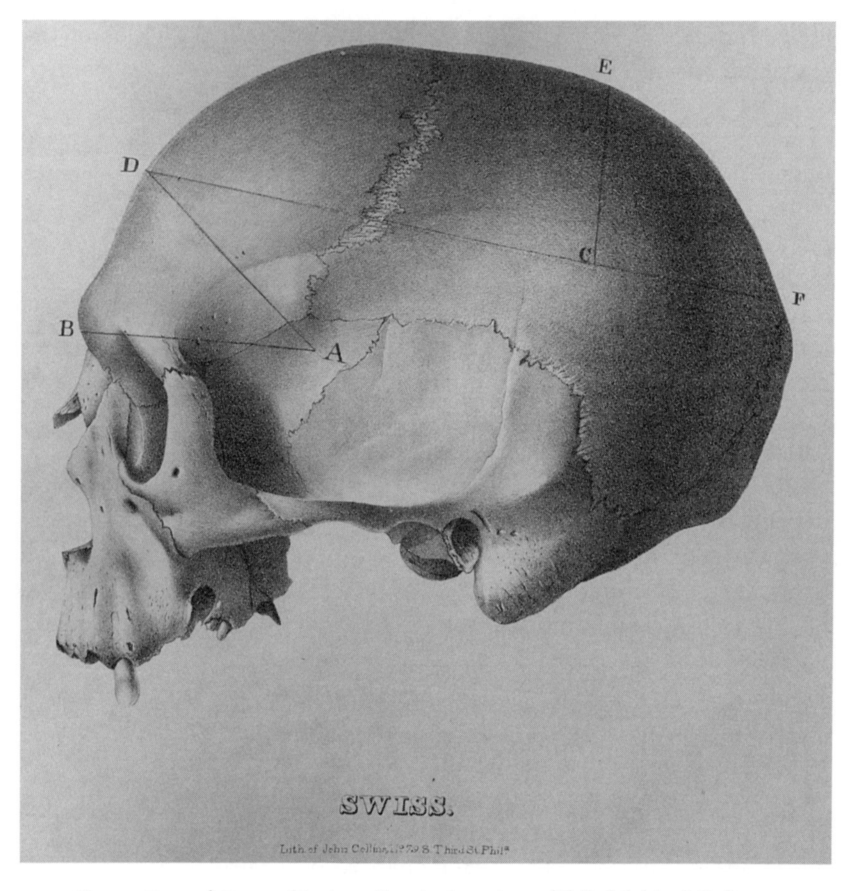

Fig. 14. Samuel George Morton, *Crania Americana* (Philadelphia: J. Dobson, 1839), pl. 71, Whipple Library, University of Cambridge.

all being drawn as large as nature, their relative proportions will become apparent."[88]

Once again, the material form of the illustrations mattered. With the lithographer John Collins charging $8.75 per commission, Morton initially hoped to lower costs by having the Swiss skull drawn at a reduced size.[89] He informed Combe that it would be "lithographed at least at half size . . . which I hope will answer the purpose."[90] For Combe, this would not do. The skulls all needed to be lithographed to the same scale in order to allow for comparison by eye. Indeed, Combe had earlier complained that many of Morton's other illustrations appeared "all foreshortened," ultimately convincing Collins to start over.[91] On seeing a proof of the Swiss lithograph while in Philadelphia, Combe eventually persuaded Morton to have it redrawn at full size.[92] He also ensured that the lithograph accurately indicated the major phrenological divisions, explaining in his appendix that "the space included in D, A, B, denotes the dimensions of the anterior lobe devoted to intellect," whereas "the space included in E, C, D, to manifest the moral sentiments."[93]

Soon after the printing of this final lithograph, word reached Combe of Prichard's success in Birmingham. The early reception of Morton's plates in Britain then fed back into the printing and publication of *Crania Americana* in Philadelphia. With this disciplinary challenge under way, the appendix took on even greater importance. Earlier in the year, Morton had informed Combe that it would need to be printed in small type: "I promised Mr Fuller it should not exceed 30 pages, & that I would keep it within the limits by reducing the type." Morton tried to account for this in terms of the lack of support from his publisher, writing, "I am not backed by some responsible & enterprising publisher: but having to meet every expense from my own purse."[94] This chimed with a narrative in which American naturalists considered it difficult to secure patronage at home: most notably, John James Audubon had found it necessary to sail to Britain in order to seek both subscribers and a publisher for his extraordinary *Birds of America*.[95] But, on hearing the news of Prichard's display at the British Association, Combe put his foot down. In the end, Morton relented, and in October 1839 he recorded that the "Appendix to my book is nearly printed in the same type as my own."[96] Combe also tried to persuade Morton to superimpose phrenological divisions on his earlier plates. However, by this time, it was too late. Morton reasonably informed Combe, "As the whole edition of every plate is already struck off, it will not be in my power to insert marks for the centres of ossification & causality."[97] Still, Morton did make one concession. On plate 40, "the Cotonay head," Collins added a small "X" (fig. 15). This indicated the point at which the parietal bone joins the sphenoid, from which the "reflecting organs" had been measured.[98]

Despite these successes in late 1839, not everything went the phrenologists' way. As soon as he returned to Bristol following the British Association meeting in August, Prichard wrote to Morton describing the favorable reception the plates had received: "I took your beautiful plates to the meeting of the British Association at Birmingham where they were exhibited publicly and much admired."[99] In particular, he informed Morton about the interest the plates had generated among his fellow ethnologists, rather than phrenologists such as Watson. In an attempt to cement this disciplinary connection, Prichard wrote, "My friend Dr Hodgkin . . .

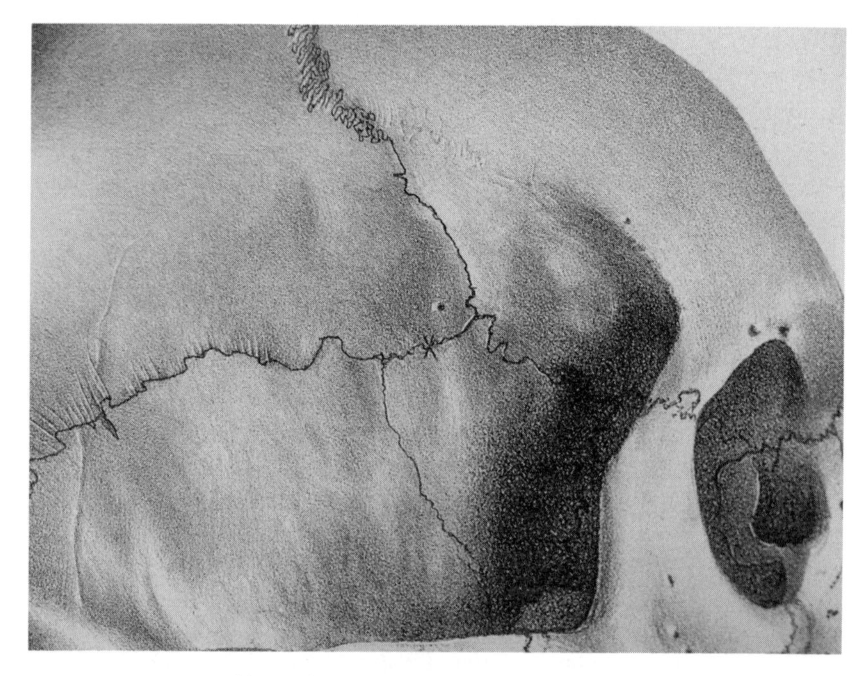

Fig. 15. Detail from pl. 40, Samuel George Morton, *Crania Americana* (Philadelphia: J. Dobson, 1839), Whipple Library, University of Cambridge.

is particularly interested in the subject of ethnography . . . I will mention your book to him."[100] Finally, Prichard highlighted his efforts in promoting *Crania Americana* in Britain, offering to help "accelerate the acquaintance of my countrymen with your work."[101]

At a time when Morton was embroiled in a troublesome working relationship with his alcoholic publisher, John Fuller, this news evidently pleased him.[102] In fact, Prichard's early use of the plates at the British Association ultimately prompted Morton to divide the entire edition of *Crania Americana* into two different states. In December 1839 Morton separated the final 500 printed copies into what he described as "two editions, the American and the Foreign."[103] The "American edition" of 400 copies featured two dedications: one to John Phillips, Morton's assistant in Philadelphia, and another acknowledging the help of William Ruschenberger, an American naval surgeon and colleague at the Academy of Natural Sciences of Philadelphia.[104] Morton had something else in mind for the remaining 100 copies. As he explained to a despairing Combe in 1840, "I dedicated my work to Dr Prichard. . . . He has shewn a great interest for the success of my work in several letters addressed to me, & his communication to my friends."[105] The "foreign edition" therefore featured three dedication pages. Phillips remained, but Ruchenberger was omitted entirely. In his place, Morton inserted a dedication to his Irish uncle James Morton, and a separate sheet praising "the learned and ingenious author" James Cowles Prichard. Even more disturbing for the phrenologists, Morton explicitly linked his own publication to Prichard's ethnological project. The dedication went on to read that *Crania Americana* was "designed to illustrate a portion of the same interesting enquiry" as Prichard's *Researches into the Physical History of Mankind* which had recently entered into a third and expanded edition (fig. 16).

Combe did his best to convince Morton that this was an error. He

TO

JAMES COWLES PRICHARD, M. D., F. R. S., &c. &c. &c.,

OF BRISTOL, ENGLAND,

THE LEARNED AND INGENIOUS AUTHOR

OF

"RESEARCHES INTO THE PHYSICAL HISTORY OF MANKIND,"

THE FOLLOWING WORK,

WHICH IS DESIGNED TO ILLUSTRATE A PORTION OF THE SAME INTERESTING INQUIRY,

IS MOST RESPECTFULLY

DEDICATED.

Fig. 16. Dedication to James Cowles Prichard in the "foreign edition" of Samuel George Morton, *Crania Americana* (Philadelphia: J. Dobson, 1839), Special Collections, University of Bristol.

warned him that "Mr Hewett Watson intended to purchase three copies of your work at his own expense & present them to public institutions . . . , when he read the dedication to Dr Prichard he abandoned his purpose!"[106] But it was too late. Early reviews in Britain were quick to spot the dedication, and many naturally assumed *Crania Americana* should be read as part of Prichard's ethnological project. The *British and Foreign Medical Review,* published quarterly in London, helpfully informed its readers that the "volume is dedicated to our illustrious countryman, Dr. Prichard." Quoting the dedication directly, the reviewer added, "We need scarcely add our own opinion that to no one could this work, 'which is designed to illustrate a portion of the same interesting enquiry,' be more appropriately inscribed." In fact, the reviewer even recommended skipping over Morton's introductory essay, "since it contains little that will be new to the readers of Dr. Prichard's elaborate treatise."[107] In bits and pieces, *Crania Americana* made its way back and forth across the Atlantic Ocean. Throughout this process, it was read within different disciplinary traditions, both old and new. By the end of 1839, Prichard looked to have secured Morton's work as an ethnological atlas. Still, the phrenologists did not give up. Back in the United States, Combe redoubled his efforts to manage this troublesome transatlantic publication.

APPOINTING A BRITISH PUBLISHER

For Morton and his supporters on the East Coast, the reception of *Crania Americana* in Europe was at least as important as its success in the United States. Earlier in the century, episodes such as the discovery of the Gloucester sea serpent had fed the prejudices of European naturalists, many of whom believed American science to be untrustworthy. When the sightings of 1817 were confirmed as a hoax, the British geologist Gideon Mantell described them as just another "Yankee lie."[108] But even Benjamin Silliman, editor of the prestigious *American Journal of Science,* had taken the sea serpent seriously. This, alongside the 1835 "Great Moon Hoax" reported in the *Sun* of New York City, seemed to confirm the misgivings of both *Chambers's Edinburgh Journal* and the *London and Foreign Quarterly Review.* The United States was considered a "country of sensations" supported by a "degrading and disgusting" press.[109] Morton recognized this was especially true of natural historical studies of mankind, noting the field was "regarded with suspicion & distrust."[110]

Promoters of American science also worried about their dependency on European texts and technologies: book imports into the United States far exceeded exports throughout the nineteenth century, while the majority of locally printed works were simply European reprints.[111] This dependency extended to binding cloth, which was routinely sourced from European suppliers.[112] Even the limestone used for Morton's lithographs was imported from Bavaria via New York at a cost of 10 cents per pound weight.[113]Nonetheless, while early Philadelphia lithographers struggled to match the quality of French imports, the situation had changed considerably by 1839.[114] The city housed six well-respected lithography firms. These included John Collins of 79 South Third Street, responsible for the majority of lithographs found in *Crania Americana*, including Combe's Swiss skull, as well as those displayed at the British Association.[115] Lithography also came to be understood as a means by which American science could establish itself as reputable on the global stage.[116] Firms such as Thomas Sinclair, which purchased Collins's business during the publication of *Crania Americana*, specialized in natural history illustration.[117] This specialization was accompanied by a move away from the realism favored in the early nineteenth century, in which specimens were depicted as active in an environment. Instead, increased European trust in American natural history coincided with the portrayal, as in *Crania Americana*, of stationary specimens presented against a plain white background.[118]

On receiving his own copy of the final bound volume in February 1840, Silliman was therefore quick to recognize the potential value of *Crania Americana* in promoting American science abroad.[119] He reassured Morton that there was "no doubt it will do you great honor in Europe."[120] For Silliman, Morton's personal reputation was also tied to the success of American national science. He went on to write that "you cannot be compensated except in reputation & in the consciousness of having added to the reputation of your country for the sacrifice of so many years."[121] *Waldie's Journal of Polite Literature*, a Philadelphia monthly, singled out Morton's plates in particular, reporting, "We have not seen any thing of the kind from any European press, either English, French, or German, which exceeds the drawings of Mr. Collins in fidelity of representation, or in the beauty of execution or delicacy of finish."[122] The *Christian Examiner* in Boston even praised the presswork itself, calling *Crania Americana*, "typographically speaking, one of the most magnificent the country has produced."[123]

Despite all this praise in his home country, Morton was well aware that getting his work noticed in Europe would not be straightforward. First and foremost he needed a British publisher. In this respect, the phrenologists had an advantage over Prichard. While Combe toured the United

States between 1838 and 1840, he remained in close contact with Morton through correspondence. He also made a number of stopovers in Philadelphia during the spring of 1839.[124] During these visits, Morton was quick to recognize Combe's international reputation as the author of *The Constitution of Man* alongside his expertise in transatlantic publishing. In October he told Combe, "Whatever publisher you recommend I will employ & I will thank you to write to him on the subject."[125] This opportunity allowed Combe to direct Morton toward a British publisher sympathetic toward phrenology. Three days later, Combe replied, recommending "Messrs Simpkin, Marshall & Co, London, as unexceptionable publishers, safe as to responsibility, punctual, & attentive."[126] By December, the decision had been set in ink. Morton sent Combe a short note: "I am extremely obliged for your communication and advice respecting the publication of my book in London, & have accordingly put the names of Simpkin, Marshall, and Co on my title page as the London publishers."[127] This firm, based at Stationers' Hall Court, published numerous editions of Combe's own books, including *Elements of Phrenology* and *Outlines of Phrenology.* Simpkin, Marshall, and Co. also published the *Phrenological Journal* from 1827. For British audiences in particular, the presence of this publisher on the title page ensured that the phrenological import of *Crania Americana* could not be ignored entirely.

Even after choosing a publisher, there was still a lot Morton did not understand about the nature of transatlantic publishing. Combe was therefore eager to highlight his own expertise, informing Morton that "the terms on which London publishers transact business will astonish you."[128] In a series of letters, Combe carefully set out the logistical challenge of shipping *Crania Americana* to Britain and getting it noticed. Morton had already suffered one calamity. On the night of Friday, 27 December 1839, the brig *J. Palmer* foundered off the coast of Cape Cod. Forty copies of *Crania Americana*, destined for the British market, were lost to a tide of "unprecedented height." The next morning, locals combing the shoreline discovered only "some pieces of boxes" and a "waistcoat, with the name of, 'S. Browne,' on the back of it."[129] As Morton lamented in a letter to Combe, the books were "not insured. . . . My subscribers to the eastward have been prevented from obtaining their copies."[130]

Prichard already had a head start, so Combe was keen to avoid further delays. As soon as *Crania Americana* was published in December 1839, Combe begged Morton to "use the speediest means" to have a copy delivered to Watson in London.[131] He also informed Morton that *Crania Americana* would be subject to import duties on arrival in Britain. The lithographs in particular would prove expensive: with seventy-eight illustrations at 1d. each, they added another 12s. 6d. per copy.[132] Additionally, while Morton at first planned the work as a quarto with separate folio lithographs, he later agreed with Combe to incorporate both illustration and text into a single folio volume.[133] Copies were then bound locally before leaving Philadelphia, making it more difficult for owners and booksellers to discard the phrenological appendix or separate out individual lithographs.[134] But this too increased the costs associated with import, as another duty needed to be paid on arrival at 5d. per pound weight.[135] Combe also warned Morton that London publishers would charge "the expenses of carriage to their warehouse, of all advertisements, and postages." What's more, as Simpkin, Marshall, and Co. paid accounts in arrears, Combe advised Morton that there was "no prospect of your realising any sum from London sales in less time than two years."[136] In short, publishing *Crania Americana* in Britain would prove expensive and required capital up front to pay duties and freight charges.

With Morton still paying back his workmen in Philadelphia, he could not easily afford the additional costs associated with a British publisher.[137] Fortunately, his Irish uncle passed away early in 1840, leaving Morton a large legacy.[138] This solved the immediate funding problem. Still, in the face of Combe's evident expertise, Morton decided to hand control over to the phrenologist. Writing in January 1840, Morton informed Combe that he planned to "put the whole business at your disposal, with a thousand thanks for this additional proof of your friendship."[139] This ultimately allowed Combe to manage the reception of *Crania Americana* much more closely, particularly in Britain. From then on, as Morton's appointed representative, Combe was able to instruct Simpkin, Marshall, and Co. on pricing, advertisements, presentation copies, and suitable reviewers.

ANCIENT PERUVIAN HEADACHE

In early 1840, Morton began to prepare a consignment of the "foreign edition" to replace that lost the previous December. By sail, the books would take between twenty-five and forty-five days to reach Liverpool, before being sent down to Simpkin, Marshall, and Co. in London.[140] With little chance of securing an extended review of *Crania Americana* in the British press during the interim, Combe did what he could to guarantee a phrenological reception back in the United States. On topics related to American natural history, British periodicals were known to pay attention to major East Coast quarterlies, such as the *North American Review*, as well as more specialized scientific publications. A favorable review in the right place, Combe understood, would feed back into the British reception the following season.[141] Once again, Combe's proximity to key figures within the American scientific establishment proved decisive. While in New Haven during the February of 1840, he met with Benjamin Silliman, editor of the *American Journal of Science* and professor at Yale University.[142] Silliman had just received his own copy of *Crania Americana* and was mulling over whom to select as a reviewer.[143] Initially, he had been considering the Philadelphia physician and botanist Benjamin Hornor Coates. But different reviewers provided different skills. Combe advised Silliman that Coates was "more of a naturalist." In contrast, Combe offered to "shew the high moral & historical interest of the work."[144] In the end, Silliman opted for the phrenologist, writing, "No man in this country—or probably in any other—can be so good a judge of the merits of this work as yourself."[145]

Combe quickly accepted. The format in particular was appealing. Silliman offered Combe a review in "any form & to any extent you choose."[146] This provided an opportunity to address a number of issues left out of the appendix because of Morton's restrictions. Indeed, the final review submitted to Silliman occupied thirty-five pages compared to the twenty-three permitted in *Crania Americana*. Additionally, Combe had written the appendix without access to Morton's finished text: he had seen only the lithographs and skulls. A review would therefore allow Combe to establish a phrenological connection with the main body of *Crania Americana*. As he reminded Morton, the review would also "shew the bearing of your work on Phrenology."[147]

Nonetheless, Combe still needed to negotiate other aspects of the format. Silliman had initially planned to print Combe's notice in the April issue under "the authority of your name."[148] But Combe knew that if his review was to be taken seriously back in Britain, it needed to be anonymous. He explained to Morton that "as I wrote the appendix for you, my name on the Review would have injured its influence. It would have made it appear like a notice written by the author." What's more, Combe specifically had the British reception in mind when he made this request, informing Morton that "the article on your work will appear Editorially. I begged for this on your account & my own. An Editorial notice has double the weight of a communicated one. It will tell more in your favour in Europe."[149] Combe understood well that decisions concerning format in the United States had the potential to shape the reception of a review back in Britain.

With these points agreed upon, Combe set to work. The review opened with the usual pleasantries, praising *Crania Americana* and situating it within a transatlantic context. Combe pointed to the "beauty and accuracy" of the lithographic plates and anticipated "a cordial reception by scientific men not only in the United States, but in Europe."[150] With this out of the way, Combe then embarked on an extended summary of Morton's introductory essay before pointing readers toward the core value of the work. "Thus far Dr. Morton has travelled over ground previously occupied by other naturalists," Combe wrote. What made *Crania Americana* exceptional, according to Combe, was its connection to phrenology. Morton "had the courage and sagacity to enter on a new path . . . with a view to elucidate the connection (if there be any) between particular regions of the brain and particular mental qualities of the American tribes."[151]

Tellingly, it is only at this point that Combe introduces a series of illustrations into his review. The review features six woodcuts in total, five of which are copies of *Crania Americana* lithographs. However, the initial woodcut depicts two human brains and is entirely new (fig. 17). The brains are presented side by side from a top-down perspective and identified by Combe as "the brain of an American Indian" and "the brain of a European." Additionally, the different lobes are labeled from A to D on each brain. Combe went on to explain, "In the American Indian, the anterior lobe, lying between AA and BB is small, and in the European it is large." Making the final link to phrenology, Combe informed his readers that the anterior lobe was responsible for "the intellectual faculties."[152]

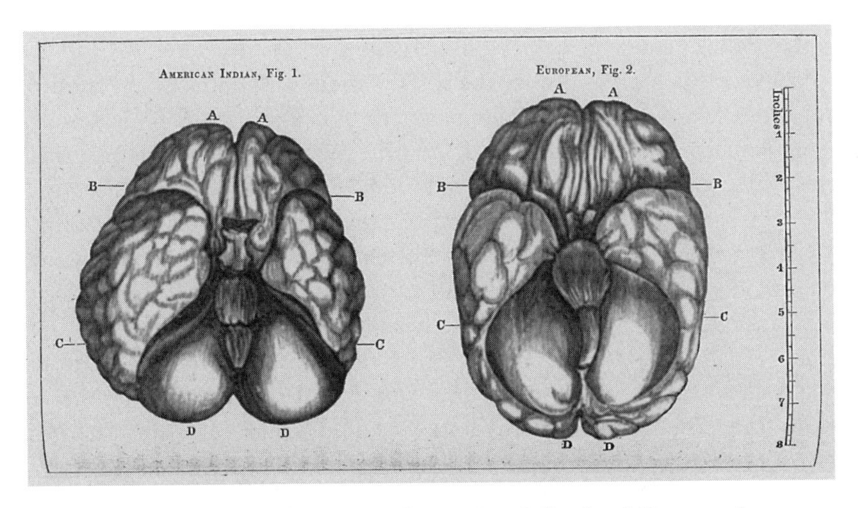

Fig. 17. Woodcut illustrations of "American Indian" and "European" brains, [Combe], "Crania Americana," *American Journal of Science* 38 (1840), Special Collections, University of Bristol.

The introduction of this particular woodcut at this particular point in the review is significant. There are no illustrations of brains in the entirety of *Crania Americana*. This woodcut therefore allowed Combe to forge a connection missing in the work itself: an inferior skull implied an inferior brain. As Combe made clear, "We use the cuts only to illustrate the fact that the native American and the European brains *differ widely in the proportions of their different parts*." It was precisely at the point in which Combe introduces phrenology that this woodcut appears. It is printed directly after the page in which Combe first claims "the necessity is very evident of taking into consideration the *relative proportions of the different parts of the brain*, in a physiological enquiry into the connection between the crania of nations and their mental qualities."[153] Combe further reinforced this link by printing a woodcut copy of the "Swiss" skull from *Crania Americana*, complete with phrenological divisions, straight after the brain illustration. He then invited the reader to compare the two images, writing, "In this figure (Fig. 3,) . . . line AB denotes the length of the anterior lobe from back to front, or the portion of brain lying between AA and BB in figures 1 and 2."[154] While Prichard certainly saw the skull as a key factor in national difference, he was unwilling to link this to a material mind.[155] Indeed, there were no brain illustrations in any of Prichard's three editions of *Researches into the Physical History of Mankind* available at this time. But seeing this connection, Combe argued, was key to reading both Morton's text and images as contributions to phrenology. This review therefore challenged Prichard's use of *Crania Americana* as emblematic of a new visual language distinctive of ethnology.

Despite his high praise for the work, Combe was not entirely uncritical of *Crania Americana*. The chapter on Peruvian skulls proved problematic, both politically and intellectually. Morton explicitly denied that these specimens showed evidence of head binding and artificial deformity. But this seemed to contradict Combe's phrenological assessment. While preparing the review in February 1840, Combe wrote to Morton setting out his misgivings: "The only part of your work which puzzles me is that which treats of the ancient Peruvian heads, & at once denies that they are compressed & yet ascribes to them high civilization."[156] Indeed, Morton had divided the Peruvian skulls into "two families": the "Modern Peruvians" and the "Ancient Peruvians."[157] The Ancient Peruvians, according to Morton, typically featured a "low facial angle" and a "sloping forehead." Yet these "seemingly brutalised crania" had been discovered amid the magnificent archaeological remains at Tiwanaku in South America, a site Morton compared to "the Theban catacombs." Morton therefore concluded that while one might assume "a people with heads so small and badly formed would occupy the lowest place in the scale of human intelligence," they had in fact "attained a considerable degree of civilization and refinement."[158] This idea of an ancient intellectual race also provided support for South American independence movements, to which the United States government was broadly sympathetic.[159] In political tracts, Agustín Gamarra, the president of independent Peru at the time, contrasted the Spanish era of backwardness with a glorious ancient past. Earlier preindependence nationalists such as Tupac Amar II even claimed descent from ancient Incas in the hope of amassing support.[160]

Combe, however, was not impressed. He later wrote in the review that "there is a discrepancy between this description of these skulls and the civilization ascribed to their possessors."[161] A number of readers soon picked up on the damning implications of Morton's analysis for phrenology. The *North American Review* complained that "whatever may be the views entertained with regard to the truth of the doctrine of the phrenologists, we are not apt to attribute a high degree of mental capacity to heads of an *anti-Caucasian* formation."[162] John Augustine Smith, a New York physician and critic of phrenology, also cited Morton's work in his *Select Discourses on the Functions of the Nervous System in Opposition to Phrenology*. He too singled out the chapter on Peruvian skulls, ridiculing the idea that these Peruvians "had less talent than those whom they ruled."[163] To make matters worse, Morton had already rejected one possible solution to this problem, writing in *Crania Americana* that "it is difficult to imagine by what complex contrivances the present shape could have been produced."[164] Combe then queried Morton on this point, writing, "If these skulls had been compressed by art, we could have understood that certain portions of the brain might have been only displaced, but not destroyed."[165]

With the review close to press and no resolution in sight, Combe privately invited Morton to reconsider his position: "How can these contradictory facts be reduced to consistency with nature? . . . I should be greatly obliged by your remarks by return of post."[166] Morton did respond quickly, stating privately that he had "been hasty in considering it the cranial type of the nation. . . . I cheerfully abandon a hypothesis which is at variation with nature & analogy."[167] However, with the book already printed, options for redress were limited. At first, Morton offered to publicly retract his entire interpretation of the ancient Peruvian skulls. At the beginning of March 1840, he sent Silliman a letter to this effect, suggesting it be printed at the end of the *American Journal of Science* review.[168] But Combe intervened. He warned Morton that the letter "was too broad an admission of your own error, too strong a condemnation of what you had written about the ancient Peruvians, and too complete an abandonment of your own opinion & inferences." With European trust in American natural history still hanging in the balance, Combe could not risk accusations of incompetence, particularly back in Britain. Combe warned Morton that this would "reduce the reliance of your readers on your general case, caution, & accuracy." Combe therefore edited the letter, informing Morton that he had "avoided, as far as possible, this evil, & at the same time made you state the essential fact that no. 4 is not the cranial type, & that you are engaged in procuring further information."[169] In the final printed review, Morton's heavily edited postscript simply read: "I wish to correct the statement, too hastily drawn, that it is *the cranial type* of their nation . . . Signed, SAMUEL GEORGE MORTON."[170]

Combe wasn't the only barrier to Morton's retraction. The transatlantic publishing context also presented further complications. Initially, Morton had hoped to issue a second edition of *Crania Americana* soon after the first. The plan was to "wholly remodel" the chapter on Peruvian skulls.[171] But there was a problem. Morton had already agreed not to publish a second edition "while any copies remain unsold in the hands of Messrs S. M. & Co." Given the value of the stock soon to be sent to London, Morton could not risk violating the terms of his agreement with Simpkin, Marshall, and Co. Furthermore, by March 1840, only a few copies had been bound and sent to reviewers. Indeed, Combe's private criticism had caused Morton to stall: both the "foreign" and the "American" edition remained at the printer's warehouse. With little prospect of a second edition any time soon, Morton therefore chose to make changes to those copies that had yet to leave Philadelphia. He redrafted the opening of his Peruvian chapter, making a few subtle changes to the wording on two of the four pages comprising sheet 25. Morton then cancelled and replaced this entire sheet in all remaining copies, writing that "in reference to my 'Ancient Peruvians' I mistrusted my inferences . . . & had cancelled two pages of that chapter after the first copies had been sent to New York." For the review, Combe had access only to Silliman's early unaltered copy, but Morton explained, "By comparing the copy you received in that city with the other sent from here, you will see that I softened down my position as much as possible."[172] Following Combe's earlier advice, the changes were not extravagant. Morton simply omitted a line in which he described an ancient Peruvian skull as "altogether peculiar."[173]

The "American edition" therefore existed in at least two different states: one with the original sheet 25 and one with the replacement. As the surviving copies of the "foreign edition" had yet to leave Philadelphia at this point, these feature only the replacement sheet. These changes therefore represent another means by which, within a transatlantic world of reviewers, printers, and publishing agreements, *Crania Americana* took on an increasingly fragmented form. Combe's hope of securing a global phrenological reading looked increasingly unlikely.

COMPRESSING AND SUPPRESSING REVIEWS

For an expensive atlas such as *Crania Americana*, limited to 500 copies, the reception of a review could prove just as significant as the reception of the book itself. Particularly in Europe, where copies were even more scarce, a greater number of readers encountered *Crania Americana* as an octavo review rather than as a complete folio volume. Combe's notice in the *American Journal of Science* proved a particularly common source on both sides of the Atlantic. Certainly, the journal itself was considered noteworthy in Europe and the United States, as evidenced by the range of international resellers. Combe's April publication was also relatively early, with most American reviews not reaching the press until July 1840 and British reviews often being delayed even further into October and November. Morton himself acknowledged this, informing Combe that the work was "selling much better in this country than I could have expected; which I attribute chiefly to your prompt & flattering review in the American Journal of Science."[174] But most importantly, Combe's six woodcuts allowed readers to experience, albeit indirectly, what was for many the most novel aspect of *Crania Americana*: Morton's lithographs.

At the beginning of March 1840, Silliman's compositors in New Haven had finished setting the *American Journal of Science* review in type. By making use of local publishing networks, Combe ensured that both his notice and the accompanying illustrations moved in tandem, at least on the East Coast. Prior to publication in April, Combe arranged for offprints to be sent down to Philadelphia.[175] There, Adam Waldie, publisher of the *American Phrenological Journal*, prepared for the review to be copied into the June number. Waldie knew that, if he wanted to use the original woodcuts, he would need to wait until after the publication of the *American Journal of Science* in April. This would have pushed his reprint back to July. He therefore decided to hire a local artist to copy the illustrations from the offprints.[176] The *American Phrenological Journal* was then able to reproduce Combe's entire notice in June, a month before most other periodicals. After this, American reviewers broadly followed Combe in acknowledging a connection between phrenology and *Crania Americana*. The *Christian Examiner* in Boston praised the appendix, describing it as the "completion of the inquires and observations contained in the body of the work."[177] John Bell's *Eclectic Journal of Medicine* in Philadelphia concurred. It singled out Morton's anatomical measurements as "the most valuable illustrations of philosophical craniology extant—the more so, indeed, as they are followed by a table of Phrenological Measurements."[178]

Still, there was a limit to the power of Combe's notice, even in the United States. The July number of the quarterly *North American Review* had likely gone to press prior to the publication of the April number of the *American Journal of Science*. Without Combe's prompt, the reviewer simply situated *Crania Americana* amid "the learned and philosophical works of Blumenbach and Prichard," much to the phrenologist's disdain.[179] Farther west, in Ohio, the *Ladies' Repository* offered another alternative reading. This Methodist monthly operated in a world both geographically and intellectually distinct from Yale University and the Academy of Natural Sciences. Here, *Crania Americana* was introduced as a work of natural theology. The journal informed its female readership that "man stands at the head of the animal kingdom. . . . He is properly styled, 'lord of the lower world.'" This biblical account of man's place in nature, the *Ladies' Repository* suggested, also applied to different races. In support of this, it copied out a "description of these several varieties or races . . . from Dr. Morton's *Crania Americana*."[180] Less than a decade earlier, the remaining Shawnee tribes in Ohio had been forcibly relocated west of the Mississippi River.[181] In the wake of Andrew Jackson's policy of Indian removal in 1830, it is perhaps unsurprising to find the "American Race" recorded in this particular journal as "averse to cultivation, and slow in acquiring knowledge; restless, revengeful, and fond of war."[182]

In Britain as in Ohio, neither *Crania Americana* nor Combe's review was always read as he would have wished. Prichard continued to promote his ethnological account wherever he could. On receiving his own copy, signed by the author, Prichard presented Morton's work at a meeting of the Royal Geographical Society. There he praised Morton's lithographs, describing the "accuracy of his delineations," before again linking *Crania Americana* to the ethnological project of the Aborigines' Protection Society, to which Morton had recently been elected an honorary member.[183] This "valuable work," Prichard told the Royal Geographical Society, contained "specimens of the skulls of all the aboriginal races in America, many of which have now become extinct."[184] Prichard also used his influence within the medical profession to arrange a favorable notice in the

British Medical and Foreign Review, edited by John Forbes in London. He later boasted that "the first review of it written in this country was made by a friend of mine."[185] Morton had made things easier, having sent Prichard's copy ahead of the shipment to Simpkin, Marshall, and Co. Sure enough, the review was dismissive of phrenology, informing its readers that "we suspect the phrenological student needs more guidance than he will find here, in order that his conclusions may be satisfactory."[186]

Other reviewers sympathetic toward the ethnologist did not have such easy access. Indeed, Combe was making things difficult. Morton had put him in charge of issuing presentation copies from Simpkin, Marshall, and Co's stock, with recipients including David Craigie, editor of the *Edinburgh Medical and Surgical Journal*, and James Kennedy, a regular for the *Medico-Chirurgical Review*.[187] Predictably, both these journals printed notices either favorable or neutral toward phrenology.[188] With access to *Crania Americana* under Combe's control, others were not so lucky. Most prominently, Robert Jameson, Regius Professor of Natural History at the University of Edinburgh and editor of the *Edinburgh New Philosophical Journal*, did not receive a presentation copy. This wasn't too much of a shock. Jameson was known to be hostile to phrenology, having lectured against the subject in the 1820s.[189] But, while *Crania Americana* itself proved difficult for him to acquire in Edinburgh, the *American Journal of Science* did not.[190] Jameson therefore chose to reprint Combe's review instead.[191] However, as the phrenologists soon discovered, the reprint had been subject to heavy editing. Combe later identified this in a letter to Morton, writing that Jameson had omitted "all notice of the direct bearing of Phrenology on the Crania Americana & all mention of my name as in any way connected with the work."[192] This was a fair assessment. Jameson had deleted most paragraphs that either mentioned or supported phrenology. These included an extract from *Crania Americana* in which Morton declared, "I am free to acknowledge . . . that there is a singular harmony between the mental character of the Indian, and his cranial developments, as explained by phrenology." Once again, Morton's plates proved a point of contention. Unlike in Philadelphia, Combe could not control the use of his woodcuts so easily. Jameson copied a number of these into his own journal but, crucially, left out the figures of the "Indian" and "European" brains, alongside the "Swiss" skull and accompanying text. The *Phrenological Journal* provided an accurate analysis when it stated that "Mr Combe's index figure . . . is omitted: thus, the *phrenological* explanation of the four figures is suppressed in that Journal."[193] Through careful editing, Jameson repackaged Combe's review in Edinburgh to promote a very different reading. With these two woodcuts in absence, along with all mention of phrenology, *Crania Americana* could simply be described as a work of "the natural history of the native inhabitants of the New World."[194]

By October 1840 the reception of Morton's work seemed to have further fragmented on both sides of the Atlantic: readers interpreted *Crania Americana* in traditions ranging from natural history and ethnology to natural theology and phrenology. Still, as Morton's appointed representative in Britain, Combe maintained a considerable amount of influence over Simpkin, Marshall, and Co. This allowed him to promote certain periodicals' reviews over others. Combe also hoped to use the cheap periodical press, rather than just the prestigious quarterlies and monthlies, to shift the balance back in his favor. Advertising was central to this strategy. In November 1840, Combe arranged for a full-page advertisement to be printed and inserted into *Chambers's Edinburgh Journal* alongside a range of other monthlies that had yet to notice the work. *Chambers's Edinburgh Journal* itself was developing an overseas connection at this time, available for a couple of pence in both the United States and Britain.[195] To reach such a broad audience, Combe needed to print 60,000 copies of the notice, each featuring six carefully selected extracts from both American and British reviews. Quotes from the sympathetic *Medico-Chirurgical Review* and the *Edinburgh Medical and Surgical Review* appeared alongside Combe's own *American Journal of Science* notice. Another extract from the *Phrenological Journal* read, "The beautiful lithographic drawings by which this publication is so copiously illustrated, render it worthy of a place by the side of the large works of Gall and Vimont."[196] Thus *Crania Americana* was presented as one among a range of phrenological atlases, from Franz Joseph Gall's *Anatomie et Physiologie du Système Nerveux* to Joseph Vimont's *Traité de Phrénologie*.

It wasn't just through the choice of extracts that Combe hoped to nudge readers toward a phrenological interpretation. He also arranged for the opposite page to feature an advertisement for his own phrenological works. The economics of binding facilitated this, with Combe pointing out that the "expense of stitching to the Journals is the same for one page as for four." This also justified Combe's decision to divide the cost of the paper, printing, and binding equally between Morton and himself.[197] Readers therefore encountered this advertisement for *Crania Americana* within a few pages of, or in some cases directly opposite, a notice for Combe's own *System of Phrenology* and the *Phrenological Journal*. The notice also promoted the "People's Edition" of *The Constitution of Man* at 1s. 6d. just across from Morton's folio volume at £6 10s. This odd juxtaposition is better understood when we appreciate the function of the advertisement to establish *Crania Americana* as a phrenological work among as broad an audience as possible.

Advertisements then did not simply act as a means to shift stock. However, this tactic was always limited by the reality of who could access the spaces in which *Crania Americana* was held. Readers of *Chambers's Edinburgh Journal* might be members of mechanics' libraries, but they would be hard-pressed to come across a copy of Morton's folio there. As the decade wore on, it became more and more evident that *Crania Americana* could be found only in libraries frequented by gentlemen of science. Certainly the work was expensive. But with a limited number of copies, both Morton and Combe also targeted the more prestigious European institutions. In London, both the Royal Society and the Linnean Society held copies.[198] The Glasgow Mechanics' Institute did not.[199] A letter from J. J. Flanders, a semiliterate man from New York, further reminds us of the difficulties working-class men and women, along with slaves and Native Americans, continued to face in accessing expensive folios. Flanders explained to Morton that he had heard of "a work caled the Crania Americana written by you acompaned with ingravins of skulls [*sic*]." However, Flanders lamented that he had "sirched Boston and sent to New York but can not find one [*sic*]."[200] Indeed, as in London, the New York Mechanics' Institute, where Flanders might have practiced his reading, did not have a copy.[201] In writing global histories, it is easy to get caught up in a narrative of movement and circulation. As this chapter has already shown, managing the transatlantic publication and reception of *Crania Americana* was by no means straightforward. But Combe's advertisement also invites us to pay greater attention to the absences of movement and the places where Morton's work did not travel. For working-class men like Flanders, Prichard and Watson's jostling at the British Association meeting might have seemed far away.

BRAZILIAN TEETH

Perhaps surprisingly, given its content, *Crania Americana* initially received little attention in South America.[202] This was partly due to the networks of academic prestige that Morton cultivated. While he corresponded with European collectors traveling to Brazil and Peru, hoping to acquire specimens, Morton showed little interest in developing a readership among South America's growing community of physicians and ethnologists. Presentation copies were targeted at the Linnean Society in London and the Académie des Sciences in Paris.[203] These were the institutions that mattered, at least in Morton's eyes, for cementing the status of American racial science. As a consequence, no copies of *Crania Americana* appear to have reached South America in the immediate years after its publication. In fact, even as late as the 1890s, the leading medical library in Brazil did not hold a copy.[204] Phrenologists might have dreamed of a global science of race, but in reality, those outside the North Atlantic world of scientific institutions and correspondence networks were often excluded from such a project.

Morton's lack of engagement with South American readers is all the more pertinent given that phrenology had just started to take root in Brazil at this time. In 1832, the emperor of Brazil, Pedro II, founded two medical schools, one in Rio de Janeiro and another in Bahia. Part of a wider reform movement, these were modeled on the Faculty of Medicine in Paris. And as in France, every student in Brazil was required to submit a thesis on completion of the medical degree.[205] A number chose to write on phrenology. In 1841, Antonio Pereira d'Araújo Pinto, a student at the Faculty of Medicine in Rio de Janeiro, submitted a dissertation on phrenology. Titled "Algumas Proposições de Phrenologia," Pinto's Portuguese treatise was one of the first to introduce Brazilian medical audiences to the work of Gall and Spurzheim.[206] A few years later, Pinto was followed by another medical student, José Manuel de Castro Santos, who in 1846 submitted a dissertation at the Faculty of Medicine in Bahia giving a phrenological account of human intelligence.[207] Yet despite their interest in the new sciences of the mind, neither Pinto nor Santos was in a position to read *Crania Americana*. The library at the Faculty of Medicine in Rio de Janeiro was well stocked with English, French, and German medical works, including some on phrenology, but it did not hold a copy of Morton's atlas.[208] The connection between phrenology and *Crania Americana* was therefore absent from the start in Brazil.

Morton certainly showed little interest in cultivating a South American readership during his own lifetime. Nonetheless, his work did find an audience in Brazil later in the century. From midcentury onward, Brazilian physicians and anthropologists were at the center of debates concerning national identity. They drew on a range of racial and mental sciences, including eugenics, criminology, and phrenology, to try to understand what it meant to be Brazilian. Much of the debate centered on the mixed nature of the Brazilian population, with a racial history incorporating Africans, Europeans, and indigenous people. For some, like the pharmacist José do Patrocínio, racial mixing was what defined the Brazilian nation. Of mixed ancestry himself and active in the Brazilian abolitionist movement, Patrocínio argued, "We have been able to fuse all races into a single native population."[209] Others were less optimistic. Raimundo Nina Rodrigues, a physician at the Faculty of Medicine in Bahia, agreed that Brazil was defined by racial mixing. However, he believed that this undermined claims to both national identity and ethnic superiority. These debates only intensified in the years running up to the abolition of slavery in 1888 and the foundation of the First Brazilian Republic in 1889.[210] And it was precisely at this time that Brazilian thinkers turned to Morton.

In 1882, the Museu Nacional in Rio de Janeiro hosted the first Brazilian Anthropological Exposition. As with similar exhibitions in Europe and the United States, the Anthropological Exposition featured staged displays of indigenous people as well as artifacts, including arrows, pottery, and musical instruments. Alongside these ethnographic exhibits, Brazil's leading anthropological and medical thinkers met and discussed the latest racial science.[211] The program included a series of talks by the physician João Baptista de Lacerda, later director of the Museu Nacional itself. It was here that Morton found his place within yet another disciplinary tradition: physical anthropology. Developing a theory that was a far cry from phrenology, Lacerda argued that racial heritage could be traced through the study of dental composition. Much like Prichard earlier in the century, Lacerda invited his audience to look at a set of lithographs taken from *Crania Americana*. However, he promoted a very different reading. "The excellent lithographical prints of Morton provided us with the elements for the comparative study of the teeth in the American races," Lacerda explained. He even suggested that the collection of skulls held at the Museu Nacional in Rio de Janeiro could be compared with the illustrations in Morton's atlas. (It seems likely, as with Prichard, that Lacerda had access only to the plates and not the complete work.) Again, rather than focusing on the cranium, Lacerda asked his audience to look at the jaw. On doing so, Lacerda argued, one would find the "sign of animality imprinted on the teeth of the American skulls." According to Lacerda, the "Indians of Brazil," the "deformed skulls of Bolivia," and the skulls of the Chippewas in the United States all shared a dental configuration characterized by thick molars and the projection of the canines. This was what united the "American race," not the width or breadth of the skull.[212]

While *Crania Americana* was originally conceived as a phrenological text by Morton in the 1830s, this connection was completely lost by the time it reached Brazil in the 1880s. Instead, Morton's atlas was positioned as part of an attempt to cement the status of Brazilian anthropology, particularly a strand grounded in the state's growing anatomical collections held at the Museu Nacional. Lacerda in particular believed that Brazilian science had fallen behind Europe and the United States. "The course of scientific inquiry in South America is still far from its northern counterpart," he told his audience at the Museu Nacional.[213] But by adopting Morton, and reading *Crania Americana* within yet another disciplinary tradition, Lacerda hoped to lay the foundation for a Brazilian science of race, one that would hold just as much weight in Philadelphia or London as in Rio de Janeiro.[214] At the same time, Lacerda used that same racial science to differentiate the indigenous Brazilian population from the white elite, reinforcing a sense of national identity through the study of mankind's physical characteristics.[215] Here then we see how the history of scientific disciplines has a geography, one bound up with nationalism. From a study of the brain in the United States to a study of teeth in Brazil, the circulation of *Crania Americana* across the Atlantic world opened up a space in which both the text and the images were always open to negotiation.

Crania Americana, despite its title, cannot be read as a straightforward product of antebellum American culture. This chapter has shown how actors on both sides of the Atlantic shaped the publication and the reception of Morton's alluring folio volume. What's more, this chapter has suggested how histories of science and histories of the book may contribute further to one another as we move beyond national studies. In this case, the material world of transatlantic exchange both helped and hindered the efforts of those looking to promote *Crania Americana* within their respective disciplines. The boundary work in establishing different scientific disciplines therefore took place over a greater range of geographies than has previously been acknowledged. Particularly for anthropology, there is a need to appreciate the material and intellectual relations between different national traditions. Furthermore, in the case of phrenology, there is a need to consider how the boundaries between science and nonscience were also forged within an increasingly globalized world.

Beyond *Crania Americana*, historians have acknowledged the importance of scientific atlases in the emergence of new disciplines. Lorraine Daston and Peter Galison rightly argue that atlases lay "the visual foundations upon which many observational disciplines rest." Atlases thus "define the working objects of disciplines." Throughout their study, Daston and Galison stress the importance of different modes of image production, from copper engravings in the eighteenth century to photography in the twentieth. They also chart the relationship between the production of an image and accompanying "epistemic virtues."[216] But in making this particular connection, Daston and Galison assume that atlases act as relatively straightforward windows onto the intentions of authors. They are also left with little to say about the complex material processes of reproduction and reception uncovered in this chapter. In fact, Morton's demands as an author were constantly moderated by a range of interested parties, from his publisher in Philadelphia to Prichard in Bristol. Crucially, this was a process that began prior to publication. The sheer heterogeneity of reader responses to *Crania Americana* also raises serious questions about the utility of charting the emergence of disciplines and epistemic virtues from the perspective of authors alone.

Atlases in fact appear as rather unlikely candidates to structure scientific activity. They are after all expensive to print, cumbersome to move about, difficult to access, and often limited to short print runs. These problems are all compounded when we consider reception on an international scale. But despite this, Morton lived through a period that saw the emergence of a range of influential examples, from Murchison's *Silurian System* to Thomas Say's *American Entomology*.[217] By studying reviews, extracts, and advertisements we can make better sense of the relative success of this format. As a prestigious volume, *Crania Americana* was reproduced and reinterpreted time and time again, whether as a loose lithograph at the British Association or as a woodcut in the *American Journal of Science*. This intense process opened up a space in which different audiences confronted one another. Disciplinary categories then emerged as a means to promote one reading over another. Despite all the merits of Daston and Galison's work, it is important to recognize that atlases did not shape scientific disciplines in isolation. Rather, they gained disciplinary purchase through the variety of ways in which they were reproduced and read.

Finally, while this chapter has concentrated in detail on the relationship between British and American protagonists, it also lays the foundation for histories of scientific atlases stretching beyond the Atlantic world. Throughout the 1840s, Morton continued to promote his work overseas, placing it within an even broader range of institutional and disciplinary settings. In 1843 Morton sent a copy of *Crania Americana* to the Asiatic Society of Bengal where it was read by British army surgeons returning from the frontier.[218] By 1844, the Société Ethnologique in Paris had also obtained a copy.[219] There it was read by the "father of ethnology in France," William Frédéric Edwards, a man Prichard had hoped to emulate through the foundation of the Ethnological Society of London in 1843.[220] Soon afterward, the Royal Society of Northern Antiquaries in Copenhagen also received a copy. This group approached Morton's plates within yet another embryonic tradition, that of folkloric studies of the Arctic.[221] Today, scholars are too quick to situate Morton within histories of anthropology. While many did read *Crania Americana* as a work of anthropology, including physicians like Lacerda in Brazil later in the century, this was not the only or indeed the most common reading.

By following *Crania Americana* to Russia, we can get one last look at just how difficult it was for nineteenth-century readers to pin down Morton's beautiful yet troubling volume. In 1845 Charles Cramer of the Imperial Mineralogical Academy in St. Petersburg wrote to Morton complaining of the troubles he had faced in obtaining a copy. Cramer explained how a shipment of boxes, possibly containing *Crania Americana*, had been sent from New York addressed to the "Theological Institute" in St. Petersburg. But no such institution existed. Cramer suggested to Morton that the addressee perhaps meant the "Geological Institute." Still, this institution was distinct from the Imperial Mineralogical Academy at which Cramer worked. By a "slight misnomer of [Morton's] agents," *Crania Americana* had been lost amid the plethora of expanding scientific institutions in Nicholas I's imperial Russia.[222] As Cramer learned to his cost, what to write on an address label was tantamount to answering how *Crania Americana* should be read. Theological or phrenological, ethnological or geological? In the nineteenth century, no one was certain.

CHAPTER 4

Letters

What was the most important reform of the nineteenth century? For Orson Fowler, editor of the *American Phrenological Journal* in New York City, it was the penny post. In an article calling on the United States to adopt a similar system to the one established in Britain in 1840, Fowler argued that "probably no reform of modern times compares with this in its destined utility." For the American phrenologist, all other reforms—from new prison systems to the abolition of slavery—relied on "communion on paper." The penny post would "promote mental discipline, and the public good." It would "bind man to man in one common brotherhood."[1] This vision of reform was grounded in a phrenological understanding of the mind. Fowler explained that "every letter, every paper, and book, becomes a teacher of some kind, and exercises, and thereby develops the mind of both producer and reader."[2] Phrenologists on the other side of the Atlantic were equally enthusiastic. Hewett Watson, later editor of the *Phrenological Journal* in Edinburgh, also championed "Cheap Postage," writing, "The free interchange of ideas . . . tends most powerfully to promote the moral improvement of mankind."[3] Rowland Hill, the pioneer of the Uniform Penny Post in Britain, was even subject to a phrenological analysis. The significance of Hill's reform was matched to a protruding forehead, illustrating a "general intellect of a very high order."[4] Phrenologists also hoped that such a system of universal postage might be expanded internationally. In a letter to an American correspondent, George Combe complained that "foreign postal regulations" were still on the "old plan." Even worse, Combe noted, "many of my foreign correspondents believe that the penny system applies to all letters which it does not."[5] In the meantime, the Edinburgh Phrenological Society subsidized Combe's international postage, paying for letters he exchanged with correspondents in the United

States, Canada, France, and India.[6] Other phrenological societies were also encouraged to put membership fees to similar purposes, exchanging noteworthy letters and papers internationally.[7]

Picking up on the phrenologists' concerns with correspondence and the politics of the mind, this chapter explores the use of letters in three major nineteenth-century reform debates: the abolition of slavery, prison discipline, and education. In doing so, this chapter develops two overarching arguments. First, phrenologists used correspondence to engage with reform as a global project. Second, phrenological language allowed reformers to present their arguments in terms of a new understanding of human character.

While political historians have started to explore the global nature of reform and revolution in this period, the existing historiography of science treats political movements within formulaic national settings.[8] In France, phrenology is associated with the July Monarchy, while in the United States it is an accessory to Jacksonian democracy.[9] In Britain, it is the Reform Act, and in Ireland it is Catholic emancipation.[10] In India, phrenology is part of British colonialism, and in the Cape Colony it is an episode in the history of race.[11] This chapter demonstrates both the merits and the limits of the existing historiography. Sciences of the mind were certainly intertwined with British colonialism in India, but phrenologists also believed that schools in Bengal could inform those in the United States. Similarly, phrenologists in Britain and Prussia looked to New South Wales prisons when considering the architecture of their own penitentiaries. By following the actual geographies occupied and imagined by nineteenth-century historical actors, this chapter brings the existing national histories of phrenology and political movements into relation with one another. In doing so, it uncovers how reform and phrenology became intertwined on a global scale.

The current historiography also raises serious questions about the relationship between science and politics more generally. John van Wyhe rightly points out just how heterogeneous even a national political context could be. In Britain phrenology found supporters ranging from Tory churchmen like the Reverend Henry Wintle to radical leaders like the Chartist William Lovett, a man who read Combe's *Constitution of Man* while serving time in Warwick Gaol. Given this diversity of political support, ranging from Tories to Chartists, van Wyhe concludes that "phrenology was not, essentially, about reform."[12] There is, however, a serious flaw in van Wyhe's argument. It arises from an extremely narrow understanding of what constitutes the history of political thought. Historians of political thought do not try to establish one-to-one relationships between social contexts and political views. Instead, they uncover how particular ideas—from liberalism to Marxism—are used in practice. Identical political language is often put to a variety of conflicting uses.[13] The very word *reform* is a case in point. In the nineteenth century, the meaning of the term ranged from "moral correction" to "institutional restructuring." This period also saw the development of a "reform programme." Previously separate projects, such as the improvement of school and prison facilities, came to be linked.[14] Most importantly, *reform* did not denote a particular party politics. Whigs, Tories, Owenites, and Chartists all presented themselves as "reformers" when it suited them, particularly after 1832.[15] Van Wyhe therefore wrongly assumes that political diversity undermines the connection between phrenology and reform. In this case, the historical work is precisely to uncover how phrenology could be adopted by both British abolitionists and American slaveholders.[16] In this chapter I treat reform as it was in the nineteenth century, as something open to contestation and appropriation. In doing so, I reestablish phrenology as a reform science.

The global scale is an important part of my argument. By following letters across traditional political geographies, we get a clearer sense of how historical actors deployed phrenology when confronted with both complementary and conflicting moral worlds. Often this tension was resolved by imagining local reform as part of a global project. So despite the obvious social, environmental, and political differences between prison colonies in the Pacific and American penitentiaries, phrenologists separated by thousands of miles came to conceive of the two as part of the same scheme. It is only by looking at letter writing as a material practice that we can fully understand how phrenology could be deployed over such an incredible range of geographies and put to such diverse political ends. Letters were above all material objects—pieces of paper, inscribed with ink, folded, stamped with wax, and transported by runners, mail coaches, and steamships.[17] Nineteenth-century writers took advantage of this materiality, but they also struggled with it. Letters could be used to think with, to jot down ideas, and to rehearse arguments that might later appear in print.[18] They allowed phrenologists in the Pacific and India to imagine they were having a conversation with a gentleman in London or Boston, rather than sweating in the tropics. But letters could also be lost or damaged, contributing to a sense of distance and separation.

Once we consider correspondence as properly materialized, we can also understand it, like the phrenologists, as a practice of reform in its own right. Composing and sending a letter, particularly when the correspondents are separated by a significant geographic distance, allowed phrenologists to connect apparently local issues to global politics. In fact, the very passage of a letter implied a material connection that might otherwise have gone unnoticed. Nineteenth-century political action relied upon establishing exactly this kind of causal connection.[19] A successful campaign for abolition in Britain was only possible once middle-class consumers accepted the link between purchasing sugar in London and the plight of the slave in the West Indies. Yet many of the same consumers did not feel any moral obligation toward slaves outside of the British Empire. Establishing and negotiating these "moral geographies" was the principal occupation of nineteenth-century reformers.[20]

As birth and property gave way to less tangible notions of respectability, character served as the other dominant concept in nineteenth-century political thought.[21] Letters in particular were understood to reveal the authentic character of an individual in a way even their public conduct could not.[22] When we consider reform as a global project, this issue becomes even more pertinent. Correspondence provided a means to assess the trustworthiness of individuals who might never have met. At the same time, character also served as the very object of reform. New prisons were designed to produce a permanent change in the character of the inmates, while abolitionists debated the relative moral qualities of slaves and slaveholders. Of course, anyone advocating for reform on this basis needed to carefully manage his or her own public image. With character playing such a central role, phrenology emerged as an effective political language. It provided a universal understanding of human character, one that could be applied in almost any situation, from reading a letter in the Pacific to managing a prison in the United States. Phrenology and correspondence together therefore offer an unparalleled opportunity to uncover how politics and science operated on the global stage.

PHRENOLOGY ON THE PLANTATION

Every spring Charles Caldwell set off from his hometown in Kentucky, traveling down the Mississippi River by paddle steamer, before finally arriving in New Orleans. Once there, he would unpack his collection of phrenological busts, ready to begin his annual lecture tour. At the New Orleans Lyceum, the governor of Louisiana listened attentively, while the local organizing committee praised Caldwell for his "highly intellectual and interesting exposition of the philosophy of the human mind." Following Caldwell's initial tours in the 1820s, white southerners took an increasing interest in phrenology.[23] Caldwell even complained of competition from itinerant lecturers in Louisiana.[24] Boston artisans were soon receiving orders for steel calipers and plaster busts to be sent down to plantations in South Carolina, while local enthusiasts in Alabama printed an account of a "negro boy" exhibiting exceptional mathematical ability.[25] Despite describing the young slave as a "living wonder," the authors proceeded to offer the boy's skull as a "valuable acquisition" to any phrenological collection.[26] One phrenologist even admitted to acquiring the skull of a slave who had been struck on the head with an axe by his master.[27] In the South, phrenology and violence went hand in hand. Born in North Carolina in the 1770s, and owning slaves himself in Kentucky, Caldwell was no stranger to this world.[28] He was also keen to defend it.

Caldwell first broached the subject in a letter of 1835, tentatively asking Combe, "What think you of my views of Africans and Caucasians?"[29] The two had been introduced ten years earlier by Combe's brother, Andrew. An American always stood out in Paris, particularly one from the South. Andrew Combe had bumped into Caldwell at Johann Spurzheim's phrenological lectures on rue de Richelieu in the summer of 1821.[30] Caldwell's starting position was relatively straightforward. For him, phrenology proved that Africans and Caucasians were fundamentally different "species." Invoking the idea of an omnipotent creator, common to southern arguments against abolition, Caldwell suggested to Combe that "by *original organization* and therefore *radically* and *irredeemably*, the African is an *inferior race*. Nothing short of the power that made them can ever raise them to an equality with the Caucasian."[31] For Caldwell, it was the large "animal organs," located toward the back of the head, that rendered Africans unfit for freedom. In a long letter to Combe on this subject, Caldwell drew repeated parallels between animals and slaves, writing that "by good pasture and feeding, you may increase the size of your horses and cows. But you cannot bestow on them the bulk of the rhinoceros or the elephant."[32] In another, Caldwell wrote that he found the difference between Africans and Caucasians to be "much greater than the difference in organization between the dog and the wolf; or between the fox and the jackal. Yet they are acknowledged to *differ in species*."[33] Abolitionists could not hope to change "either the Ethiopian's skin or the leopard's spots." In conclusion, Caldwell argued, "The Africans must have a master."[34]

While these views were no doubt popular at the New Orleans Lyceum, Caldwell initially preferred to keep them out of print. He wrote to Combe explaining that it was "more than probable that I shall never write any thing for the press on the subject of slavery."[35] Caldwell's caution reflects his uneasy position at the intersection of northern and southern American society. He certainly enjoyed his time lecturing in the slave states, but Caldwell also derived much of his status from having published in respectable northern periodicals. These included *Annals of Phrenology*, the official publication of the Boston Phrenological Society.[36] The secretary of the society, the abolitionist Samuel Gridley Howe, would not have been pleased to learn that one of its contributors was in fact a defender of slavery. For both Caldwell and Combe, correspondence provided a relatively safe space in which to test out their ideas and identities.

Caldwell's long letter on the "animal organs" reached Combe in September 1839. At the time, Combe was staying in Portland, Maine, having left Edinburgh in 1838 to conduct his own two-year lecture tour of the United States. Relaxing in his hotel by Cape Elizabeth, Combe spent some time preparing his response. The material form of the letter helped (fig. 18). Combe jotted down his initial thoughts in the margins, writing, "I think more of Africans than he. They are inferior to white, but this is not the question." It did not matter, as most whites agreed, that Africans possessed inferior brains. What mattered was whether they could be entrusted with freedom. Filling the space to the side of Caldwell's first sheet, Combe compared the character of Africans and Native Americans. Drawing on Caldwell's own animal comparison, he wrote that Native Americans were "indomitable, ferocious savages. They are not tameable. They are not slaves because they are not tameable." With the Second Seminole War still raging in Florida, this argument appealed directly to southern fears of the Native American population. In contrast, Combe argued, "Africans are mild, docile & intelligent, compared with them. They are slaves because they are tameable." Here Combe first expressed an idea in note form that he would later return to in print. Stories of violent slave rebellions in Virginia and Jamaica fueled white fears of immediate abolition throughout the 1830s. Caldwell himself argued that "strife and blood-shed would soon become the daily occupation" of the free African. Combe responded to these concerns, arguing that phrenology in fact showed African character to be essentially placid, writing that "the qualities which make them submit to slavery are a guarantee that if emancipated & justly dealt with, they wd not shed blood."[37] Combe later repeated this argument in his *Notes on the United States of North America* as well as in the fifth and expanded edition of his *System of Phrenology*.[38] But he first worked it out in the margins of Caldwell's letter.

Still, Caldwell's defense of slavery rested on much more than his dim

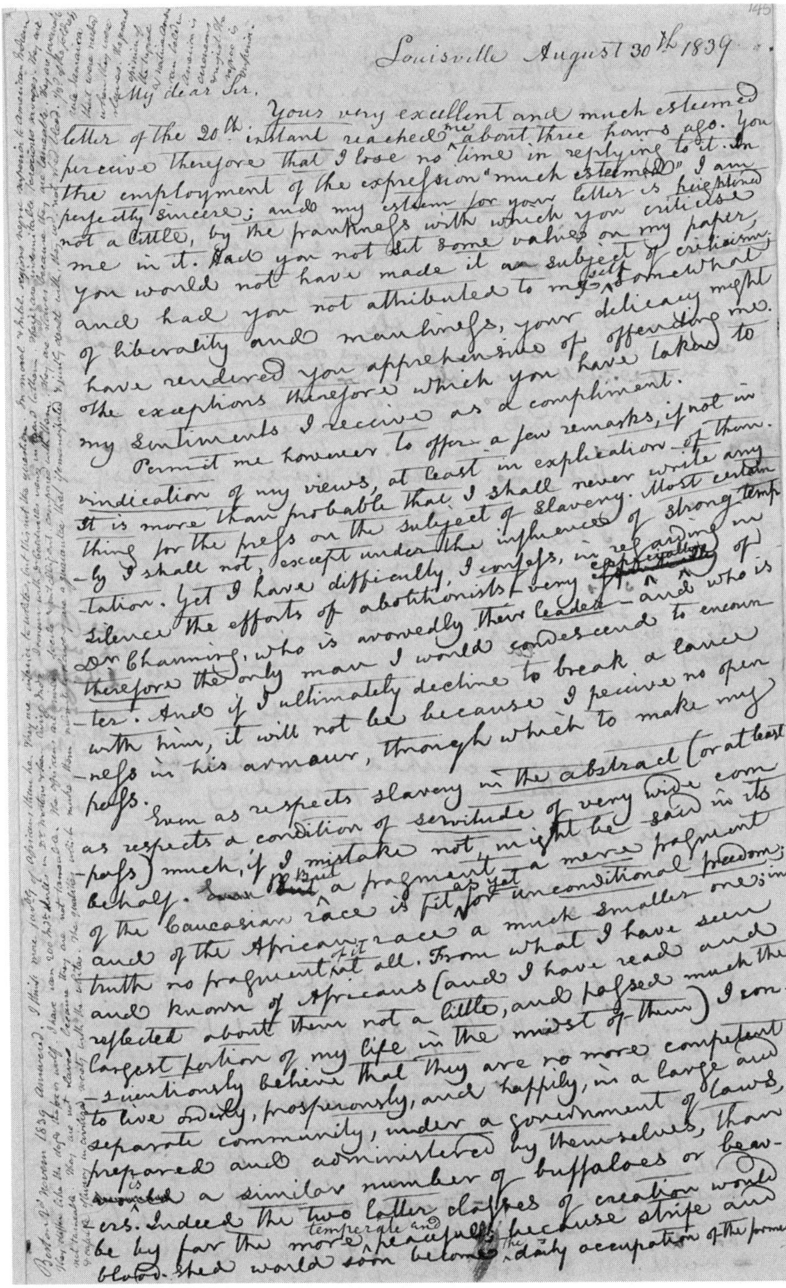

145

Louisville August 30th 1839.

My dear Sir.

Your very excellent and much esteemed letter of the 20th instant reached me about three hours ago. You perceive therefore that I lose no time in replying to it. In the employment of the expression "much esteemed" I am perfectly sincere; and my esteem for your letter is heightened not a little, by the frankness with which you criticise me in it. Had you not set some value on my paper, you would not have made it a subject of criticism; and had you not attributed to myself somewhat of liberality and manliness, your delicacy might have rendered you apprehensive of offending me. The exceptions therefore which you have taken to my sentiments I receive as a compliment.

Permit me however to offer a few remarks, if not in vindication of my views, at least in explication of them. It is more than probable that I shall never write any thing for the press on the subject of slavery. Most certainly I shall not, except under the influence of strong temptation. Yet I have difficulty, I confess, in regarding in silence the efforts of abolitionists, especially of Dr Channing, who is avowedly their leader—and who is therefore the only man I would condescend to encounter. And if I ultimately decline to break a lance with him, it will not be because I perceive no openness in his armour, through which to make my pass.

Even as respects slavery in the abstract (or at least as respects a condition of servitude of very wide compass) much, if I mistake not, might be said in its behalf. But a fragment, and as yet a mere fragment of the Caucasian race is fit for unconditional freedom; and of the African race a much smaller one; in truth no fragment at all. From what I have seen and known of Africans (and I have read and reflected about them not a little, and passed much the largest portion of my life in the midst of them) I conscientiously believe that they are no more competent to live orderly, prosperously, and happily, in a large and separate community, under a government of laws, prepared and administered by themselves, than is a similar number of buffaloes or beavers. Indeed the two latter classes of creation would be by far the more temperate and peaceful; because strife and blood-shed would soon become the daily occupation of the former.

Fig. 18. Combe's notes on slavery in the left margin of Charles Caldwell's letter, Caldwell to Combe, 30 August 1839, MS7249, f. 145, Combe Papers, National Library of Scotland.

view of African character. He was adamant that European abolitionists could not understand the conditions of slavery without having visited the southern states themselves. On arriving in the United States, Combe received a letter from Caldwell to this effect: "You do not like the slave-holding states of the Union. My good friend, shall I tell you why? You have never seen them."[39] Well aware that Combe planned only to visit the northern states on his lecture tour, Caldwell attempted to change his mind, writing, "Without sojourning some time among us, you can never depict us aright; because you can never know us aright."[40] Here, Caldwell showed his sensitivity to the moral geography of antislavery arguments.[41] He also understood the significance of another aspect of abolitionist rhetoric, one that phrenologists were particularly well placed to engage with. African character was important. But for many, the character of the slaveholders themselves was of greater relevance.[42] Opponents of abolition often argued that slavery could still be maintained with good conscience, so long as masters acted with care and diligence toward their slaves. Caldwell himself invoked this argument in a letter to Combe, writing, "We profess to be as virtuous, moral, and religious as our brethren of the north. . . . We have much humanity, benevolence, and magnanimity. . . . And those attributes all plead, and plead successfully, for kindness towards our slaves."[43] In a letter of Christmas Eve 1838, Caldwell even made the outrageous claim that "there is nothing appalling in this aspect of slavery. My servants live much more comfortably than I do."[44]

After consulting his friends in Boston, including the abolitionist William Ellery Channing, Combe finally took up Caldwell's offer and decided to visit Kentucky and Virginia in April 1840.[45] He was not impressed with what he found. Writing in his *Notes on the United States of North America*, Combe complained that "nothing can exceed the fertility and beauty of Kentucky; yet slavery makes it languish."[46] Phrenology itself provided a language of character in which Combe could dismantle proslavery arguments. In the run-up to the abolition of slavery in the West Indies in 1833, the *Phrenological Journal* in Edinburgh had printed an article entitled "Colonial Slavery Tested by Phrenology." The author of the article argued that "phrenologically, we know enough of the human faculties of the average endowment . . . to come to the conclusion that the Negro slave cannot be humanely treated; that he *must* be over-worked, under-cared for, and cruelly punished by so many slave-holders."[47] Combe returned to these ideas in correspondence. He was polite enough not to identify Caldwell by name, but he was nonetheless disparaging when it came to slaveholding character. In a letter to the American abolitionist Maria Weston Chapman, he wrote, "*How is it possible* for a people so moral, religious, enlightened, and free to defend and practice slavery?"[48] In another letter he even criticized the former president John Quincy Adams for trying to placate slaveholders in the District of Colombia. Adams had claimed that abolition would be unconstitutional. Combe had never met Adams, but he had seen a bust: "I write this from his *head*. . . . If Mr Adams had had a larger Conscientiousness, he would have *felt* that there was a jarring between his principle & his conclusion."[49]

Caldwell knew that his character was under intense scrutiny. He was also acutely aware that Combe would read his letters for markers of respectability. On one occasion Caldwell even excused himself for poor handwriting, explaining, "Should you discover any unusual scrawling in this letter, attribute it to the tremor of the boat."[50] Nonetheless, Caldwell sometimes had trouble reining in his emotions. He was the sort of man who would underline every other word in a letter, leading Combe to tactfully describe him as a "severe" and "powerful" writer.[51] But Caldwell also understood that abolitionists were just as much subject to the rules of good character as he was. With this in mind, Caldwell advanced another argument concerning the moral geography of slavery. He challenged abolitionists to explain how they could care so much about the fate of African slaves when they did little to improve the condition of the white working class, particularly in Europe. Once again, this argument was restricted to correspondence. In a letter to Combe in February 1838, Caldwell wrote, "Believe me the condition of our slaves is immeasurably more comfortable, than that of the peasantry of Holland and Belgium, and much more so, than that of the labouring community in many parts of England and Ireland."[52]

This argument was not unique to slaveholders like Caldwell. In fact, it followed an earlier tradition of radical working-class criticism in Britain. In Norwich, local Chartists broke up an abolitionist meeting with shouts of "Put slavery down at home," while John Wade, a campaigner for labor reform, claimed in his *Black Book* that "the horrors of the middle passage" did not "transcend those of the infernal factory system."[53] Other more conservative phrenologists also took up these ideas. Spurzheim himself had initially been sympathetic toward abolition. Like Combe and Caldwell, he also used letters to think with. During a lecture tour of Britain, Spurzheim wrote regularly to his future wife in Paris, Honorine Périer. He asked Périer to collect all his letters so that he could consult them at a later date.[54] In one letter sent from London, Spurzheim praised the "English who, against their interest, are the only people who cry publicly against the injustice of the abominable slave trade." This he attributed to the character of local abolitionists, writing, "I found nowhere the moral sentiments stronger. . . . They have four, 10, 12, 13, 14, hope and justice, hence a very serious character."[55] But the following year, having traveled to Dublin, Spurzheim was shocked by what he saw. Attending the annual Donnybrook Fair, Spurzheim wrote again to his fiancée, explaining, "I saw what I never had seen—such a degree of wretchedness. . . . I would rather see a herd of animals, at least there would be more decency in seeing them." He found "the organisations of the children inferior" to anything he had seen before. This encounter with the Irish poor radically changed Spurzheim's view on abolition. He concluded his description with the line "When I come again to London, how I shall cry against those who take care of the Africans and do not think of a better situation for the inhabitants of their own country."[56]

IMAGINING ABOLITION

The American abolitionist Lucretia Mott understood the materiality of correspondence better than most. On finishing a letter, she would turn it over, fold it, and stamp it with her own wax seal. Once dried, the impression revealed a kneeling female slave, chained and surrounded by the words "Am I Not a Woman and a Sister?" (fig. 19). This image had been adapted in the 1820s from the original Josiah Wedgwood antislavery medallion to better reflect the concerns of female abolitionists.[57] As a wax seal, it allowed women like Mott to push the boundaries of polite female correspondence.[58] In contrast to Caldwell, Mott first made contact with Combe in person rather than on paper. The two met in Philadelphia in 1839 at a Quaker meeting house.[59] Combe's wife, Cecilia, was also traveling with him, and the two stayed with Mott and her husband, James, while in the city. They spent the time discussing phrenology and getting to know the Motts' children. Mott had read Combe's *Constitution of Man* and soon received an invitation to join the Philadelphia Phrenological Society, one of the few to admit women.[60] Later in the century, she even allowed Lorenzo Fowler to perform a phrenological analysis on her children.[61] And unlike Caldwell, Mott's letters are full of the details of family life, including the state of her health and the progress of her husband's textile business. George and Cecilia Combe tended to reply on the same piece of paper: a display of familial unity, but one that also saved on transatlantic postage.[62]

Phrenology and family life might have been deemed suitable subjects

Fig. 19. "Am I Not a Woman and a Sister?" glass seal, Collection of the Massachusetts Historical Society, Boston.

of correspondence for middle-class women, but the case was not so clear with slavery. Caldwell himself had been scathing regarding female abolitionists. In a letter to Combe of October 1838, he complained of Harriet Martineau's attack on slavery in her *Society in America*, writing:

> Her misrepresentations are numerous, grievous, and gross. . . . She condemns with the bitterness of a disappointed old maid, every thing in which she does not fully concur. That she has mind, nobody doubts,

> but alas! her temper is fully a match for it. Her Destructiveness would make a high-wayman murder.

Caldwell recognized that these were "perhaps harsh terms to apply to a woman."[63] But that was the point. For him, antislavery was unladylike.

Within this climate, Mott's wax seal provided a means to turn even routine familial correspondence into political expression. Combe saw this figure each time he opened one of Mott's letters, no matter what the contents. Soon enough, he became comfortable with the idea of discussing abolition with her, albeit in gendered terms. In a letter to Channing, Combe praised Mott, describing her "strong, bold, independent intellect," which was "combined with exquisite gentleness, delicacy & taste." He also noted that that she had "embraced Phrenology strongly & sees its moral power & practical tendencies."[64] Indeed, phrenology itself provided a language in which Combe and Mott could safely discuss abolition without descending into the aggressively male political arena. In May 1838 the newly constructed Pennsylvania Hall had been burned to the ground by an antiabolitionist mob.[65] Mott witnessed the violence firsthand. Reflecting on the unrest in Philadelphia, Mott explained to Combe that she wished to avoid the "intemperate zeal" associated with the phrenological organs of "Combativeness and Destructiveness." Instead, according to Mott, "the cause of abolition should be pleaded by the moral sentiments and addressed to the moral sentiments."[66]

Combe and Mott agreed on the basic principle that slavery was a sin and ought to be abolished. However, they did not agree on the means. Before leaving Edinburgh, Combe had overseen the publication of an article in the *Phrenological Journal* entitled "On the American Scheme of Establishing Colonies of Free Negro Emigrants on the Coast of Africa."[67] This presented a positive view of the American colony of Liberia situated in West Africa. Liberia had been founded by the American Colonization Society (ACS) as part of its plans to abolish slavery by freeing slaves and sending them to Africa. The scheme received support from a number of influential southern politicians, including the Kentucky senator Henry Clay, a man Combe later described as possessing a "mediocre" brain, and the Virginia congressman Charles Mercer.[68] Advocates of colonization believed slavery was inhibiting the industrial development of the United States. They also believed that African slaves were not capable of integrating into white civilized society.[69] On the other side of the Atlantic, colonization appealed to Combe and his colleagues in Edinburgh. It provided a means to reconcile a belief in the inferiority of the African mind with antislavery sentiment. In 1833, the *Phrenological Journal* declared that "we heartily approve [of] the American Colonization Society." The article also praised Liberia as "a community of Africans, without a white to claim the white's ascendancy." For the author, phrenology also explained why it was so important to separate the white and black populations: "When mixed, the white and Negro *must* stand to each other in relation of a superior and inferior race, with all the injurious effects of such a relation on both."[70]

Following the end of slavery in the West Indies, British abolitionist efforts increasingly turned toward the United States throughout the 1830s. Given the growing international interest in American slavery, the ACS also began to seek support from major British antislavery figures. Indeed, the *Phrenological Journal* article noted that the "venerable [Thomas] Clarkson . . . has lived to see and applaud it in the strongest terms."[71] The ACS also attempted to align itself with similar British efforts, including Thomas Buxton's Society for the Civilisation of Africa. With the British government still yet to formally recognize Liberia as a legitimate colony, the reception of the ACS in Britain had the potential to influence the path taken by American antislavery efforts.[72] And Combe, as a best-selling author and prolific writer on American society, certainly held sway over a large number of British readers.

Mott knew all of this well. In fact, she suggested that Combe might wish to publish a selection of her correspondence on slavery in Britain.[73] For Mott, as for many other northern antislavery campaigners, particularly Quakers, colonization was not abolition. Mott wanted British abolitionists to recognize this as well. She was therefore dismayed to learn of Combe's support for colonization and, lamenting the limits of the medium, wrote a long letter in June 1839 beginning, "I wish, in this space, I could make thee understand why colonizationists and abolitionists cannot harmonize." Mott believed the problem lay partly in Combe's access to the appropriate publications, with the ACS flooding the British market. She recommended that while he was in the United States, Combe should visit the "Anti-Slavery Office in New York" and pick up a copy of "Jay's Inquiry, & J.G. Berney's Letter on Colonization." She also suggested that Combe should read the annual reports of the ACS itself. He would then understand that there could be "no affinity" between "colonization and abolition societies."[74]

Combe received Mott's letter while staying in the White Mountains, New Hampshire. He was not enjoying himself. Living "amidst clouds of mosquitoes," he and Cecilia had been "awake all night," and they spent "all day in the most ignoble of exercises with our nails on our skin." With no stagecoach due for another week, there was "no consolation except that which I now enjoy of writing to you." Combe recognized that Mott's letter required an equally lengthy response. He certainly had the time to spare, beginning, "I forsee I will fill my sheet." He also suggested that he could potentially give Mott what she wanted, writing, "If I shall publish any work on America on my return, I should like very much to print your letter in it. . . . It contains a great deal of interesting information, well stated, & in a short space." Still, Combe was unwilling to change his own opinion. He recognized the significance of the reports Mott had quoted, and began by attempting to reassure her that he was not a covert supporter of slavery himself, writing, "I embrace without reservation, your fundamental principle that 'man cannot rightfully hold property in man,' and I am satisfied, from your quotation, that many members of the Colonization Society, hold the opposite opinion." Combe also admitted that the ACS "may have erroneous views & even bad motives at the bottom." "Nevertheless," he argued, "their scheme . . . may be capable of being turned to good account in virtue's cause, & it appears to me that it will be so turned."[75]

Why was he being so stubborn? In his letter to Mott, Combe joked that "a large mosquito has bitten me on the organ of Firmness, & has raised, what was always an abundant elevation, into promontory."[76] But later, in his *Notes on the United States of North America*, Combe was more serious. He wrote:

> In the United States the free negroes suffer many evils from the climate and from their degraded social condition, and they also encounter great obstacles to their advancement, from being forced to compete in all branches of industry with a race superior to themselves in native energy of mind.

For a phrenologist who believed simultaneously in African mental inferiority and the evils of slavery, Liberia represented "an asylum for such of the American negroes as could find no satisfactory resting-place in the Union."[77]

In response to Combe's letter from the White Mountains, Mott replied that she sympathized with his "annoyances with mosquitos." She did not, however, sympathize with the views expressed in the letter. In a final and ultimately fruitless effort to get Combe to change his mind on the subject of colonization, Mott pointed out that "the coloured citizens of the free states have proselytised against such removal." In fact, she noted, free African Americans had opposed colonization "long before

Wm L Garrison denounced it, and before an anti-slavery society was found in this country."[78] Mott was right. Her good friend James Forten, a wealthy African American businessman based in Philadelphia, had challenged the ACS throughout the early 1830s prior to the foundation of the American Anti-Slavery Society in 1833. Forten wrote a number of articles for the abolitionist newspaper *The Liberator* citing the high death rate in Liberia as evidence that African character was not especially "adapted" to the African environment.[79] Frederick Douglass, himself a former Maryland slave, also argued that "colonization is out of the question." Douglass was equally critical of phrenology's role in bolstering slavery. In a later lecture at the Western Reserve College in Ohio, Douglass argued that "it is fashionable now, in our land, to exaggerate the differences between the negro and the European." He went on to explain that "if, for instance, a phrenologist . . . undertakes to represent in portraits, the difference between the two races—the negro and the European—he will invariably present the *highest* type of the European, and the *lowest* type of the negro." Douglass linked this misrepresentation of African mental capacity directly to slavery, concluding, "By making the enslaved a character fit only for slavery, they excuse themselves for refusing to make the slave a freeman."[80]

For African American writers, correspondence also provided a significant literary space in which these debates could take place. James McCune Smith, an African American physician, wrote a series of letters addressed to Douglass on the subject of phrenology. Smith had taken a medical degree at the University of Glasgow in the early 1830s and returned to his home in New York City to lecture on the "fallacy of phrenology" in 1837.[81] His letters to Douglass on the subject were then published in *Frederick Douglass' Paper* in Rochester as part of a series entitled "The Heads of Colored People."[82] Smith's sketches of African American character on the streets of New York parodied the language of phrenology. Smith also invoked a nascent form of black nationalism, arguing that "the ideal of the American nation, *is not a white man. . . .* His complexion is swarthy, between a mulatto and a quadroon." Mimicking Combe's own phrenological vocabulary, Smith wrote that "the greatest height of skull is just before the ear: he has not what Phrenologists call concentration or self esteem."[83]

African American attitudes toward racial character were not homogeneous. While Douglass praised Smith's critique of phrenology, he still worried that this account of working-class life, from boot polishers to washerwomen, painted a negative picture of African American achievement. In an open letter Douglass emphasized the geographic limits of Smith's understanding of African American character, writing, "Very little can be learned of the colored people as a whole by merely seeing them in the streets of this or any other city." Instead, "one must see them at their homes, in their places of business, at their churches, and in their literary and benevolent societies to get an idea of their character and condition." Douglass concluded by asking, "Why will not my able New York correspondent bring some of the real *'heads of the colored people'* before our readers?"[84]

Douglass's and Smith's critiques of phrenology provide a brief glimpse into the world of black counter-narratives opposing European racial science.[85] But this world was carefully segregated, not just physically but also intellectually. In fact, the "color line" separating whites from blacks was sometimes stronger in correspondence than in daily life. Combe met a number of African Americans during his tour of the United States. These included a free hotel manager in Philadelphia, whom Combe described as possessing "a brain that would do no discredit to an European," alongside enslaved laborers in Kentucky.[86] And when Douglass visited Britain in the 1840s, Combe welcomed him into his home in Edinburgh. The two ate breakfast together alongside the British abolitionist George Thompson and the American abolitionist William Lloyd Garrison. Douglass recorded the meeting in his *Life and Times of Frederick Douglass*, writing, "Mr. Combe did the most of the talking, and did it so well that nobody felt like interposing a word, except so far as to draw him on." Phrenology was the chief topic of conversation. Douglass recalled that Combe "looked at all political and social questions through his peculiar mental science. . . . Phrenology explained everything to him, from the finite to the infinite." Douglass was impressed by what he heard, having recently read Combe's *Constitution of Man*. In fact, Douglass singled out this book as one of the few phrenological works containing a fair portrayal of African character.[87]

Yet despite these convivial meetings, Combe and Douglass did not exchange a single letter. There was certainly an institutional element to this. In the United States, African American use of the postal service was carefully regulated. By law, only a "free white person" was allowed to carry the mail. And if you were black, just going to the local post office could be an ordeal. Racial abuse and stolen correspondence were common. Douglass himself complained that postmasters often "refuse to hand a black man a letter . . . because he is of the hated color."[88] More broadly, Combe managed a correspondence network that stretched across the United States, Africa, India, and the Pacific. It took in factory workers and aristocracy as well as women and children. Combe even corresponded with a group of Bengalis from Calcutta. Yet he never wrote to or received a letter from an African American. In a medium in which character remained the final arbiter, Combe's belief that the black mind was ultimately "inferior to white" proved too difficult to overcome.[89]

THE GLOBAL PENITENTIARY

Alexander Maconochie watched as the ship carrying the bundle of letters he had written over the previous six months dipped below the horizon. Stationed at the penal settlement of Norfolk Island, a small volcanic rock in the Pacific Ocean, he was over 1,000 miles northeast of the New South Wales coast.[90] Once they arrived in Sydney, the letters addressed to Britain would then travel a further 14,000 miles around the Cape of Good Hope, taking over six months to reach Bristol.[91] It would be at least another year before he received a reply. And so, despite living among over 1,500 people, Maconochie felt isolated. On the island he had no one with which to discuss his developing regime of "penal science."[92] Correspondence provided the outlet Maconochie was looking for. He could forget the alien landscape of coral reefs and Malayan convicts, replacing it with a gentlemanly world of polite discussion.

On first writing to Combe from Norfolk Island in June 1841, Maconochie imagined the two were having a conversation, writing, "I note down now to talk to you of all these matters."[93] With little prospect of a reply in the near future, if at all, the notion of "talking" with Combe was largely rhetorical. The letter allowed Maconochie to conjure up a respectable and familiar context in which to work through his ideas. He could look back to the early 1820s when he and Combe had both been living in Edinburgh, meeting up to discuss phrenology and prison reform.[94] Back on Norfolk Island, the idea of conversing with Combe once again seemed appealing. Maconochie opened his letter by announcing that "out of the fullness of satisfaction . . . my pen writes." He then set down the details of his "experiment" in prison discipline, suggesting that Combe and his fellow phrenologists would take interest in the "great moral questions so involved." Maconochie began by describing the problems Norfolk Island posed. The settlement housed "the refuse of other colonies" and, prior to his arrival in 1840, had been organized under a "horrid system of brute coercion."[95] Under the former superintendent, convicts labored knee-deep in the sea, cutting stone in the "wet quarry" by the coral reef.[96] If they refused or acted with insubordination, they would be savagely beaten. For this reason, the governor of New South Wales, Ralph Darling, considered Norfolk Island "the extremest punishment short of death."[97] Only the worst reoffenders, those already sentenced to transportation to New South Wales or Van Diemen's Land, would be transferred to Norfolk Island. Maconochie explained to Combe that nowhere on earth were there "individuals less fitted than any others possibly could be" for his "system of moral influence."[98] The implication was simple but powerful: if Maconochie's prison reform could work on Norfolk Island, it could work anywhere.

At the center of Maconochie's vision was the "mark system." Under this regime, sentences were measured by the accumulation of marks rather than by time. Instead of a seven-year stint on Norfolk Island, a convict would need to earn approximately 6,000 marks before being released. Marks could be gained through labor and good behavior. Convicts could also opt to form groups and accumulate marks together.[99] But they also needed to spend marks in order to acquire basic provisions and, if they desired, further luxuries.[100] Maconochie even made the convicts pay for access to a library of "Moral and Religious works" purchased for £50 in Sydney. These books were to be "improving" and to "open the mind to a perception of the value of minute distinction."[101] Combe's *Constitution of Man*, available in New South Wales at the time, no doubt fitted the bill. Finally, Maconochie abolished all physical punishment on the island. Instead, convicts would forfeit marks if they misbehaved.[102] And so, in common with many other penal reformers of his generation, Maconochie believed the prison should mirror an ideal capitalist society.[103] He explained to Combe:

> I have nowhere heard it said that a poor family in England ever comes to distress . . . that had over accumulated by its own exertions £5 in a Savings Bank: and that is the great principle that I desire to apply to Prisoners.[104]

For Maconochie, criminal minds could not be reformed through fear of physical punishment. Instead, convicts needed to learn the value of good work discipline and money management. The prisoners he encountered on Norfolk Island were "hostile . . . to the function of society." Yet, as he explained to Combe, "it is with their *aid*, not merely their submission or neutrality, that I may *succeed*." Maconochie did however recognize that many believed the Norfolk Island convicts to be "irredeemably bad." Was every criminal really capable of being reformed? If not, then his system would be deemed a failure from the start. Phrenology allowed Maconochie to challenge the idea of an incorrigible mind. He explained to Combe that his system was based on "Phrenological principles" and that he believed "the general intellect of the criminal population is for the most part greatly under-stated." On Norfolk Island, Maconochie observed, "there are many excellent heads among them."[105]

Still, it wasn't just the character of the convicts Maconochie needed to worry about. As with abolition, the mental and moral qualities of the prison officer played a significant role in reform debates. Maconochie himself was not a popular man in the colonies. Prior to his appointment to Norfolk Island, Maconochie penned a damning report based on his experience at the Hobart penal settlement on Van Diemen's Land in 1837. He forwarded the report, describing the system there as "cruel, uncertain, prodigal," directly to the Parliamentary Select Committee on Transportation in Westminster.[106] Maconochie was soon unwelcome on Van Diemen's Land and accepted a transfer to Norfolk Island. Still, the colonial officials in New South Wales did their best to get their own back. Rumors began to circulate in the Sydney press that Maconochie's eldest daughter, Mary Ann, had been seduced by a convict and fallen pregnant. And when Maconochie held an extravagant party for the prisoners complete with rum and fireworks, albeit in honor of Queen Victoria's birthday, he was ridiculed. The colonial secretary, Edward Deas Thomson, dispatched a short letter calling on Maconochie to rein in his own "sanguine temperament."[107] In writing to Combe, Maconochie therefore hoped to project a favorable impression of both his scheme and his own character back in Britain. Maconochie admitted in his letter that "you must have heard of my great controversy in this hemisphere," but he wished Combe to know that he was "for now successful." Even the "old mutineers" on the island were displaying "earnest, energetic & trustworthy tempers."[108]

When Combe finally received the letter in December 1841, he too felt a sense of distance. The parcel had clearly come a long way, having been stamped with four postmarks, including "New South Wales" and "Norfolk Island." The conversation Maconochie had imagined over six months ago in the Pacific would have to wait. Combe chose not to reply until Maconochie had returned to Britain in 1844.[109] By that time, Combe had completed lecture tours of both the United States and northern Europe. In the course of these he had visited the Eastern State Penitentiary in Philadelphia, Auburn Prison in New York, and the House of Correction in Hamburg.[110] He had also received a letter from Henry Spry of the Bengal Medical Service detailing the imprisonment, trial, and execution of a group of thieves in northern India.[111] And so looking over his correspondence again in the mid-1840s, Maconochie's letter in hand, Combe was now in a position to imagine a system of prison discipline stretching across North America, Europe, India, and the Pacific.

The result was a pair of articles entitled "Norfolk Island" and "Penal Colonies" published in the *Phrenological Journal.* Combe acknowledged the problematic geography of penal reform in the first article, writing, "What is transacted in a remote isle of the Pacific Ocean, without a newspaper, is never seen, and not heard of."[112] But Combe wanted to do much more than bring the mark system to Britain. Reflecting on the similarity between Norfolk Island and the penitentiaries he had seen in the United States, Combe advanced the idea of a regime of prison discipline capable of operating in any environment. The mark system was "a practical embodiment of phrenological principles . . . calculated to work certainly and easily, at any distance, and under any executive power." The language of phrenology allowed Combe to identify a global criminal class, irrespective of locality. The "thugs" in northern India, the African American convicts in the Eastern State Penitentiary, and the mutineers on Norfolk Island were all united by cerebral organization. Criminals the world over possessed "large animal, active and powerful intellectual, and very deficient moral organs." Maconochie's scheme was no longer an "experiment" confined to the Pacific. Rather, it was "applicable to prisons and penitentiaries at home as well as abroad."[113]

With readers in Britain, France, Prussia, India, and the United States, the *Phrenological Journal* was certainly a powerful medium through which to advance the idea of a global regime of prison discipline. But Combe also reinforced this through correspondence. In the United States, reformers continued to debate the relative merits of two different systems of prison discipline. Under the separate system at the Eastern State Penitentiary in Philadelphia, convicts spent their entire sentence in isolation. In contrast, under the silent system at Auburn Prison in New York, convicts spent the days laboring together but were forbidden to communicate. Supporters of both systems advanced the ideal of reforming the soul rather than merely punishing the body.[114] Samuel Gridley Howe, secretary of the Boston Phrenological Society, was one of those in favor of the separate system. Like many others, he believed that Auburn Prison relied too much on the whip to enforce a regime of silence.[115] However, Combe had not been impressed with the separate system during his own visit to the Eastern State Penitentiary on a freezing cold January morning in 1839. As he explained in his *Notes on the United States of North America*, "This system is not founded on, nor in harmony with, a sound knowledge of the physiology of the brain." In fact, "the system of entire solitude . . . leaves the moral faculties still in a passive state, and without means of vigorous active exertion."[116] In short, long-term solitary confinement was too extreme, tending to produce mental deficiency rather than reform. Three years later Charles Dickens reached a similar conclusion during his own visit to the United States, writing, "I hold this slow and daily tampering with the mysteries of the brain, to be immeasurably worse than any torture of the body."[117]

After reading *Notes on the United States of North America*, Howe wrote to Combe in 1846 explaining that he was "surprised to find that I come to a different conclusion from you on the advantages of the separate system."[118] But Combe was not going to change his mind. Earlier that year he had received favorable reports of the influence of phrenology at Auburn and Sing Sing Prison in New York, also conducted under the silent system. As Combe himself observed, all books except for the Bible were forbidden at the Eastern State Penitentiary.[119] In contrast, reading was much more widely permitted at Auburn and Sing Sing. Combe already earnestly believed that reading improved the criminal mind. He had even received a letter from an inmate who had read *The Constitution of Man* while incarcerated at Rochester Prison in upstate New York.[120] But Combe probably couldn't have imagined just how far his ideas had penetrated. In February 1846 he received a letter from Elisha Hurlbut, a lawyer in New York City. Hurlbut explained that he had met the female prison warden at Sing Sing. Incredibly, she was "a phrenologist—and teaches phrenology to the unfortunate beings under her charge." Not only that, Hurlbut explained, but "she often takes the 'Constitution of Man' in hand and lectures from it to the female prisoners in the chapel of their prison."[121] John Morrison, the superintendent at Auburn, had also shown a great interest in phrenology, characterizing the typical inmate as "sloping from the forehead."[122] With converts on the inside, and a greater emphasis on actively improving the mind, there was really no contest between the silent system in New York and the separate system in Pennsylvania. Combe replied to Howe, politely dismissing the merits of the Eastern State Penitentiary, writing, "I regret to observe that you & I differ in regard to the advantages of the solitary system of prison discipline." He did, however, recommend that Howe read the latest number of the *Phrenological Journal*, as it included an "article on Norfolk Island."[123]

From then on, Maconochie's mark system became an effective means for Combe to express his ideal of a global system of prison discipline. He forwarded copies of Maconochie's privately printed pamphlet, *The Man-*

agement of Prisons in the Australian Colonies, to both Hurlbut and Horace Mann, the American politician and educational reformer.[124] Mann, also a subscriber to the *Phrenological Journal*, had recently completed his own tour of the Prussian schools and prisons.[125] He thanked Combe for the pamphlet and suggested that the application of phrenology "to Criminal Legislation, Jurisprudence, &c &c, will, in time, I have no doubt, work revolutions in those departments."[126] Other correspondents actively sought Combe out for his knowledge of American and colonial prisons. The German legal scholar Karl Joseph Anton von Mittermaier wrote to Combe after having attended his lectures on phrenology in Heidelberg. Mittermaier described how he had been "persuaded that Phrenologie is the best way to form a very sound foundation for Psychologie."[127] Combe had actually introduced the basics of the mark system and the separate system in his first lecture.[128] Mittermaier wanted to know if Combe still believed that "entire solitude . . . would be prejudicial to the physical and mental state of the prisoners."[129] It seemed to Mittermaier, following Combe's introduction to Maconochie, that "by forming the habit of accumulating spare money in savings banks" the mind "may be properly directed by early training to diligence and industry." Combe responded in agreement, noting that there was "a radical defect in both of these systems of prison discipline, as administered in America." The problem was that neither Auburn nor the Eastern State Penitentiary provided "sufficient means for strengthening the moral and intellectual faculties of the prisoners." Only Maconochie's mark system actively promoted those phrenological organs found to be deficient in criminals. Mittermaier and Combe's exchange was then printed in both the *Phrenological Journal* in Edinburgh and the *Sächsische Vaterlands-Blätter* in Leipzig.[130] Combe wrote to Maconochie again in 1846 informing him of the great reach his scheme now enjoyed: "I have commended your mark system in letters to my American friends, as an admirable practical realisation of the principles which they already approve of."[131] And so, through correspondence, Maconochie's mark system found a global audience of phrenological reformers.

CALIPERS IN THE CLASSROOM

Bathing in the Hooghly River, just to the north of Calcutta, a small group of Bengali boys could hear the school bell ringing. At seven o'clock each morning they dried themselves and walked the short distance from the river to the modest bamboo schoolhouse at Monirampore. Once there, they were taught English and Sanskrit by a local Brahmin, no doubt hoping to proceed to Hindu College or Sanskrit College in the city. From the outside, this school could easily have been mistaken for any one of the hundreds of traditional *pathshala* operating in Bengal at the time.[132] But in early nineteenth-century India, another phrenological "experiment" was under way. In June 1825 George Murray Paterson founded what he described as "nothing less than a Phrenological School on the banks of the Hooghly."[133] It was the first of its kind in the world.

Paterson had joined the East India Company as an assistant surgeon in 1818, and, like Maconochie, he knew Combe from his time as a young man in Edinburgh.[134] He was also an early and avid supporter of phrenology, having read Combe's *Essays on Phrenology* shortly before leaving Britain.[135] Thankfully for Paterson, Calcutta was a lot better connected than Norfolk Island. Attached to the 16th Bengal Infantry at Barrackpore from December 1824 onward, he maintained a steady correspondence with phrenologists in both Britain and the United States. Every morning a Bengali postal runner, or *dak*, arrived at the officers' headquarters. With Paterson's letters packed into a satchel, the runner would then travel the fifteen miles south to the Calcutta General Post Office on the Chowringhee Road. Without regular steam shipping from Bengal in this period, the letters were sent by sail, arriving in Britain between four and six months later. Those addressed to the United States would then take another month to cross the Atlantic.[136]

In his first letter to Combe from Bengal, Paterson described himself as "the Champion of Phrenology in the East."[137] He went on to give an account of the meetings of the Calcutta Phrenological Society. Founded in March 1825, the society sought to provide a regular course of phrenological lectures in the city as well as to undertake a comprehensive "Oriental anthropology."[138] Paterson told Combe that they had recently held a debate on "the use of this System in improving Education." He had also delivered "a course of six lectures on the principles & practice of phrenology" at the Asiatic Society in Calcutta.[139] These lectures were illustrated by a collection of over ninety phrenological busts purchased from the London cast manufacturer James De Ville. According to Paterson, "flocks of visitors" came to see the busts, each individually labeled and arranged on the tables at the sides of the Asiatic Society's lecture hall.[140] In another letter to the American physician John Bell, corresponding secretary of the Philadelphia Phrenological Society, Paterson described how he had nurtured a "nascent appetite for phrenological studies in Calcutta, by distributing elementary books & busts."[141] Like other voluntary associations in the city at the time, the Calcutta Phrenological Society maintained strong ties with the British colonial elite. The president was Clarke Abel, physician to the governor-general of Bengal, while a number of East India Company merchants attended the meetings held each month in the library of the *Bengal Hurkaru* newspaper on Tank Square. John Grant, a surgeon at the Calcutta General Hospital and a man Paterson described as possessing an "extremely loft" forehead, took the role of vice president.[142] Paterson also corresponded with Horace Wilson at the Asiatic Society, forwarding notes on "the True Origin of the Earth" as well as articles on "Electricity and Galvanisation."[143] With the support of the colonial medical establishment, the phrenologists found their place amid a broad range of scientific lecturers operating in early nineteenth-century Calcutta, from James Dinwiddie's spectacle of galvanism on Cossitollah Street to David Ross's chemical experiments at Hindu College.[144]

But Barrackpore, where Paterson lived and worked most days, was not urban Calcutta. Writing to Bell again in the summer of 1825, Paterson was taking a break from lecturing in the metropolis, "the weather being so intensely hot for two months to come." While most of the other members of the Calcutta Phrenological Society lived in the center of the city near Fort William, Barrackpore and neighboring Monirampore were much more rural. In his letter to Bell, Paterson described how he had "taken an opportunity while living in quiet cantonments, to try an experiment which proved eminently successful beyond my most sanguine expectations." The "Phrenological School of Munerampoor" was the result of that experiment. Paterson suggested that, given the American phrenologists were advocating for their own educational reforms, "it will be perhaps interesting . . . to learn something regarding the progress of the true & plenary System of anthropology in Asia."[145]

The school was modeled on a traditional *pathshala*, "built of bamboo & brick" on a plot provided rent free by a local Bengali landlord, or *zamindar*.[146] Paterson began by emphasizing the need for a phrenological analysis of both teachers and pupils. He described seeking out "a native schoolmaster in the village—who came & had his head examined." Eventually, Paterson selected "a Brahmin" with "a fine configuration of a head," paying him "one gold Mohur a month," or approximately 15 rupees. The pupils, twenty-five Bengali boys between two and twelve years old, were then arranged into classes based on Paterson's own phrenological analysis of their heads. He described how "small classes were formed of those whose heads manifested the greatest similitude in the three regions." It was also "necessary to insulate boys from the dissimilitude of their developments." The school day was then divided into three as well, representing the major divisions of the phrenological organs. From 7:00 a.m. to 9:00 a.m. the boys exercised the "animal region," from 11:00 am to 1:00 pm the "intellectual region," and from 3:00 pm to 5:00 pm the "moral region." Paterson then personally measured the progress of the boys with a pair of calipers. He described this practice in his letter to Bell:

> On entrance every lad's head was manipulated, measured & registered in a book kept at the school for the purpose, and every month afterwards it was regularly measured again, so that any slightest alteration might be noted with extreme accuracy.

Looking over the entries in the ledger at the end of 1825, Paterson concluded that the school had proved successful in demonstrating "the fruits of a Phrenological course of instruction." In fact, Paterson told Bell, "most of the lads had so improved in appearance that the uncles & aunts & cousins who lived at some distance & had not seen them during these 4 months, could scarcely recognize them."[147]

On the opposite side of the river from Monirampore stood Serampore College, a missionary school established by William Carey and his fellow Baptists in 1818. Paterson's school represented a phrenological analogue to the growing number of missionary establishments in early nineteenth-century Bengal. Like the missionaries, Paterson was skeptical of Indian religious practices, referring to the "Brahminical labyrinth of superstition" in a letter to Combe.[148] But he was not opposed to Indian educational traditions. Paterson himself had studied Indian languages, completing a course in Hindustani with the noted Orientalist John Gilchrist in London.[149] Paterson's school therefore provided for the teaching of Sanskrit, which he described in a letter to Combe as the "classical tongue . . . the same as we learn Latin in Britain."[150] This would be taught in the middle of the day while exercising the "intellectual region." Paterson also encouraged the teaching of English. However, he ultimately believed that the Bengali mind was not suited to learning European languages because of a small organ of Form.[151] For both Paterson and the missionaries at Serampore College, the "Hindoo mind" needed to be improved within its own traditions before it could fully appreciate European learning.[152]

In common with many other British colonialists, particularly military officers, Paterson found Bengali men to be weak, lazy, and effeminate.[153] The 47th Bengal Native Infantry was stationed at Barrackpore at the time, preparing to fight in the First Anglo-Burmese War. In a letter to Combe, Paterson compared a "Hindoo" skull with a "Burmese" skull, claiming "the Burmese would laugh at a Hindoo army, were not they led and ordered by British officers."[154] Paterson had also composed an article entitled "On the Phrenology of Hindostan" during his first posting at Fort Monghyr, an East India Company garrison on the Ganges to the west of Bengal. Later printed in the *Transactions of the Phrenological Society*, Paterson's article described the "Hindoo skull" as "delicate and brittle" and attributed the "national timidity of this people" to a large Cautiousness.[155] In light of this, Paterson placed great emphasis on physical education, instigating a regime of "bodily exercises of wrestling, skipping, mugdoo (a species of long dumb-bell)." This was intended to "alter that listless & supine appearance which native lads often possess."[156] Whether evangelical or phrenological, the function of education was ultimately to provide for an obedient and productive colonial workforce. At Serampore, the Baptists imagined a legion of "Christian pundits," while on the other side of the river, Paterson hoped that the "Hindoos of British India, will change in cerebral organization, and consequently in mental manifestation."[157] He concluded his letter to Bell, writing, "This is the means; the *end* is obvious—to rear up useful members of society."[158]

As with prisons and plantations, the character of a school also depended upon the character of its proprietor. But while the intellectual and moral qualities of the Bengali pupils had apparently improved, the same could not be said for Paterson. David Drummond, founder of the Durrumtollah Academy in Calcutta, argued that the phrenological schoolmaster was characterized by "childishness" and "credulity."[159] Another of Paterson's opponents went even further. James Beatson, an East India Company officer, wrote to Combe declaring, "Dr. P. is not in a fit state to have any thing to say to such matters. He drinks like a fish! And his brain is considered to be . . . affected by the liberal potations of brandy to which he has habituated himself."[160] (Given Paterson admitted to poor health—"my liver having suffered much during my stay in Sumatra"—this might have been a fair description.)[161] Beatson recommended that Combe should send no further books to either Paterson or the phrenological school. Paterson retaliated, attacking Drummond as "puerile and erratic" and describing another opponent, John Adam of the Medical and Physical Society of Calcutta, as "unstable as water."[162] But the damage was done, and Combe agreed it was better for Beatson to sell off the remaining copies of his books rather than give them to the alcoholic phrenologist.[163]

Whatever Paterson's problems, his school nonetheless generated interest far beyond rural India. Correspondence connected these reformers materially, but it was phrenology itself that provided a language through which to imagine a universal system of instruction. Whether in Boston or Bengal, phrenologists established educational reform on two basic principles: first, the threefold division of the brain into intellectual, moral, and animal faculties; second, the materiality of the mind. For phrenologists, all education was ultimately physical education. The brain was just another muscle. So while Paterson's school was certainly influenced by his belief in the weakness of the Bengali mind and body, it also reflected a broader phrenological attitude toward education, one reinforced through transnational correspondence. Paterson was in fact writing to Bell in Philadelphia and Combe in Edinburgh precisely because he understood his scheme as more generally applicable. He summed up his view on education to Bell as follows:

> I am certain the true & plenary System of Education is by addressing our labours to every organ in particular. *Exercising* those that are too weak in the balance of the whole; and *Quieting* those that are too strong for the same balance.[164]

When Paterson wrote to Combe, he also explicitly conceived of his work as part of a global movement, writing, "I hope now the Societies in Europe, Asia, and America may illuminate each other. They have only now to found a Society at the Cape of Good Hope, to have Phrenology in the four quarters of the globe."[165]

Paterson's dream of a phrenological society in Africa never came to fruition. But by midcentury, phrenology had penetrated into some Cape Colony schools. Among these was a government school run by James Rait at the Colesberg frontier settlement. Rait initially sought Combe out for advice concerning a medical problem, but the two also discussed education and phrenology, with Combe recommending an "abundance of bodily exercise in the open air."[166] Rait soon committed to both phrenology and physical education, ordering a series of school textbooks from the phrenological publisher William and Robert Chambers in Edinburgh.[167] The teachers' handbook accompanying the series echoed Combe's and Paterson's advice, suggesting that "physical" instruction should not be viewed as secondary to "intellectual."[168] Printed in the tiny octodecimo format, the other books in the series reflected the organization of phrenological schools the world over. The series was "constructed upon a theory which acknowledges that the human being possesses a physical, a moral, and an intellectual nature, each of which calls . . . for its appropriate exercise, training, and instruction."[169] Before long the pupils at Colesberg were undertaking gymnastic routines in the newly built playground and reading from *Rudiments of Knowledge*, a book that even included a chapter entitled "Faculties and Emotions of the Mind."[170]

Phrenological schools also began to pop up in the United States. Horace Mann had just finished reading the Allen and Ticknor 1834 edition of Combe's *Constitution of Man* when he was appointed secretary of education of the state of Massachusetts.[171] Combe and Mann later met in a train carriage on the way to a school convention in New England.[172] This proved to be the beginning of an enduring friendship, one sustained over the years through regular transatlantic correspondence. Combe came to consider Mann "another brother to me" and found his character to be beyond reproach.[173] Mann was both manly and righteous, "a perfect moral Hercules" in Combe's eyes.[174] The geographic distance actually worked to increase the bond Combe felt. He told Mann it was "delightful to find kindred spirits, full of great thoughts & feelings of world-embracing philanthropy, in all countries."[175] Combe only wished the penny post would extend to the United States.[176] In any case, Mann returned the compliment, describing *The Constitution of Man* as "the greatest book that has been written for centuries" and "the only practical basis for education."[177] His good friend and fellow educational campaigner, Samuel Gridley Howe, also found Combe's books to be exemplary. He corresponded with the phrenologist in March 1839, seven months before the two met in Boston, writing that "no words can express the obligation I feel to the great author of the System of Phrenology."[178]

Mann and Howe were serious about the role of phrenology in school reform. The threefold division of the brain shaped the very meaning of the project, with Howe suggesting that education should be directed to the "physical, intellectual, & moral nature" of students.[179] When Mann became president of Antioch College in the 1850s he put these ideas into practice, explaining to Combe that the tutors and managers were all "avowed phrenologists."[180] As at Paterson's school in India, the day was divided into physical, intellectual, and moral instruction. Each morning the students would rise at 6:00 a.m. to eat breakfast and attend chapel before exercising with their teachers for at least three hours. This would then be followed by five hours of lectures and another four hours of independent study, the latter intended to promote self-discipline and moral reflection.[181] As director of the Perkins School for the Blind in Boston, Howe also structured education around phrenology. He even made sure it was taught, writing to Combe in 1839 to inform him that the "upper classes are all instructed in the general principles of intellectual philosophy" and "I have known many who have taken a deep interest in the philosophy of phrenology."[182]

But while Mann and Howe were able to enact phrenological reforms within their own institutions, the Massachusetts Board of Education itself was not particularly powerful. It did not have any formal authority to direct school policy. Instead, it simply produced annual reports in the hope of influencing local school managers in the state.[183] The situation was similar in Britain. With no national educational directive, school reform could be advanced only through setting up model institutions and publishing regularly about the results.[184] Writing letters and exchanging printed material were therefore practices of reform in and of themselves. Securing favorable reports in the foreign press could significantly enhance the credibility of a local school movement. When Mann sent Combe copies of *The Annual Report of the Board of Education*, the phrenologist thanked him and sent over the latest edition of his *Moral Philosophy* alongside a copy of *The Annual Report of the Glasgow Normal School*.[185] Howe too sent copies of *The Annual Report of Perkins School for the Blind* and asked Combe to distribute them during his phrenological lecture tour of the German states.[186] There was, however, some transatlantic postal etiquette to follow. Sending a package without proper payment could result in the recipient footing the bill. Writing to Mann in 1842, Combe explained:

> I am often annoyed by my friends on your side putting up a pamphlet . . . on board of a packet ship addressed to me, & never dreaming of its farther fate. It is put into the post office bag & reaches me as a foreign letter at an enormous postage.

Combe was aware that Mann probably faced similar problems and, eager to preserve his own good character, asked "if my packet to you costs you so much."[187] A gentleman always paid the correct postage.

Whatever the expense, there was a lot to be gained. Before he arrived in the United States, Combe learned that his books had already found their way into a number of village schools in western New York. With this in mind, he asked Mann whether it might be worth preparing a new book for this market with the title "Phrenology for Schools."[188] The reply was encouraging. Mann informed Combe that there was also a campaign under way for *The Constitution of Man* to be included in the Massachusetts School Library.[189] This, however, proved controversial. As in Britain, schools were segregated by religious denomination with education organized locally by different Christian groups. Centralized control of the curriculum, particularly the school library, was often opposed on the grounds it favored one religious sect over another. Mann knew all too well that the Board of Education needed to present itself as nonpartisan.[190] In 1840, the Democrats had actually attempted to abolish the board precisely on the basis that it was imposing Unitarian teaching.[191] Mann was dismayed, and complained privately to Combe of "the bigotry of the religionists . . . in resisting all measures which do not emanate from or cannot be controlled by them."[192] Combe told Mann they had the same problem in Britain, writing "religious sects strengthen & give permanency to themselves by teaching their doctrines to the young."[193] However, Mann hoped that phrenology might cut across religious differences. It certainly wasn't associated with any particular denomination. The only risk was that it would be considered "materialist" or "secular." In light of this, Mann informed Combe that the Boston publisher William Ticknor was preparing a new edition of *The Constitution of Man* "containing an appendix, the object of which is to show a harmony between the doctrines of Phrenology & those of the Bible."[194] Mann hoped that by repackaging Combe's book he could convince the mixed denominational committee in charge of the Massachusetts School Library to accept phrenology.

Mann's campaign for phrenological textbooks in Massachusetts schools was ultimately unsuccessful, and the committee of Baptists, Unitarians, and Congregationalists rejected the proposal.[195] However, Mann did exert direct control over the state teacher training colleges: the normal schools. As was the case for Paterson in India, phrenology allowed reformers to identify suitable teachers and principals. After all, as Combe explained to Howe, the "personal qualities of the head of an Establishment have a great influence on its life & being."[196] Mann wanted "intelligent gentlemanly teachers" while Combe explicitly recommended that normal schools "consider the temperament & brains of the candidates."[197] Through the Board of Education, Mann also ensured that *The Constitution of Man* remained a staple of the Massachusetts normal school curriculum throughout the first half of the nineteenth century.[198] Mann even personally recommended Combe's book to a number of young men and women who were thinking of training as teachers. After reading *The Constitution of Man* twice in two weeks, one "highly orthodox" trainee reported that he was still concerned that phrenology contradicted Christian teaching. Mann told Combe how he had "endeavoured to explain to him, that your system contained all there is of truth in orthodoxy." The time spent with the young apprentice was apparently not wasted, and Mann reported that he "adopted my views of the subject," quipping that it had been a "case of conversion."[199]

Combe was impressed by the progress of phrenology in American schools and promised Mann that he would "call public attention both in Europe & America to your bright example." Combe kept to his word and, in April 1841, informed Mann that he had "written an account of your system & doings & have actually got it adopted by the Edinburgh Review!"[200] This was a coup, given that the same periodical had published a damaging critique of phrenology in the 1820s. But it was precisely because phrenology had traveled to the United States and returned in the form of Mann's annual reports that Combe was afforded such an opportunity back in Britain. The article was still published anonymously, which was not unusual for the *Edinburgh Review*, but Combe did manage to emphasize the role of phrenology in American educational reform. He quoted directly from Mann's reports, insisting that schoolmasters should have "a knowledge of the human mind as the subject of improvement."[201] By sending phrenology on a circuit, beginning in Calcutta and then taking in the Atlantic world, the phrenologists ultimately enhanced both the reach and the credibility of their respective campaigns. Local reform gained momentum through global exchange.

Moving from the slave plantations of the United States to the rural schools of Bengal, this chapter has developed two overarching arguments. First, correspondence allowed phrenologists to promote reform as a global project. Second, phrenology itself provided a means to express these ideas in terms of a universal understanding of human character. The materiality of correspondence proved central to this argument. For phrenologists, the passage of a letter implied a connection between disparate locales, reinforcing a sense of moral responsibility and common purpose. Correspondence was ultimately a material practice of communication to match a materialist philosophy of the mind. When Fowler argued that reading and writing letters could "exercise" the mind, he meant it literally.

It was also through correspondence that phrenologists came to see the relationship between different reform projects. Slavery was a particularly common point of comparison. When Maconochie learned of the interest his scheme had generated in the United States, he wrote to Combe expressing his belief that "the cause I advocate seems to me even more important to humanity . . . than that of the abolition of Black Slavery."[202] In fact, like many other campaigners opposing transportation, Maconochie compared the existing penal settlements in New South Wales to the slave plantations in the United States. According to Maconochie, only the mark system "would remove that taint of slavery which, at present, corrupts every portion of it."[203] Similarly, while Mott began her correspondence with Combe on the basis of a shared interest in slavery, she too was prompted to consider the relationship between abolition and other reform agendas. Combe suggested that the Philadelphia race riots of the 1840s could partly be explained by an ineffective school system, writing "the riots were much promoted by large boys who have been expelled from the public schools . . . , cast loose on society with all their propensities untamed, and reckless of all authority human or divine."[204] In light of this, he sent a copy of his *Remarks on National Education* to Mott.[205] She concurred with Combe, expressing her belief that "the efforts to spread education . . . will prove one of the successful means, to bring about the desired result."[206] In the course of a couple of letters, the abolition of slavery suddenly depended upon enacting phrenological school reforms. William Lloyd Garrison's cousin was even reported to have started teaching phrenology to freed slaves in Jamaica.[207]

These developments were not confined to abolitionists. Caldwell also came to see the relation between slavery and reform more generally. In the 1830s he published an article in the *Phrenological Journal* entitled "New Views of Penitentiary Discipline" in which he discussed both the silent system at Auburn and the transportation of convicts to New South Wales. Adopting the same language he used to describe the slave population, Caldwell declared that crime was "the product of mere *animal propensity*" and that "to reform a criminal, then, you must make him less of *an animal*, and more of a *human being*."[208] This was a period in which the African American convict population weighed heavily on the minds of penal reformers, with the Boston Prison Discipline Society describing "the degraded character of the colored population" in its first annual report. The Massachusetts State Prison even suggested sending African American inmates to Liberia.[209] For Caldwell, phrenology explained the relationship between freedom and mental character both in the prison and on the plantation: the "animal organs" dominated in both slaves and convicts. Phrenology ultimately provided a universal language of mental character through which to bring together these diverse institutions—from schools to slavery—under a single political banner.[210] It was therefore through global exchange that the very notion of "reform" came into focus.[211]

With individuals and society connected through a common understanding of human character, it didn't take long for reform to transform into something more dramatic. In July 1848, Combe received his first letter from a revolutionary.[212] Gustav Struve, former editor of *Zeitschrift für Phrenologie* in Heidelberg, had already hinted at a change in his political outlook. In doing so, he revealed how phrenology could be tied to a new political language of revolution, one that many started to contrast with reform. In a letter sent just before the revolutions of 1848, he had explained that "phrenology is at the bottom of all my doings . . . , it has only made clear, what before was unclear, firm and cast iron, what before was wavering and uncertain."[213] Struve stuck to his convictions and in April 1848 joined the Hecker Uprising in Baden as part of the unrest that was sweeping across Europe. Although the uprising ended in defeat, Struve managed to escape to Switzerland. Unsurprisingly, news of revolution in Europe was greeted with condemnation in Britain. Holed up in the Swiss border town of Rheinfelden, Struve read the damning editorials published in *The Times*. He was therefore keen to emphasize both the legitimacy and the moral value of his action to his British correspondents and penned a letter to Combe in which he observed:

> Very soon all men of character and courage were convinced that nothing could be expected, neither from the several German princes, nor from the assembly of Frankfort. . . . It became clear to every thinking man that the old system of slavery would be kept up unless the princes should be driven away by force.

Appropriately enough, Struve composed this letter on what can only be described as revolutionary notepaper (fig. 20).[214] The intricate printed design featured an image of his fellow republican Friedrich Hecker and was headed with the revolutionary slogan "Freiheit, Gleichheit, Verbrüderung" (Liberty, Equality, Fraternity).

Despite Struve's choice of stationery, Combe was not convinced. Once again, the meaning of reform was open to contestation. For Struve, reform needed to be won from below, through the wholesale rejection of aristocratic society. But for Combe, reform had to come from above, through legislation and government. Like many others in Britain at the time, he labeled the alternative "revolution."[215] What's more, for Combe, phrenology explained exactly why political change could never legitimately come about through

117

Freiheit
Gleichheit
Verbrüderung

Muttenz

Hecker

Rheinfelden the 9 August 1848.

My dear friend!

In consequence of the wish you expressed to me in your letter of the 31 July I give you a short account of the republican movements which took place in the month of April of this year. You know that there was a state of warfare since many years in the Grand Duchy of Baden between the people and the government. Our german princes had lost all the love and all the respect the nation had bestowed upon them in former times, in consequence of the barefaced breach of all the promises they had given in the times of the wars with Napoleon, and the abominable system of slavery which they had adopted. Already the 12 September 1847 a meeting of the most eminent liberals of the Grand Duchy of Baden was held in the town of Offenburg, where the claims of the people were put together. The consequence of this

Fig. 20. Struve to Combe, 9 August 1848, MS7297, f. 117, Combe Papers, National Library of Scotland.

revolution. Like the slave plantation or the eighteenth-century jail, revolution provided no check on man's animal nature. He explained to Struve:

> I desire a moral freedom, in which the selfish animal propensities shall be restrained, & the higher faculties predominate. There *must* be a power capable of restraining & directing the propensities, *somewhere.*

A stable society could be achieved only through "a long apprenticeship to freedom under a representative government, with a monarchical head"[216]—in short, the kind of social relations already in operation in Britain. On receiving Combe's reply, Struve acknowledged that little could be said to change his mind. He therefore simply expressed his wish that, whatever their political differences, the two phrenologists might remain on good terms, writing, "I hope that our friendship is still the same and that the storm which blew around us has not weakened its strength."[217] This was Struve's final letter to Combe. They never spoke again.

CHAPTER 5

Periodicals

"Is Phrenology True?" Readers in the Caribbean were keen to know the answer. In August 1868, the *Barbados Globe* printed an advertisement for a series of lectures delivered by "W.D. Maxwell." Described as a "coloured interpreter of the science," Maxwell was a "native of Barbados." Despite the racial tensions still at play on the island, Maxwell's lectures proved "a success." Local papers explained that Maxwell "succeeded in winning the attention of his audience and in giving some of them a just idea of the principles of Phrenology." One reader of the Barbados *Times* even wrote in, reporting on his experience. "P.C. Brewster" of Bridgetown had permitted Maxwell "to manipulate my cranium, in order fully to test the validity of the science." Brewster was convinced, writing, "I was never more astonished than to find him so fully and truthfully detailing every circumstance, which the various organs were capable of performing." Brewster was compelled to admit "Phrenology to be a science." Another correspondent in Barbados noted that phrenological lectures were "becoming popular with us."[1] Before long, word of Maxwell's lectures spread. Readers of the *Daily Chronicle* in British Guyana were told that Maxwell intended to "favour the people of your colony with a lecture also." True to his word, the Barbadian phrenologist boarded a ship and delivered a "Phrenological Lecture" in Georgetown. The *Daily Chronicle* reported that the audience was "exceedingly small, but very appreciative." Maxwell, readers learned, described "the varieties of mankind" and examined the heads of two audience members. Later in the century, Maxwell made a much longer journey, crossing the Atlantic to deliver a series of lectures in Britain. Revealing the racialized reception of this black phrenologist, the *Gentleman's Journal* in London stated that Maxwell's lecture "would have done credit to any European of the highest mental calibre." Volunteers for examination were placed "under the searching glance of the dark man's piercing eye."[2]

Maxwell's career illustrates the close connection between imperialism, popular science, and the periodical press in the nineteenth century. The "marketplace" for nineteenth-century science was not confined to Britain.[3] In the United States, the *American Phrenological Journal* was propped up by the profits of slaveholders. In the Cape Colony, settlers advertised phrenological lectures in frontier newspapers. And in Calcutta, a group of Bengalis started a phrenological journal in the hope of reforming colonial society. Following these cases in further detail, this chapter argues that the periodical press played a key role in the creation of new audiences for science. Traditionally, however, historians have largely confined their studies to audiences within the British national context. Jonathan Topham and others have demonstrated how the nineteenth century saw the emergence of scientific readerships among artisans, women, and children across the British Isles.[4] This chapter expands this historiography, examining the emergence of new reading audiences for science among settlers, frontiersmen, and colonized subjects.[5] Additionally, while the existing historiography often takes national contexts for granted, this chapter examines the role played by phrenological journals in the formation of national identities. Titles like *American Phrenological Journal, Canadian Phrenological and Psychological Magazine,* and *Illustrated Australian Phrenological, Physiognomical and Hygienic Magazine* point toward the nationalism that underpinned the growth of the periodical press—a theme global historians of science should not ignore.

Even as periodicals addressed national audiences, they also looked outward. This chapter therefore examines how phrenologists used the periodical press to simultaneously imagine themselves as part of national and global communities in print.[6] This was made possible by the growing import and export markets for books, newspapers, and journals. In the colonial world, local supporters of phrenology regularly extracted articles from periodicals imported from Europe. Maxwell's lectures in London, for example, were reported back in British Guyana.[7] *The Australian,* a liberal newspaper in New South Wales, reprinted articles from the *Phrenological Journal* in Edinburgh, including an account of a criminal skull. The editors, one of whom was the son of a Norfolk Island convict, added that they would "always be glad to receive any communication on the subject."[8] Periodicals in Europe and the United States carefully fostered a reading experience that gave the impression of a global movement. *Annals of Phrenology* in Boston advertised itself specifically as an anthology of articles extracted "from the Edinburgh, Paris, and London Phrenological Journals."[9] By publishing these articles alongside commentaries, *Annals of Phrenology* acted as a guide on how to read foreign periodicals, cultivating new reading audiences in the process. One article, "On the Life, Character, Opinions and Cerebral Developments of Rammohun Roy," extracted from the eighth volume of the *Phrenological Journal*, was identified as "particularly interesting to general readers." Another, "On the Education of a Civil or Mechanical Engineer," was described as "very good" but not sufficiently adapted to an American audience.[10] Back in Britain, the *Phrenological Journal* itself regularly featured an "Intelligences" section containing extracts and news from foreign periodicals. Readers were presented with an account of "Phrenology in the Sandwich Islands" taken from the *Hawaiian Spectator*, news of the "Phrenological Society of Sidney [*sic*]" extracted from the *Hobart Town Courier*, and "a phrenological description of the character of the Hydriots" from *Minerva* in Athens.[11] In one case, news from New South Wales, Great Yarmouth, Paris, and Göttingen all featured on the same page.

Phrenological journals both presented these communities and cultivated them. Readers were invited to become active participants. The *Phrenological Journal* in Edinburgh explained that "we rely on our subscribers exerting their influence each in his own circle to extend the diffusion of the science."[12] Letters pages were central to this strategy, simultaneously generating national and international phrenological audiences. *Annals of Phrenology* informed its readers that it had received letters "from London, Edinburgh, Dublin, and Paris."[13] These included a letter from Joseph Vimont, one of the leading phrenologists in France, in which he praised the efforts of his American counterparts. Vimont's letter was printed in the same volume as correspondence from Kentucky informing readers that phrenology was "making rapid strides in the West."[14] Similarly, the *Journal de la Société Phrénologique de Paris* listed corresponding members across Europe and the colonial world. These included Benoît Trompeo, in northern Italy; François Aulagnier, a military doctor living in Marseille; and Dr. Richy, who was traveling between Madagascar and India.[15]

While periodicals were certainly a powerful means through which to promote phrenology as a global science they also had their limits. The geography of phrenology in the periodical press mirrored the uneven distribution of printing technologies and materials. In Europe and the United States, publishing emerged as a major commercial industry during the early decades of the nineteenth century. Phrenological societies in Boston and Paris found publishers with free capital willing to finance the publication of new journals. Jean-Baptise Ballière, publisher of the *Journal de la Société Phrénologique de Paris*, operated profitable offices in London and Paris selling a range of scientific and medical texts. His own business had received a considerable boost thanks to a 40,000 franc loan from the French government following the July Revolution.[16] Similarly, *Annals of Phrenology* in Boston formed part of the commercial publishing strategy of Marsh, Capen and Lyon of 133 Washington Street. The first number of *Annals of Phrenology* even featured advertisements for Marsh, Capen and Lyon's other phrenological publications, including a new American edition of Johann Spurzheim's *Phrenology, or the Doctrine of Mental Phenomena.*[17]

The economies of print in the colonial world could not have been more different. Raw materials such as paper, ink, and type were in short supply, with printers reliant on imports from Britain and France.[18] Printing presses themselves were also hard to come by, with colonial governments hesitant to grant unlimited freedom of the press.[19] Differences in local printing conditions were then reinforced by the haphazard nature of exchange networks. The latest issue of a newspaper or journal would often arrive late or not at all. In Europe, the cholera pandemic of the early 1830s significantly disrupted the cross-circulation of journals between phrenological societies across the Continent. The 1836 volume of the *Phrenological Journal* in Edinburgh noted that it was "in arrear with our notices of the French Phrenological Journal, partly from the later numbers not having been received till some months after they were due."[20] George Combe complained in private correspondence that he was also having difficulty acquiring copies of *Zeitschrift für Phrenologie* from Heidelberg and *Tidsskrift for Phrenologien* from Copenhagen during the same period.[21] Delays in receiving publications could significantly alter their relevance. *Annals of Phrenology* noted disapprovingly that many phrenological publications in the United States still featured a phrenological chart based on "the old classification and marking, which Spurzheim finding incorrect, rejected."[22] Import duties and postal costs added further to the expense and difficulty of acquiring foreign journals. The American politician Horace Mann grumbled at having to pay $1 duty on a single monthly number of the *Phrenological Journal* imported from Britain. Mann could afford the cost, joking that "no doubt I shall get my money's worth when I read it," but an entire annual subscription of twelve issues only came to $2.50 before taxes were added.[23] Similarly, Combe complained that American booksellers were sending periodicals without sufficient postage, often covering only the transatlantic voyage and leaving him to personally pick up the bill for packages forwarded on from Liverpool to Edinburgh.[24]

Examining the United States, the Cape Colony, and India, this chapter charts the power of the periodical press, as well as its limits, in making new audiences for science. By studying three different print cultures in turn, this chapter reveals the variety of ways in which phrenology was connected and disconnected from the wider world. This chapter also examines the interplay between material and print culture, uncovering how illustrations from China and transcripts of lectures from London found their way into periodicals in the United States and India. More broadly, this chapter suggests that we cannot fully understand the emergence of popular science without attention to the global history of this category. Popular audiences for nineteenth-century science were a product of both nationalism and imperialism, in Britain and beyond.

A PHRENOLOGICAL BUSINESS EMPIRE

For Lorenzo and Orson Fowler, phrenology was a business.[25] And at the center of that business was the *American Phrenological Journal*. Published monthly between 1838 and 1911, this proved to be one of the most successful periodicals of the nineteenth century. The *American Phrenological Journal* reached miners in California, slaveholders in South Carolina, Congressmen on Capitol Hill, and preachers in Boston. By the 1860s, the Fowlers boasted that "we are now sending our publications to . . . Japan, Africa, New Zealand, Australia." Missionaries even reported taking copies of the *American Phrenological Journal* to China. Through the periodical press, the Fowlers hoped to transform their modest phrenological cabinet in New York City into the center of a national as well as international business empire. However, before they were publishers, Lorenzo and Orson had been popular lecturers. The brothers had learned about phrenology while studying at Amherst College in the 1830s, reading American reprints of Johann Spurzheim's books and acquiring busts and skulls for the college natural history society. At Amherst, Orson gave phrenological examinations to his classmates for 2 cents, honing the skills by which he would make a living for the rest of his life. After graduating, Lorenzo and Orson traveled together across the United States, as far west as Missouri, down to Tennessee, and up to Massachusetts, lecturing and offering readings in town halls and lyceums.[26] The brothers were just one example of the thousands of popular scientific lecturers operating in the United States at this time.[27] What made the Fowlers different, however, is that they successfully translated this world of popular phrenological lecturing into print. By midcentury the brothers were reaching over 30,000 subscribers a month, running the *American Phrenological Journal* through Edward Jenkins's steam press in New York City.[28] In doing so, the Fowlers created new reading audiences for science in the United States and beyond.

Using the money they had earned from lecturing, the Fowlers launched the *American Phrenological Journal* in 1838. The first few volumes were printed in Philadelphia, and then the brothers moved their entire operation to Clinton Hall, just off Broadway in the heart of New York City. Once there, the young phrenological enthusiast Samuel Wells joined the firm as a partner. (Wells was also in love with the Fowlers' sister, Charlotte, who worked at the business, organizing the subscriptions.)[29] The walls at Clinton Hall were lined with skulls and busts "from all quarters of the Globe": "Cannibals," "Caribs," "Pirates," and "Indians."[30] The cabinet even featured an Egyptian mummy. Entry was free, but the brothers hoped visitors would buy something once they were inside: a phrenological bust for $1.25 or a cheap reprint of George Combe's *Constitution of Man* for 30 cents. Many also came and paid to have their character read.[31]

For the Fowlers, the periodical press provided a way to extend all this to a national audience of phrenological readers.[32] In some cases, the pages were actually presented as a recreation of Clinton Hall.[33] In one monthly series, entitled "A Shelf of Our Cabinet," the *American Phrenological Journal* described exactly what it was like to visit the collection in New York City. Readers in distant towns and cities were invited to "contemplate a single shelf of busts in our collection." The articles, accompanied by wood engravings, were designed as a kind of virtual tour, directing the reader's internal eye: "Here is the head of Canova. . . . Alongside him is the mask of Bacon, author of the *Novum Organum*. . . . Next to him, in odd proximity, is the bust of an ourang-outang." This needed a bit more explanation. What was Francis Bacon doing next to an ape? The *American Phrenological Journal* clarified, informing its readers that "our busts are not classified according to character and social standing, but according to their size or height, so that in arranging them on the shelves some strange combinations are made."[34] Whether this was a sensible arrangement or not, the article made it clear to the reader that the exact organization of the cabinet had been preserved in print.

As well as recreating the cabinet, the *American Phrenological Journal* helped to advertise the Fowlers' goods and services. These notices were typically featured on the inside cover of each monthly issue (fig. 21). The May 1849 issue, for example, invited its readers to purchase copies of the Fowlers' busts, explaining, "We can forward, from the Phrenological Cabinet, packages, boxes, or trunks, to any place on all the regular routes in the

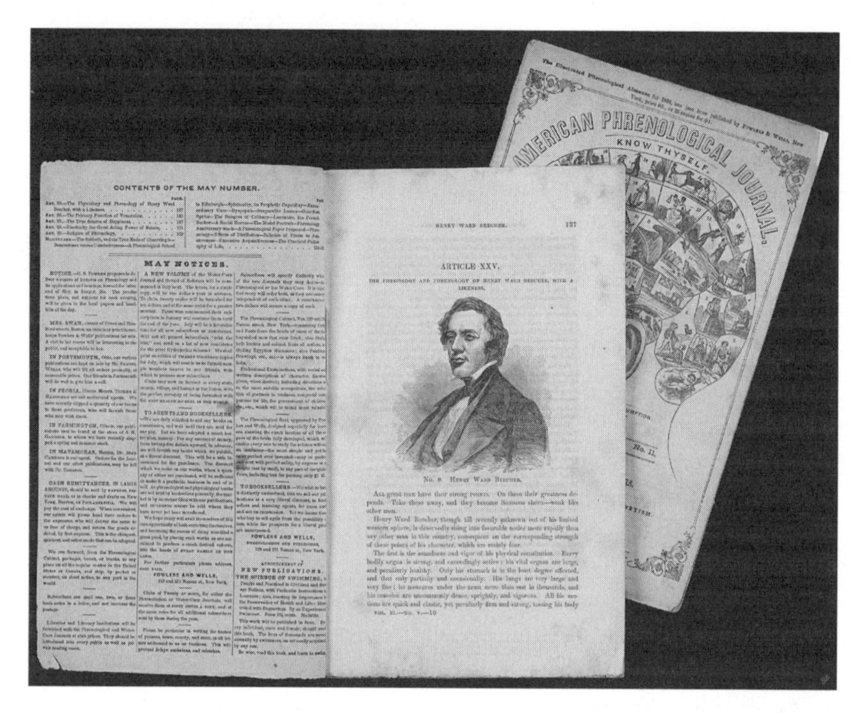

Fig. 21. Opening of the May 1849 issue of the *American Phrenological Journal*, with advertisements and monthly notices on the inside cover, Rare Book Division, Department of Rare Books and Special Collections, Princeton University Library.

United States or Canada." For those closer to New York City, or willing to make the journey, the same issue advertised "Professional Examinations, with verbal; and written descriptions of character . . . including directions as to the most suitable occupations."[35] Finally, the *American Phrenological Journal* advertised the Fowlers' popular lecture business, which they still ran on the side. For "ten dollars per lecture, or six lectures for fifty dollars," Lorenzo or Orson would visit a local lyceum or town hall and give an introductory account of phrenology, hoping to generate some more subscribers for the journal in the process.[36]

Publishing the *American Phrenological Journal* was one thing, but getting people to actually subscribe was another. Outside of New England and New York, where Spurzheim and Combe had spent the majority of their time lecturing, there was little sustained interest in phrenology in the United States in the 1830s. The idea that a settler in Oregon or a rancher in Texas might want to read about phrenology was a novelty. The Fowlers, therefore, needed to create the national audience that they hoped to address. This was not going to be easy. Very few periodicals in the United States at this time reached a truly national readership. And those that did tended to be the big quarterlies, like the *North American Review*, or otherwise addressed existing professional or religious communities. A monthly journal, on a specialist topic, and from an unknown publisher was going to have to work hard to generate subscribers.[37]

Things got off to a rocky start. Although the first volume generated 1,500 subscribers, by the third volume that number had dropped to 400. The Fowlers complained in 1842 that "the present volume has scarcely paid for its printing and paper."[38] Under the pressure of mounting debt, the Fowlers developed a range of business and publishing practices in order to generate new subscribers. First and foremost, they started sending copies of the *American Phrenological Journal* to local papers all over the country. This was facilitated by the fact that the Post Office allowed editors to exchange periodicals free of charge.[39] In 1842 the Fowlers advertised that "newspapers inserting the prospectus on the cover four times, and sending a paper containing it, shall receive a copy of the Journal."[40] This seemed to work. Before long, the *American Phrenological Journal* had been noticed in the *Baltimore Leader*, the *Western Farmer*, and the *Georgia Citizen*.[41] At least now, potential readers in the South and West could learn about the *American Phrenological Journal*. However, even if readers were interested, they needed an accessible way to actually purchase the journal. So alongside the local papers, the Fowlers started developing close relationships with booksellers in different states, offering wholesale prices for large orders. These approved booksellers were then advertised on the inside cover of the *American Phrenological Journal* itself. For instance, the same May 1849 issue described earlier listed resellers in Illinois, Mexico, and Ohio.[42] Once again, changes in the regulation of the Post Office helped. By 1845, distance was no longer factored into the price of postage, allowing the Fowlers to send their publications anywhere within the United States without additional cost.[43]

With local newspapers and booksellers on board, the Fowlers were in a much stronger position. But what really changed things was the development of a system of voluntary agents. The *American Phrenological Journal*, more than any other periodical, relied on its readers to generate new subscriptions.[44] The Fowlers directly appealed to the agency of their readership, writing, "Any and every subscriber or reader is requested to act in behalf of the Journal."[45] The idea was simple. Existing readers were asked to collect the names of local subscribers and return them to the Fowlers in New York City. In this way, the *American Phrenological Journal* could expand into a truly national publication. This is exactly how the Fowlers imagined it, writing:

> There are thousands of friends from Maine to Oregon, and from the Gulf of St. Lawrence to the Gulf of Mexico, who would gladly aid in the circulation of these Journals, and we now offer them not only facilities, but our friendly invitation to co-operate in this great work.[46]

Crucially, although other publishers attempted similar strategies, the Fowlers presented the use of voluntary agents as something particularly appropriate to phrenology. The improving power of phrenology, for the individual and the nation, made it every reader's duty to spread this new knowledge within their local community: "You have it in your power to confer a special favor upon your neighbor, by inducing him to become a reader of one or all of our journals. Will you do it?"[47] The Fowlers even claimed that "those who have read one volume of the Journal will find little difficulty in convincing any *reasonable* mind of the paramount advantages of Phrenology and Physiology in self-improvement."[48] Ultimately, the aim was "to make hundreds of new phrenological missionaries."[49] In doing so, the Fowlers would secure the financial future of their business.

To make things easier, at the end of each year existing subscribers were supplied with a printed "Pictorial Prospectus" (fig. 22). This, the Fowlers suggested, should be "posted in a conspicuous public place, where all may see and read it." Perhaps a "post-office," "factory," "steamboat," or "school-house." The prospectus, illustrated with phrenological busts, set out all the advantages of phrenology as well as giving an overview of the content that would appear in the following volume. Attempting to appeal to a broad audience, the prospectus claimed that "the Mechanic, the Farmer, the Professional Man, the Student, the Teacher, and the Mother will find each number of the Journal an instructive and valuable companion." In addition to acting as an advertisement for the journal, the prospectus doubled as a place to record new subscribers. It featured four columns in which anyone interested in signing up could fill in their name, post office, county, and state. Once filled in, the voluntary agent could then simply post the sheet back to the Fowlers, who would complete the order.[50]

While some agents were content to post the prospectus in their local town hall, others took on a more active role. Many itinerant phrenological lecturers simultaneously acted as voluntary agents, soliciting names of subscribers before they left each town. In doing so, they brought the *American Phrenological Journal* into more and more remote locations, re-

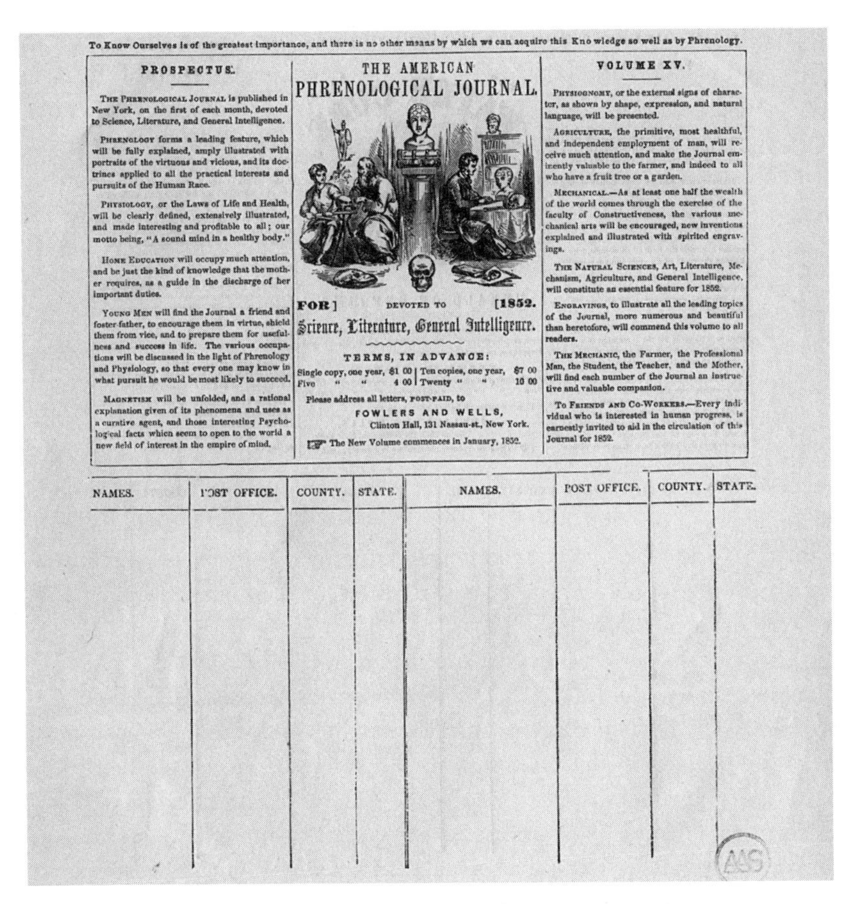

To Know Ourselves is of the greatest importance, and there is no other means by which we can acquire this Knowledge so well as by Phrenology.

PROSPECTUS.

THE PHRENOLOGICAL JOURNAL is published in New York, on the first of each month, devoted to Science, Literature, and General Intelligence.

PHRENOLOGY forms a leading feature, which will be fully explained, amply illustrated with portraits of the virtuous and vicious, and its doctrines applied to all the practical interests and pursuits of the Human Race.

PHYSIOLOGY, or the Laws of Life and Health, will be clearly defined, extensively illustrated, and made interesting and profitable to all; our motto being, "A sound mind in a healthy body."

HOME EDUCATION will occupy much attention, and be just the kind of knowledge that the mother requires, as a guide in the discharge of her important duties.

YOUNG MEN will find the Journal a friend and foster-father, to encourage them in virtue, shield them from vice, and to prepare them for usefulness and success in life. The various occupations will be discussed in the light of Phrenology and Physiology, so that every one may know in what pursuit he would be most likely to succeed.

MAGNETISM will be unfolded, and a rational explanation given of its phenomena and uses as a curative agent, and those interesting Psychological facts which seem to open to the world a new field of interest in the empire of mind.

THE AMERICAN

PHRENOLOGICAL JOURNAL.

FOR] DEVOTED TO [1852.

Science, Literature, General Intelligence.

TERMS, IN ADVANCE:

Single copy, one year, $1 00 | Ten copies, one year, $7 00
Five " " 4 00 | Twenty " " 10 00

Please address all letters, POST-PAID, to

FOWLERS AND WELLS,
Clinton Hall, 131 Nassau-st., New York.

☞ The New Volume commences in January, 1852.

VOLUME XV.

PHYSIOGNOMY, or the external signs of character, as shown by shape, expression, and natural language, will be presented.

AGRICULTURE, the primitive, most healthful, and independent employment of man, will receive much attention, and make the Journal eminently valuable to the farmer, and indeed to all who have a fruit tree or a garden.

MECHANICAL.—As at least one half the wealth of the world comes through the exercise of the faculty of Constructiveness, the various mechanical arts will be encouraged, new inventions explained and illustrated with spirited engravings.

THE NATURAL SCIENCES, Art, Literature, Mechanism, Agriculture, and General Intelligence, will constitute an essential feature for 1852.

ENGRAVINGS, to illustrate all the leading topics of the Journal, more numerous and beautiful than heretofore, will commend this volume to all readers.

THE MECHANIC, the Farmer, the Professional Man, the Student, the Teacher, and the Mother, will find each number of the Journal an instructive and valuable companion.

TO FRIENDS AND CO-WORKERS.—Every individual who is interested in human progress, is earnestly invited to aid in the circulation of this Journal for 1852.

NAMES.	POST OFFICE.	COUNTY.	STATE.	NAMES.	POST OFFICE.	COUNTY.	STATE.

Fig. 22. Loose printed "Pictorial Prospectus" for the *American Phrenological Journal*, BDSDS.1851. Courtesy, American Antiquarian Society, Worcester, MA.

inforcing the link between phrenology as print culture and phrenology as popular spectacle. Combe himself was an early pioneer of this approach, agreeing to distribute copies of the prospectus of the *American Phrenological Journal* after his lectures in New England.[51] Farther afield, "H. Wisner in Illinois," "Dr. Trotter in Georgia," and "Dr. J. Anton in South Carolina" advertised themselves as phrenological lecturers, collecting subscriptions as they went.[52] One voluntary agent, Nelson Sizer, was so successful at generating subscribers after his lectures, that the Fowlers later invited him to join their business in New York City. In Connecticut, Sizer remembered making "more than thirty examinations, and in this place I obtained thirty-six subscribers to the *Phrenological Journal*."[53]

While the Fowlers appealed to the improving power of phrenology to turn their readers into lecturers and agents, they also recognized the need for a financial incentive. They therefore offered reduced rates for large orders of the *American Phrenological Journal*. Agents could then make a profit, benefiting from what were effectively wholesale prices. For example, an individual yearly subscription to the *American Phrenological Journal* cost $1. However, a voluntary agent could acquire twenty yearly subscriptions for just $10. In addition to this, the Fowlers offered "club rates" and "prizes" to groups and libraries making large orders. For $50 the Fowlers would supply "fifty subscriptions" to the journal as well as "a Phrenological Cabinet comprising . . . busts, masks, and skulls, in plaster."[54] In short, for $50, the Fowlers would provide everything you needed to start a local phrenological society or become an itinerant lecturer. The success of the *American Phrenological Journal* ultimately depended upon this reciprocal relationship, whereby the Fowlers created a national community in print through connections to provincial cultures of popular science. In doing so, they generated new markets for their business in phrenological books and busts.

The word *American* in the title of the Fowlers' journal was no accident. Nationalism was central to the marketing of the *American Phrenological Journal*.[55] The Fowlers presented the United States as uniquely positioned to both benefit from and contribute to phrenology. Contrasting the United States with Europe, the first issue of the *American Phrenological Journal* claimed that "this country presents some facilities for the cultivation of practical phrenology, to which the old world is almost entirely a stranger." Readers were told that the United States was exceptional. "We find here a greater variety of character and of talents than in any other single nation upon earth," explained the *American Phrenological Journal*. In particular, the United States was presented as an especially rich place for the study of racial science, with the journal informing its readers that "there are identified with this country three of the five varieties of the human race."[56] American readers were therefore invited to take part in a peculiarly American project, documenting the character and races of the New World. Once again, these ideas fed back into the Fowlers' cabinet, with readers collecting and forwarding skulls from newly acquired territories. This nationalist ethos continued throughout the history of the *American Phrenological Journal*, growing in tandem with the expansion of the United States. Over the years, the Fowlers even started to develop the idea that there was something particular about the "American mind," meaning white settlers. The contrast was with despotic Europe:

> And all because our institutions, by paying every man for his labor, stimulate every faculty of man to its highest pitch of exertion. It is this stimulus to effort given by our country to its citizens in which her present and prospective greatness consists.[57]

Here, phrenology itself explained the link between American democracy and the development of the mind. In the United States, "unfettered by the clogs of monarchism," the brain was guaranteed to be exercised appropriately by hard work and fair reward.[58] For phrenologists, the American mind was a product of the American Dream.

In the United States, nationalism and imperialism went hand in hand.[59] The annexation of Texas in 1845 and the subsequent Mexican-American War opened up new territories such as Utah and California to the west.[60] For the Fowlers, American imperialism was good for business.[61] New territories represented new markets for phrenological books, busts, and lectures. More than ever before, the success of the *American Phrenological Journal* depended upon forging close links with local agents. As the Fowlers recognized, many potential subscribers were now "three thousand miles west of New York." The Rocky Mountain Book Store, run by George Ottinger and Charles Savage in Salt Lake City, was an early and important outpost. Ottinger and Savage stocked "phrenological books, and every variety of *useful* books," posting advertisements in Utah's newspapers. Ottinger, a talented artist, also helped to bring the west back to New York City. In return for periodicals and books, Ottinger forwarded illustrations to the Fowlers that were then reproduced in the *American Phrenological Journal*.[62] These included a photograph of a "group of Moquis Indians" as well as a sketch of Salt Lake City. In addition to these more formal links, the Fowlers hoped to entice individual settlers into becoming phrenological pioneers. One article, entitled "Phrenology in California," suggested that there was plenty of money to be made from lecturing in the West, explaining that "an able phrenologist would find hearing ears and practical inquiring minds . . . and that lectures upon Phrenology would be as remunerative to the lecturer, in a pecuniary point of view." When the gold rush ended, a number of pioneers in California did turn to phrenology. The Fowlers received reports from a "Mr. D. A. Shaw" who had been "lecturing upon the Science of Phrenology, in the interior towns of California, during the past winter, with great success." Shaw happily reported that "the Phrenological Journal is highly appreciated here."[63]

With a growing number of lecturers and agents in the West, the Fowlers decided to expand their rewards system, offering new "prizes" suitable for settlers. In doing so, the brothers found themselves in the unique position of peddling both phrenology and guns. For $80 an agent would receive forty subscriptions to the *American Phrenological Journal* as well as "one of the New Breech-Loading Rifles." This, the advertisement claimed, would be most useful for those in "the new Territories or on the frontier." Alongside rifles, the Fowlers promoted books designed to encourage and prepare prospective settlers. Titles such as *Iowa As It Was in 1855* and *Minnesota As It Is in 1856* featured in advertisements in the *American Phrenological Journal* alongside notices for plaster casts and reprints of Combe's works. The Fowlers even ran a business in helping customers make "bounty land" claims, whereby those serving in the United States Army could apply for a plot in the new territories.[64]

Alongside the promise of financial reward, the Fowlers believed that phrenology would aid in the intellectual development of frontier society. New states and territories like California required "the application of the principles of Phrenology, and the observance of its truths, to aid in the proper organization and elevation of society," explained the *American Phrenological Journal*.[65] At the center of this vision was the periodical press. The Fowlers consistently promoted the idea that reading would bind distant communities together.[66] Before the opening of the Union Pacific Railroad in the 1860s, those in the West were isolated, at risk of descending into "an inane, dissatisfied, torpid state." According to the Fowlers, poor settlers were "literally starving for communion of mind." The solution, of course, was to subscribe to the *American Phrenological Journal*. The Fowlers even suggested that pioneers give up "tea, coffee and tobacco" to save money for a subscription. It was "by reading" that those newly arrived in California or Utah could ensure that they stayed connected to the civilized world. From a phrenological point of view, periodicals were imagined to have a special effect on the mind. The *American Phrenological Journal* explained that "periodical reading is best, because, coming at intervals, every new arrival arrests attention, awakens curiosity."[67]

Expansion to the west and south opened up new markets for the Fowlers' business. However, the growth of the United States brought the problem of slavery into sharper focus. Every time a new state joined the Union, such as Texas in 1845, slaveholders and abolitionists battled over whether the federal government could and should regulate slavery. The Fowlers in New York City, as the chapter on Eustache Belin illustrated, were broadly sympathetic to the antislavery cause. The *American Phrenological Journal* was even advertised in the abolitionist press, with William Lloyd Garrison's *Liberator* carrying copies of the prospectus.[68] Yet, as was the case for many northern businesses, the Fowlers recognized the importance of southern states as a market. Especially in its early years, the *American Phrenological Journal* depended upon southern subscribers, a number of whom pledged large sums of money to keep the periodical going.[69] Lecturers like Charles Caldwell in Kentucky and Joseph Buchanan in Florida helped to promote phrenology in the growing number of southern debating societies and lyceums.[70] The Charleston Conversation Club held debates on phrenology in the 1840s, and, in the 1850s, Lorenzo himself delivered a series of lectures in New Orleans, drumming up subscribers as he went. With readers in Mississippi, Texas, and the Carolinas, the *American Phrenological Journal* rarely addressed slavery head-on. When Lorenzo visited the South he, like the journal, chose to stick to safe topics, such as the "Intellectual, Moral, and Bodily Development of Man." Slavery was off the agenda.[71]

The outbreak of the American Civil War in 1861 brought this uneasy relationship to an end. The nationalism that the Fowlers had been promoting now seemed less secure. When it came to it, the *American Phrenological Journal* was on the side of the Union. At one point, the Fowlers' phrenological cabinet was even turned into a temporary headquarters for troops marching south.[72] At least the Fowlers could now be more forthright about their abolitionist stance. "As a nation, we have sinned," exclaimed the *American Phrenological Journal*. "Negro slavery and the free white labor system could not well coexist."[73] Yet with the outbreak of war, the Fowlers had to contend with a fractured publishing environment.[74] For one, the *American Phrenological Journal* lost many of its southern subscribers. The separation of currencies added to the difficulties, with the Fowlers making it clear that they would not accept "Confederate notes, or other worthless trash" as payment.[75] Even in the North, readers were disrupted by the war. With armies skirmishing up and the down the country, the postal service became much more unreliable. Copies of the *American Phrenological Journal* regularly failed to arrive, with the Fowlers speculating that "rebels interfere and confiscate it."[76] Still, the brothers adapted. They recognized that a new network of subscribers was needed. In an article entitled "Our Journal in the Army," the Fowlers claimed that "many of the soldiers new in the field miss their favorite A. P. J." Once again, the *American Phrenological Journal* appealed to its readers for help, hoping to replicate the system of "agents" and "clubs" under conditions of war:

> We beg our friends "under arms" to take the matter into their own hands. Let an agency be established in every camp. . . . Large numbers

> of the Journal are sent regularly to the officers and men in the navy, on all the stations, and it should be the same in the army. Soldiers, let us hear from you *direct*.[77]

The army was now a market for popular science. Some certainly did subscribe. John Williamson, stationed in South Carolina, wrote in 1862 that he was "determined that my subscription shall not cease. I inclose one dollar for the Journal."[78] Others, such as James Andrew, fighting in Louisiana in 1863, wrote letters to the journal, reporting on the progress of the war. Andrew saw the cotton fields burning. He worried for the future of the United States as well as its black population, writing, "The negroes are in a sad state. . . . I fear that their suffering will be greatly increased by our occupation of the country."[79] War, in the end, was not good for business.[80] When the conflict came to a close, the Fowlers promoted reconstruction, underpinned by a renewed sense of national identity. "The close of the war will introduce a new era," claimed the *American Phrenological Journal*. "There will be in sentiment no North, no South, but we shall be one in rights and privileges, in education, politics, and religion."[81] The Fowlers even put a phrenological spin on the differences between North and South, contrasting "a representative man from Massachusetts, and another from South Carolina." According to the *American Phrenological Journal*, the two were not so different.[82]

The success of the *American Phrenological Journal* mirrored the making and unmaking of the nation. It followed pioneers west to California as well as troops south into the Confederacy. At the same time, the United States transformed into an imperial power, stretching farther and farther beyond the North American continent. California opened a gateway to the Pacific and Asia. Following the end of the First Opium War in 1842, the United States negotiated the Treaty of Wanghia, securing trading rights with the Qing and granting missionaries access to new port cities, such as Fuzhou and Ningbo.[83] Hawaii and the Philippines later came under American control, as the United States jostled with European powers for control of the Pacific.[84] In parallel with these political developments, the *American Phrenological Journal* began to develop connections to a much wider world. In doing so, the Fowlers hoped to transform their national business into an international empire.

The brothers looked west, to the Pacific and Asia, for both content and subscribers. Missionaries played a major role in this respect. In 1869, the Fowlers received a letter from Justus Doolittle, a missionary with the American Board in Fuzhou. Doolittle had traveled to China in the 1850s, packing copies of the *American Phrenological Journal* next to his Bible. However, after having spent nearly twenty years in the "Flowery Kingdom," he complained that "I have not seen a copy of your Journal for some time past." Doolittle therefore offered to send the Fowlers "something relating to the Chinese, with illustrations" in exchange for a subscription to the *American Phrenological Journal*.[85] In making this agreement, the Fowlers replicated the strategy that had proved so successful in the United States. Doolittle, an American missionary in China, was transformed from a passive reader into an active contributor. True to his word, Doolittle sent four articles entitled "Street Sights in China." In addition to these, Doolittle forwarded a series of sketches depicting life in Fuzhou (fig. 23). These illustrations had been made, not by the American missionary, but by local Chinese artists. In this way, the *American Phrenological Journal* connected readers in the United States to increasingly distant visual and material cultures. The sketches, copied as engravings into the journal, depicted a "Chinese Servant Boy," complete with "the tobacco pipe of his master," as well as a "Woman Making Needles." Doolittle was broadly positive concerning Chinese character, approving of the skills of local artisans. However, when it came to religion, Doolittle betrayed his missionary zeal. He was dismissive of the "idols" favored by the "heathen Chinese" and lamented the worship of "ancestral tablets." Doolittle's account was bolstered by a phrenological reading. "The Chinese skull has a larger proportion of its bulk back of the opening of the ear, and less prominence and elevation of the forehead than that of the Caucasian," explained the *American Phrenological Journal*. This was "the organic cause" of the principal characteristics of the Chinese. They were said to excel at "persistent mechanical activity" yet, without a well-developed forehead, were "destitute of religious feeling and belief."[86]

American missionaries were also active in South Asia during this period, particularly in northeastern India.[87] In the 1860s, the *American Phrenological Journal* printed an article based on an account by the Reverend W. W. Hicks, a missionary with the Methodist Episcopal Church in India. Incredibly, Hicks also forwarded two Indian miniatures, painted on ivory, and depicting "The King of Oude" as well as "Nena Sahib, a Hindoo chieftain and the leader of the Sepoy rebellion in 1857" (fig. 24). These "exquisite works of art" were reproduced as wood engravings for readers in the United States. From these material artifacts, the Fowlers produced a phrenological account of Indian character, one heavily influenced by missionary accounts of "Hindoo" immorality. "There are few redeeming qualities in the character of this Indian king," explained the *American*

238 *PHRENOLOGICAL JOURNAL.* [APRIL,

Either one roll, or three, or a larger quantity are burned at once, always accompanied by the burning of at least two candles, in private houses and in temples.

CHINESE SERVANT BOY.

CHINESE SERVANTS.

The engravings on this and the following page represent the servants of a well-to-do Chinese family in their best attire. The servant boy has the tobacco pipe of his master, near one end of which is appended the pouch of tobacco in true Chinese style. The servant girl is in the act of carrying to her mistress a cup of hot tea. She belongs to the class which has large feet, and probably has been bought from her parents, or a responsible party, for service as a servant.

Many rich families have several servant girls and servant boys for doing the menial work of their households. Very frequently girls are bought when eight or ten years of age, or less, to be employed as servants, on the understanding that soon after arriving at a marriageable age they are to be provided with husbands, and then be no longer servants, unless they are hired. Such are usually sold to men who desire wives, who pay generally more than the first cost of the girls. In this way their services are obtained by their mistresses, at the cost of only their food and clothing, during the period of service.

Few boys are now-a-days sold by their parents or guardians to be servants. They are usually hired, receiving but a very small sum besides their food and clothing. It is the business of the favorite boy servant to wait upon his master at home, and when he makes calls, or leaves home on business, to accompany him, carrying his pipe and pouch of tobacco and visiting cards. It is the custom for Chinese gentlemen to smoke a whiff of tobacco at very short intervals.

Very frequently the kind of pipe used is of a style different from the one exhibited in the Picture, being made of brass in such a manner that the fumes pass through a little water in the pipe, *en route* from the bowl of the pipe to the mouth of the smoker. The kind of tobacco used in such pipes is yellowish, and is in the form of a powder, in which, it is said, a small quantity of arsenic

Fig. 23. Sketch of a "Chinese Servant Boy," sent by Justus Doolittle from Fuzhou, *Phrenological Journal and Science of Health* 50 (1870): 238, Wellcome Collection, CC-BY.

1864.] AMERICAN PHRENOLOGICAL JOURNAL. 57

PORTRAIT OF NENA SAHIB.

Johnny always performs his share of the solemn ceremonies.

Just six years old last March they say,
But a manly heart has he;
He's rattled his drumsticks many a day,
His height scarce up to your knee.

Why work so hard? say, Johnny dear,
Why do you drum so loud?
"I drum," he said, "for the Volunteers,
And I want to make them proud.
I'll rattle my sticks for the Seneca Boys—
They're going to fight the foe."

And the soldiers are proud of their drummer boy
As on to the war they go;
But many an eye now bright with joy,
May soon be quenched in woe.

"Come, go with us," the soldiers said,
"And drum on the battle-field;"
With youthful bravery John replied,
"I'll go if my mother will."

TWO HUMAN MONSTERS.

Having been permitted to examine two beautiful miniature portraits brought from India by Rev. W. W. Hicks, of the Methodist Episcopal Church, and now in the possession of Dr. E. H. Dixon, editor of the *Scalpel*, we begged permission to have copies taken of them for publication in the Phrenological Journal. The originals were painted on ivory, in the most perfect style, and are really exquisite works of art. They have since been photographed, album size, by the Messrs. Anthony, of Broadway, and from their copies our engravings have been made. They are now before the reader, with the remarks which they have suggested, and a few of the facts found recorded concerning their subjects. We will first introduce to you

NENA SAHIB.

That this man has a strongly marked character no one who looks upon his likeness will be induced to question; nor is it less evident that he is more developed in the sensual than in the spiritual part of his nature. The base of the brain predominates largely, and the temperament—vital and motive rather than mental—is one that must give force and energy to the action of his predominant organs. His hair and beard are thick, bushy, and black; his skin coarse and swarthy; and his whole organization evinces bodily power rather than mental or spiritual force.

Without particularizing more fully, we may say that the perceptive intellect is largely developed; the reflective organs full; and the top-head high in the crown, especially in the region of Self-Esteem and Firmness. We do not observe the physiognomical indications of Conscientiousness, Veneration, or Spirituality in any marked degree of development. The head is long—projecting far back—rather than high or broad, and we should not expect to find so much steady propelling power as dogged obstinacy in his character. His vital functions are evidently highly efficient and active, and he has been a good liver, though not an epicure.

Such a man as this would attract attention in any community, and would be likely to aspire to leadership. He would be more great than good, under the most favorable circumstances; with his passions fully aroused and unrestrained, he might become a demon in human form.

Mrs. Dixon sends us the following interesting note in reference to Nena Sahib. Parents will do well to heed the moral drawn from the touching little incident related of the child and the picture:

I have understood that in early life this monster became an orphan, and that harsh or cruel treatment from strangers, who knew not what they did, hardened his heart and developed a cruelty which the beasts of the forest could not equal. He was finally adopted by a family of distinction and wealth, who gave him every facility for education and improvement, and having a fine person, as well as a brilliant intellect, he availed himself of his opportunity, and to the great astonishment of his friends made himself master of varied accomplishments and many languages. His adopted parents finally died, and he who had looked forward to state and station was driven from his home as an interloper. The lady of his love scorned him; and thus, frenzied with rage and disappointment, he took revenge upon her nation, for she was an Englishwoman. The torturing hours of his own childhood returned to him, and the history of his crime is the result.

While looking upon his miniature, which had been brought from India, I exclaimed: "Well, after all, this man is handsome." But a little child of only five summers, who was at my side, said: "No, no, he is naughty! he is naughty in his eyes! he is naughty!"

If the impressible nature of childhood be thus instinctively conscious of good and evil, how great must be the effect of example upon the infant character, and how earnestly should we strive to surround children with all loving influences! For whatever the temperament may be, this alone will guide it aright. A frown or a blow is *hate* to them, and as such is engraven upon the heart and the brain, which time has not yet matured. Then take care that this spirit of hate be driven from your nurseries and replaced by that of patience, charity, and loving kindness. Look into the innocent eyes of your babes, mother, and "get knowledge, get understanding," that we may have no more Nena Sahibs.

Dhundoo Punt, Nena Sahib (the latter being his title), was a Hindoo chieftain and the leader of the Sepoy rebellion in 1857. He was the son of a Brahmin of Deccan, and was born in 1824 or 1825. When a little more than a year old he was brought to Bittoor, where he was soon after adopted by Bagee Row, the chief of the Mahrattas. On the death of Bagee, without natural heirs, the East India Company refused to acknowledge the right of his adopted child to his principal estate, which had been conditionally bestowed on the former by the company. The Nena sent a agent to England to advocate his claims, but without success. This wrong he never forgave. He had still much wealth and influence, and when the insurrection broke out was ready to devote both to the cause of the rebels, and to put himself at their head. Of his terrible cruelties perpetrated during the war which followed everybody has heard. A single instance will be sufficient to put on record here:

On the 27th of June, 1857, the English at Cawnpoor, after an obstinate defense, surrendered to the Nena, on his promising to send them safe to Allahabad. They were permitted to embark, but immediately afterward fired upon and many of them killed. The rest were brought back to land, where the men were at once put to death. The women and children, after surviving nameless outrages, were finally all massacred on the 15th of July, and their bodies thrown into a well.

Long after all the other rebel leaders had submitted, the Nena continued with about 10,000 rebels to infest the northern parts of Central India and the frontiers of Nepaul. It was reported that he died of a fever in 1859, but the report was not generally credited, and it is now considered uncertain whether he is dead or alive.

PORTRAIT OF THE KING OF OUDE.

THE KING OF OUDE.

Voluptuous! The most charitable construction we can put upon this character is implied in this term. To gratify his propensities would be the first impulse of the man; the second—one degree higher—would be to indulge his vanity and love

Fig. 24. Portraits of "Nena Sahib" and "The King of Oude," woodcuts copied from Indian miniatures on ivory, *American Phrenological Journal* 39 (1864): 57, Francis A. Countway Library of Medicine, Harvard University.

Phrenological Journal. The king of Oudh, Wajid Ali Shah, was presented as a despotic ruler, enjoying a "luxurious mode of living." He was even compared to "King Henry VIII of England . . . , the vilest of the many vile ones who have occupied the English throne." The portrait, possibly the work of the court painter Muhammad Azam, was deployed as further evidence. The reader was invited to look upon Wajid Ali's features: "His head is round and his physiology coarse"; "a gross, vain, indolent, sensual and animal organization." If the reader found the king of Oudh sensational, then they were in for another shock when they heard about Nena Sahib. This man was a "monster," accused of murdering European women and children, as well as committing other "nameless outrages," during the Rebellion of 1857. The *American Phrenological Journal* described Nana Sahib's character, noting that "the base of the brain predominates largely, and the temperament—vital and motive rather than mental—is one that must give force and energy to the action of his predominant organs." It was men like this that explained the need for missionaries in India. The article concluded by warning that "with his passions fully aroused and unrestrained, he might become a demon in human form." Christian morality would provide a check on Indian passions.[88]

By the 1880s, the *American Phrenological Journal* claimed that "no other monthly magazine has a so widely distributed constituency. . . . Packages go to New Zealand, Tasmania, and other far-off lands in the East."[89] In the opposite direction, missionaries and merchants from across the American empire sent letters, skulls, and artifacts to New York City. The Fowlers sought to transform individual readers, whether at home or abroad, into active agents of a global phrenological movement. The periodical press would "phrenologize our nation, for thereby it will reform the world."[90] In many ways, the Fowlers were right. Across the world, the periodical press helped to create new audiences for science, including phrenology. However, the technologies and politics of publishing were quite different outside of the United States. In the following two sections, I examine the fate of phrenology and the periodical press in the Cape Colony and India. In each case, the kind of audiences that phrenology appealed to were quite different. In the Cape Colony, phrenology found an audience among frontier settlers, whereas in India, phrenology was taken up by colonized elites. Nonetheless, whether they were publishing in Cape Town or Calcutta, phrenologists presented their local periodicals as part of a broader movement.

PHRENOLOGICAL SETTLERS IN THE CAPE COLONY

The *South African Quarterly Journal* kept readers in Cape Town abreast of all the latest scientific news. The "Intelligence" section, located at the end of each number, often featured reports on local lectures and scientific meetings. Readers were reminded of papers given on "the Geology of

South Africa" as well as "the Culture of Exotic Vegetables." In 1830, the *South African Quarterly Journal* also printed an advertisement for a series of phrenological lectures. It read: "Mr. Macartney, M.R.C.S. begs to announce his intention of delivering, during the ensuing month of May, six Lectures on Phrenology." The advertisement, which took up an entire page of the journal, described the content of the six lectures. The first would set out the basic "Principles of Phrenology," the second "the five external Senses," followed by a third on "the Brain: its Structure and Physiology." The fourth and fifth lectures would enumerate the various "sentiments," "propensities," and "faculties," while the final lecture would set out "the practical utility" of phrenology. Readers of the *South African Quarterly Review* were promised an entertaining series in which "the Seats of the various Organs" would be "shewn and pointed out on a model and drawings."[91]

These six lectures marked the beginning of Henry Macartney's phrenological career in the Cape. Trained as a surgeon in London, Macartney spent his early years working for the East India Company in Bengal before settling in Cape Town in 1829.[92] Over the next twenty years, he traveled between the Western and Eastern Cape, bringing phrenology to a new audience of settlers and frontiersmen. And during this period, he developed a close connection with the colonial periodical press, writing columns on frontier policy and generating notices of his lectures. However, unlike the Fowlers in the United States, Macartney was in no position to start his own phrenological journal. The material and political culture of print in the Cape Colony placed serious restrictions on the development of reading audiences for science. For a start, limited freedom of the press was only granted in 1829. Before then, the British and Dutch authorities had forbidden all private printing operations, with the exception of a few missionary presses on the frontier. While the British settlers who arrived in the Cape in 1820 saw the freedom of the press as central to the development of colonial society, the autocratic governor, Lord Charles Somerset, worried about the spread of radicalism from Europe. Phrenology and associated "materialist" doctrines were exactly the sort of thing he hated. Nonetheless, with the rise of the reform movement in Britain, alongside campaigns in Cape Town, colonial policy shifted. Somerset was kicked out, and, in 1829, the new governor, Sir Galbraith Lowry Cole, passed Ordinance 60, promising not to interfere with the press.[93] Even with these new freedoms, printing a periodical in the Cape presented plenty of challenges. Equipment was hard to come by, with just a few wooden hand presses available in the 1820s. Iron presses began arriving in the 1840s, but the first steam press only reached Cape Town in 1854. Paper, ink, and type all had to be imported from London at great expense. Shortages of supplies were common, with newspapers sometimes resorting to printing on blotting or wrapping paper.[94]

Macartney, therefore, could not realistically hope to start his own phrenological journal. He certainly was not trying to bring phrenology to a mass audience. Instead, Macartney looked to insert phrenology into the Cape's small but growing scientific and print culture. The *South African Quarterly Journal* was part of this world. Andrew Smith, the editor, was also secretary of the South African Institution as well as the founder of the South African Museum.[95] By the 1830s, Cape Town housed a botanical garden, public library, and astronomical observatory. Both Charles Darwin and John Herschel were in Cape Town at the same time as Macartney, while new journals like the *Cape of Good Hope Literary Gazette* reported on the latest scientific news from Europe.[96] From his arrival in 1829, to his death in 1846, Macartney worked hard, co-opting editors, printers, and politicians in an effort to promote mental and racial science in the Cape. In doing so, he brought phrenology to a new audience of colonial settlers.

Macartney delivered his lectures at the Commercial Exchange in the center of Cape Town. The choice of venue was important. The Commercial Exchange was at the heart of respectable settler society, acting as an auction house, storeroom, and social space. The large room in which Macartney delivered his lectures doubled as a ballroom, while the north wing housed the South African Public Library.[97] The Commercial Exchange also represented a material connection to Europe, in terms of both trade and knowledge. Alongside the wool, wine, and brandy in the warehouse, the South African Public Library held copies of European periodicals and books.[98] These included a copy of Combe's *Elements of Phrenology* as well as George Mackenzie's *Illustrations of Phrenology*, both imported from Edinburgh.[99] By speaking at the Commercial Exchange, Macartney presented phrenology as part of the liberal world of middle-class settlers and merchants in Cape Town. He hoped to appeal to men like John Fairbairn, a leading member of the Commercial Exchange and editor of the liberal *South African Commercial Advertiser*.[100]

In addition to his discussion of phrenology, Macartney described the principles of "Physics" and "Physiology," linking this more general scientific knowledge to the constitution of man. He announced that "the Laws of Physics and Life . . . will necessarily lead to a consideration of the Nature and History of Man." Presenting a story of continual improvement, Macartney explained "the progress of the human mind" could be "traced from the commencement of life up to the period of manhood, when the intellectual faculties are supposed to attain their full perfection." He even invoked "the history of science," giving an account of "the first rude beginnings . . . to the most profound productions or speculations of modern times." Significantly, Macartney linked the improvement of the human mind to the development of popular science and liberalism, particularly in the colonial world. "In preparing these lectures, I was solely actuated by a desire to promote the intellectual advancement of the Colony," explained Macartney. Comparing the Cape to Europe, he noted that "looking to other countries which stand high in the scale of civilisation, it will be found, that public Lectures and Literary productions are the characteristics that stamp the character or mark their superior advancement." Macartney then directed his attention to the freedom of the press, something he relied on to get phrenology into the Cape Colony papers. He argued that the popularity of his lecture series "rebuts the ridiculous insinuation in respect of the Inhabitants of this Colony, not being yet sufficiently advanced in Intellect to enjoy Free Institutions." This kind of rhetoric would certainly have appealed to Fairbairn, who had personally been involved in the campaign for the freedom of the Cape Colony press in the 1820s. Linking freedom, liberalism, and science, Macartney concluded by stating that "the flood of Knowledge is bearing down the ancient landmarks of Ignorance, and forcing the still swelling tide of Wisdom into almost every corner of the Globe."[101]

Despite Macartney's appeals to liberalism, his lectures received a mixed reaction in Cape Town's periodical press. The *South African Commercial Advertiser* had been skeptical concerning phrenology even before the arrival of Macartney.[102] Fairbairn might have been a liberal, but he still worried about "materialism and infidelity." The *South African Commercial Advertiser* also noted that phrenologists faced "serious difficulties . . . in protecting their theories against some stubborn facts that threatened the whole science with annihilation." The same article described "the case of John Thurtell," a murderer. This was a "cold-blooded, savage, treacherous and cowardly" individual, yet phrenologists had found Thurtell to possess "organs of Benevolence very large." In the end, rather than celebrating the intellectual improvement of mankind promoted by Macartney, the *South African Commercial Advertiser* simply suggested it was best for "the reader to decide for himself."[103] A similarly lukewarm reaction greeted phrenology in the pages of the *Cape of Good Hope Literary Gazette*, edited by the liberal abolitionist Alexander Johnstone Jardine. Shortly after Macartney's lecture, the *Cape of Good Hope Literary*

Gazette featured a review of *A Catechism of Phrenology*, printed in Edinburgh in 1831.[104] The review admitted that the book was "calculated to inform and enlighten." However, the *Cape of Good Hope Literary Gazette* did not count itself "amongst the admirers or supporters of what is called the science of Phrenology." Rather, it suggested that *A Catechism of Phrenology* was best read as "an excellent text-book for the student, by means of which he may ascertain both the merits and defects of *the science*."[105]

After five years in Cape Town, Macartney decided to move on. Toward the end of 1835, he traveled to Grahamstown on the eastern frontier. Arriving in the aftermath of the 1834–35 Xhosa War, Macartney was shocked by what he saw. Frontier life was violent. The recent war was just the latest in a string of conflicts between the Xhosa and European settlers. Following the abolition of slavery in 1833, and the move toward free labor, frontier farmers had sought to expand into Xhosa territories to the east. Arson, murder, and hangings were common.[106] Grahamstown was also relatively disconnected, not just from Europe, but from Cape Town. Prior to 1846, there was no regular postal service in the Cape Colony and only a few poorly maintained roads.[107] By coach it took eight uncomfortable days to travel from Cape Town to Grahamstown.[108] All of a sudden, the liberal world of the Commercial Exchange seemed a long way away. Enlisting the support of Robert Godlonton at the *Graham's Town Journal*, Macartney began to rework phrenology for a new audience of frontiersmen.[109]

Over the next six years, Godlonton promoted and reported on Macartney's phrenological activities. In November 1835, the *Graham's Town Journal* printed an advertisement for a new series of lectures. Godlonton even printed the cards of admission and sold them at his offices (3 rix-dollars for a single lecture or 7 rix-dollars for the complete course). This time, each of Macartney's lectures was delivered in John Mandy's house, a local lodging room for travelers and settlers freshly arrived on the frontier.[110] The audience, therefore, was very different from that of Cape Town. Rather than wealthy merchants and men of science, Macartney was now presenting phrenology to poor farmers recovering from the recent war. With no library and limited access to Cape Town, local newspapers like the *Graham's Town Journal* were the only means by which frontier settlers could read about phrenology.

This close connection between Macartney, Godlonton, and the *Graham's Town Journal* had a profound effect on the way phrenology developed on the eastern frontier. Founded in 1831, the *Graham's Town Journal* was presented as a counter to the *South African Commercial Advertiser*. Godlonton was fiercely critical of the liberal intelligentsia in Cape Town.

For him, Fairbairn's liberalism appeared hopelessly naive, a product of a world far removed from the daily struggles of frontier life.[111] The culture of print on the eastern frontier was at the center of a society engaged in constant violence. Louis Henri Meurant, the co-owner of the *Graham's Town Journal*, had actually served in the 1834–35 Xhosa War.[112] During the fighting, presses were destroyed, and type was melted down to manufacture ammunition.[113] Throughout the conflict, the *Graham's Town Journal* reported on the "Kaffir Depredations," whipping the local population up into a frenzy. In the wake of the latest violence, many farmers felt that previous treaties with the Xhosa had been far too lenient. The *Graham's Town Journal* supported this campaign, arguing that the territory between the Kat River and the Keishkamma River should be brought under direct colonial control. (The treaty following the previous Xhosa War in 1819 had designated this area as "neutral land," forbidding farmers to settle there.)[114] This call for land appropriation was tied to the construction of African racial identities.[115] According to the *Graham's Town Journal*, "Our present disasters have been caused by mistaken philanthropy, and by erroneous estimates of the Kafir character." Contrary to reports from liberals in Cape Town and London, the Xhosa were in fact "a few degrees above zero" in the scale of civilization. Their character was marked by "cruelty and treachery"; they were a "ferocious and wretched people." The *Graham's Town Journal* concluded that the only way to secure "peace, safety, or protection of property" was "to declare war against the whole Kafir nation, and carry it on with vigour."[116]

In the Eastern Cape, these attitudes fed into how phrenology was presented in both the lecture halls and the periodical press. Macartney's lectures now focused much more strongly on race. This was made easier thanks to the close connection between phrenology and skull collecting during times of war. In the 1820s, Thomas Pringle had sent a "Bushman or Bosjesman's scull" to the Edinburgh Phrenological Society. Pringle explained that the man had been "shot by some Boors of my acquaintance on the Caffer Frontier in the act of stealing cattle."[117] Now, following the most recent conflict, Macartney's lecture on "feelings and moral sentiments" was illustrated with "some Kafir skulls, lately received."[118] Another lecture, on the "Perceptive & Reflective Faculties," was accompanied by "various casts and skulls of native tribes." Macartney then explained to his audience "how much these powers depend on the size and shape of the forehead."[119]

Life on the eastern frontier radicalized Macartney. Abandoning the liberalism of his time in Cape Town, this settler phrenologist now used the periodical press to make an explicit connection between Xhosa character and land appropriation.[120] In a long article for the *Graham's Town Journal* entitled "A Few Words on the Treaty System," Macartney directly attacked Fairbairn's *South African Commercial Advertiser*. The "Advertiser," Macartney wrote, believed "the Kafirs" to be a "highly virtuous people . . . in a state of honest tranquillity." Those living on the eastern frontier knew better. According to Macartney, the Xhosa were a "set of savages" living in the "lowest barbarism and ignorance." Macartney then linked Xhosa character to frontier policy. He questioned whether "the course pursued in upholding by British arms a set of savages in the invasion of the Colony, and the perpetration of the most savage acts, be compatible with any known or established principles of justice." Macartney even thought the existing treaty system damaged the mental constitution of settlers and Xhosa alike, writing "the insane, ridiculous, and absurd system . . . far from repressing crime has a direct tendency to brutalize the mind."[121] Macartney concluded that the only way to deal with the Xhosa was to abandon the existing treaties and annex the "neutral land" between the Kat River and the Keishkamma River.[122]

Godlonton reinforced this link between phrenology, Xhosa character, and frontier policy through careful editing. A number of Macartney's lectures were printed alongside news from the Xhosa War. Most strikingly, the lecture featuring "Kafir skulls" appeared directly below a "Government Notice" reporting the "Treaty of Agreement entered into by order and in the name of his Excellency the Governor and Commander in Chief, with the Tambookie Chief Mapassa" (fig. 25). Readers therefore encountered an advertisement for a phrenological lecture on Xhosa character immediately after reading that "the Kafirs of the Tribes of Gaika, Eno, Botma, and T'Slambie, have submitted to the English Government." This treaty effectively granted many frontier farmers their wishes, with the new governor Benjamin D'Urban declaring that the "neutral land" would now be brought under colonial control as the District of Victoria.[123] Another treaty, printed in the *Graham's Town Journal* above Macartney's lecture on the "skulls of native tribes," ominously warned the chief Maphasa that "without British protection he is to-day a dead man."[124]

This link between phrenology, land appropriation, and the periodical press continued with the expansion of the Cape Colony. In 1843, the British annexed the short-lived Boor Republic of Natalia to the northeast of Grahamstown.[125] And with annexation, came phrenology. Macartney didn't travel to Natal, but his cause was taken up by a series of local lecturers, voluntary societies, and newspaper editors. In the summer of 1850,

THE

Graham's Town Journal

OR,

CAPE OF GOOD HOPE EASTERN PROVINCE REGISTER.

Published in Graham's Town every Thursday, 3 o'clock P.M. and forwarded to all parts of the Colony immediately.
TERMS OF SUBSCRIPTION:
In Graham's Town, 6s. 6d. per Quarter.—Every other part of the Colony, POSTAGE INCLUDED, 8s. per Quarter.

VOL. IV. THURSDAY, DECEMBER 10, 1835. No. 207.

TERMS FOR ADVERTISEMENTS:
For Eight Lines and under 3s. 6d.—above 8 to 16 lines 3d. per line,—above 16 lines 2d. per line. A reduction of One-third after the First insertion.
SUBSCRIPTIONS AND ADVERTISEMENTS to be paid IN ADVANCE.

GOVERNMENT NOTICE.

THE following Treaty of Agreement entered into by order, and in the name of His Excellency the Governor and Commander in Chief, with the Tambookie Chief MAPASSA, is hereby published for general information.

By His Excellency's Command,
(Signed) W. SMITH,
Acting Goverment Secretary,
in attendance.

Uitenhage, 8th December, 1835.

THE Tambookie Chief, Mapassa, being present at Shilo, on the Klip Plaats, this 23d Oct., 1835, was acquainted as follows:

1st.—That the Kafir Tribes belonging to CHELI, are confined to the limits of the left Bank of the Great Kie.

2d.—That the Kafirs of the Tribes of GAIKA, Eno, BOTMA, and T'SLAMBIE, have submitted to the English Government; and have placed them selves under its protection.

3d.—That having received from the Missionaries of the Moravian Institution of Klip Plaats a favourable account of his conduct, the Governor is willing to receive him, and his Tribes under the protection of the British Government, within the limits of the Colonial Territory as now defined. Always provided that they abstain from depredations on the Colonists of all denominations; that they obey the orders of the Governor, and of those acting under his authority; that they abstain from all connexion and alliance against us, with the Kafir Tribes, lately our enemies, and from listening to all proposals of war, plunder, and depredations upon the Colony, that may be made on the part of these people.

4th.—MAPASSA, in reply, states, that he is under the Government, and that his extreme wish is to leave peaceably under the Government, and that without British protection he is to-day a dead man.

5th.—He promises to be faithful to the British Government, and states, that he has abstained entirely from all connexion with the Kafirs in their late depredations upon the colony. He further promises to do all in his power, not only to prevent depredations, but to punish all or any of his people whom he may convict of committing depredations upon the colony; and to restore to the colony all cattle or horses which he may ascertain have been brought into the kraals of his people.

6. This done, Colonel SOMERSET has assured him, in the name of His Excellency the Governor, of the protection of the British Government, so long as he, his Chiefs, and their Tribes, conduct themselves according to the terms of this Agreement.

(Signed) H. SOMERSET, Lieut Col.,
Commandant of the Frontier of Kaffraria
✕ Sign of MAPASSA.
(Signed) AD BONATZ, Missionary.
(Signed) R. FAWKES, Capt. 27th Regt.

THE THIRD LECTURE,
On PHRENOLOGY,
TO-MORROW EVENING, (FRIDAY)

THE Third Lecture will be delivered TO MORROW EVENING (Friday) the 11th inst., at half-past 7 o'clock. The subject of the Lecture will be the "*Feelings and Moral Sentiments.*" The seats of these organs will be pointed out on a variety of casts, and familiarly explained. It will be shewn how they act on the Intellectual Faculties described last Lecture; and as they materially influence the character at all periods, the necessity of giving them a proper direction in early life will be pointed out. To parents and guardians of youth this part of the Lecture will be found interesting, as a variety of intertaining matter will be brought forward. Some Kafir Skulls, lately received, will be examined in the course of the evening, and the character described according to the principles already detailed.

Cards of admission at Messrs. MEURANT & GODLONTON's, High Street.

THE Undersigned being about to leave the Frontier, all persons indebted to him are requested to pay the same to Mr. THOMAS STUBBS before the 1st January, 1836. Also those persons having Claims, are desired to send in the same previous to that date.

WILLIAM M'GRAW.

MARRIED at her Father's Residence, in Graham's Town, on Monday the 7th of December, by the Rev. J. HEAVYSIDE, Acting Colonial Chaplain, Mr. GEORGE SOUTHEY, to ELEANOR DIXIE, Second Daughter of R. H. RUBIDGE, Esquire, of Gletwyn, Albany.

CAPE OF GOOD HOPE
FIRE INSURANCE COMPANY.
TO BE ESTABLISHED ON THE 1ST OF DEC. 1835.
CAPITAL £20,000.
IN 400 SHARES OF £50 EACH.
A moiety of the Capital to be paid by Instalments.

Directors,
Honorable J. B. EBDEN, Chairman,
Honorable HAMILTON ROSS,
WILLIAM GADNEY, Esq.,
THOMAS SUTHERLAND, Esq.,
HARRISON WATSON, Esq.,
SAMUEL BLACKALLER VENNING, Esq.,
THOMAS TENNANT, Esq.,
BARON VON LUDWIG.
HOWSON EDW. RUTHERFOORD, Esq.

Auditors.
THOMAS ANSDELL, Esq.,
GEORGE HODGSKIN, Esq.

Secretary,
Mr. RICHARD WEBBER EATON.

Rates of Premiums.

First Class,	2s. 6d. per £100
Second Class,	4s. 6d. per 100
Third Class,	6s. 8d. per 100
Fourth Class,	
Fifth Class,	17s. 6d. per 100

The advantages which are offered to Insurers in this Office are, ECONOMY with regard to Rates of Premium, and SECURITY in respect to the number of Proprietors,—the former by being established much below those hitherto existing in the Colony will virtually afford to the INSURER a participation in the Profits of Insurance; while the distribution of Shares among so large a Proprietory proportionally enlarges the field of responsibility, and consequently extends the security to the Public.

The Office will be opened for the transaction of business at 22, Heeregracht, on TUESDAY, the 1st of December next.

Agent in Graham's Town, Mr. E. NORTON, to whom all applications for Insurance must be made.

R. W. EATON, Secretary.

COMMISSARIAT.

OAT HAY,

On the Eastern Frontier, during 1836.

NOTICE is hereby given to such Persons as may be willing to supply Oat Hay for the Service of this Department, commencing on the 1st January next, and ending 31st December 1836, that Tenders marked on the Envelope "Tender for Oat Hay" will be received by the undersigned at his Office in Graham's Town, until 12 o'Clock on TUESDAY the 15th December next, for the Supply of such quantities as Individuals may be willing to furnish at the following Stations respectively, as may be required by the Commissariat, viz:

Graham's Town, and Cantonments of Rifle Corps, at Sypher Fontien,
Fort Brown, (Hermanus Kraal,)
Bathurst,
Kafir Drift, (Cawoods,)
Fort Beaufort, and
Fort Willshire,

The Tenders to express in words at length, the price per 100 lbs English Weight in Sterling Money, and the quantity Tendered, and as Security will be required Two responsible Persons must be named either in the Tender, or by letter, who are willing to become bound for the due fulfilment of the contract.

In Graham's Town the Hay will be received direct into the Commissariat Magazine by the contents of one or more Stacks at a time, as may be arranged, and the lowest priced Tenders will have the advantage of the earliest deliveries at each Post until the quantities Tendered shall have been completed.

Any further information may be obtained on application at this Office, or to the Officers in charge at the several Posts.

CHARLES PALMER,
Act. Dep. Com. General.

Commissariat Office,
Graham's Town, 11th Nov. 1835.

SALE OF LANDED PROPERTY.

ON TUESDAY the 15th inst. will be sold on the respective situations, one piece of Land, situate in New Street, two do. on erf No. 39, in New Street, and one in African Street.

J. D. NORDEN, Auctioneer.

NOTICE TO SHEEP FARMERS AND OTHERS.

PUBLIC Sale of pure Saxon Merino Sheep, at the residence of Lieut. GRIFFITH, Prospect House, Graham's Town.

An extensive and peremptory Sale will be holden on SATURDAY, the 19th inst., of 175 pure Saxon Merino Rams, positively without reserve, bred by W. H. Warland, Esq., of Hunter's River, selected with great care and attention, and consigned direct to the Frontier, per Bark *Leda* Captain G. ROBB.

The above sheep need no comment as to their purity of breed and fine fleeces, the following certificates with many others, which will be produced at the time of Sale, will be a sufficient guarantee that they are a well selected lot.

Persons residing at a distance, wishing to become purchasers at the above Sale, may attend with a perfect reliance that no sale of the above sheep will take place previous to the 19th inst.

J. D. NORDEN, Auctioneer.

"Sydney, 28th February, 1835.

"I hereby certify that I imported into the colony of New South Wales in the year one thousand eight hundred and twenty eight, seventy-seven pure Merino Ewes, from the celebrated flocks of WILLIAM CLAPCOTT, Esq., of Down-House, in the county of Southampton, which ewes were considered the best flock ever imported; since which, until the year 1832, they were improved by the best imported rams from the equally celebrated flocks of Mr TRIMMER; at which time I delivered them over to W. H. WARLAND, Esq., of Hunter's River, who has still farther improved them by crossing with the best Saxon rams, from the flocks of W. E. RILEY, Esq. of Raby. These sheep have lately been inspected by some of the best judges in the colony, who pronounce them equal (if not superior) to any in New South Wales.

J. VERGE."

"Sydney, New South Wales, 28th Feb. 1835.
"I certify that I have examined the sheep shipped by Mr WARLAND, on the bark *Leda*, for the Cape of Good Hope, and that they are in my opinion equal to the general stock from the Raby Estate. WILLIAM SHERWIN."

"Sydney, Feb. 23, 1835.
"DEAR SIR,—If any opinion of the rams you are taking to the Cape with you is of any use, I assure you I am very happy to be able to say, that I am quite surprised at the appearance of them in general. It is certainly difficult to form a decided opinion of rams so lately shorn as they have been; but I will say that they appear to me to be generally very fine, and I did not think that so fine a lot of rams could be collected in the colony. Great attention appears to have been paid in their breeding, and I should say that the fleeces of the greater part of them are very even in fineness.

Believe, my dear Sir, your's
WILLIAM RILEY, of Raby."

RDS. 50 REWARD.

SOME evil disposed person or persons having destroyed certain Bills posted up, publishing a Sale of Sheep,—Notice is hereby given that if any person will give information to the Undersigned that shall convict any person or persons of doing the same, the above Reward will be given.

J. D. NORDEN,
Auctioneer.

In the Insolvent Estate of P. J. GAUGAIN.

A Special Meeting of Creditors under this Estate, will be held before the Resident Magistrate on the 9th of January next, at 10 o'clock, for the purpose of Revising and Altering the Conditions of Sale of the immovable Property, and agreeing upon new conditions on Sale to be then submitted by the Trustees.

HENRY NOURSE,
W. J. EARLE } Trustees.

ON MONDAY next, the 14th inst. will be Sold on the premises of Mr WILLIAM M'GRAW, a large quantity of Blacksmith's Tools, Smith's Bellows, a superior set of Farrier's Implements, and a quantity of Household Furniture, also two or three good Horses.

J. D. NORDEN, Auctioneer.

THE under-mentioned Horses, confined in the Pound at Somerset, will be sold by Public Sale on FRIDAY the 18th inst., unless previously released:

1 brown Gelding,
1 red roan Mare.

THOS DRY, Pound Master.

THE Undersigned being about to leave the Frontier, has for Private Sale the undermentioned Articles, viz.:—

A Billiard Table complete; 12 Lamps, for use of do.; 5 floor Carpets, 2 Fenders and 2 sets of Fire Irons, 1 plate Warmer and a large passage Lamp; Wagons and Oxen, at present employed by the Commissariat.

The above may be seen on application to Mr. JOHN MCKENNY, or the undersigned.

JAMES M'NAMARA.

NOTICE TO DEBTORS & CREDITORS.

THE Undersigned being about to leave the Frontier, requests all to whom he may be indebted to send their respective Accounts to Mr. MANDY, Free-mason's Tavern, for immediate payment; and those indebted to him are requested to settle their Accounts without delay, or will subject themselves to be prosecuted according to Law—without distinction of persons.

JAMES MACNAMARA.

NOTICE TO SHEEP BREEDERS.

SEVERAL applications having been made to the Undersigned to part with some of his valuable young Stock of Saxon Merino Ewes, previous to the Sale to be held on the 19th inst., of the Australian Rams imported by Capt. ROBB; to encourage this Sale, and the circulation of so valuable an importation of the large, healthy, pure breed of Rams, from the sister colony, into the district of Albany,—100 Saxon Merino Ewes, dropped last February, will be for private Sale on the Morning of the Public Auction, in small lots, not less than five, at £3 each Ewe, as they run out. Further, for the Butchers, 150 fat Merino Wethers, and some young Saxon bred Rams, from Burnt Kraal, will be for disposal on the day of Sale.

CHARLES GRIFFITH.

Prospect House, Graham's Town.

NOTICE.

A MEETING of the Creditors of the Estate of the late Mr. JAMES WEBBS, will be held at the Stores of Messrs. W. COCK & Co., on MONDAY the 14th inst., at 4 o'clock in the afternoon.

W. COCK,
J. WALKER,
W. M. BOWKER } Executors.

THE Undersigned begs to inform his friends and the public in general, that he has commenced business as TAILOR and DRAPER, in the house lately occupied by Mr. A. ANDERSON, Solicitor, High Street, and hopes by strict attention and punctuality to merit a share of public patronage.

JAMES M'MASTER.

PUBLIC SALE.

ON TUELDAY, the 15th December, at the Commission Rooms of Undersigned, a Public Sale will be held, when a large variety of Goods will be offered, and about twelve barrels of English Ale, ten bags of good Sugar, three bags of Wool, and a large quantity of other goods will be positively sold without reserve.

THOS. JARMAN, Auctioneer.

WOOL FOR SALE.

TENDERS will be received at the Store of Messrs. C. & H. MAYNARD, until the 23d December next, at three o'clock, for about 12,000 lbs. of Merino Wool, from the flock of Lieut. DANIELL, at per pound English, to take the clip.

The Wool will be washed, sorted, well packed, and delivered at Port Elizabeth by the 15th January next, purchaser paying for the bags.

The Undersigned having expended the sum of £370 within the last two years for pure Saxon and Merino Rams, he can confidently recommend his Wool to Exporters as an article of the first quality. A credit (on approved Bills) of Two and Four months will be given.

RICHARD DANIELL.

Sidbury Park, Albany,
16th Nov. 1835.

STEAM-MILL FLOUR.

THE Undersigned is receiving per *Conch* and *Knysna*, Fifteen Thousand Pounds Flour.

THOS. NELSON.

WANTED, in a Store, at Somerset, an active Lad, about 14 years of age. Apply to Mr. THOS. BUTLER, at Somerset, or at the office of this paper.

Fig. 25. *Graham's Town Journal,* 10 December 1835, featuring "Government Notice" following the 1834–1835 Xhosa War (*top left*) directly above an advertisement for Henry Macartney's phrenological lecture (*bottom left*), The National Archives, London.

the Pietermaritzburg Debating Society held a series of meetings to decide the motion "Is phrenology a science or not?" According to a correspondent for the local paper, the *Natal Witness*, "The subject appeared to have been well studied by the contending parties." After three weeks, a vote was held. The Pietermaritzburg Debating Society decided that phrenology was a science, although only "by a majority vote of *one*."[126] Around the same time, itinerant lecturers started bringing phrenology to a new audience of displaced Boors and British settlers. "Mr. Jenkins" explained the principles of "Mesmerism and Phrenology" to "a numerous and interested audience" in Pietermaritzburg. He even examined a soldier from the local British garrison, attempting "to excite different powers of the mind through the medium of different organs of the brain."[127]

With readers aware of the basic principles of phrenology, the editor of the *Natal Witness*, David Dale Buchanan, began linking mental science more closely to frontier wars.[128] Like the *Graham's Town Journal*, Buchanan's *Natal Witness* provided its readers with updates on the latest conflicts. In 1850, the *Natal Witness* reported that "a force proceeded from Bloem Fontein . . . to punish Molitzani for carrying war into the peaceful mission station of Umpukani." Moletsane, chief of the Taung, was accused of having "destroyed the settlement of Makansi, compelling the Wesleyan Missionaries whom he found established there, to abandon it." Buchanan then deployed a phrenological reading of Moletsane's character in order to justify the British reprisal. The *Natal Witness* noted that "such of our readers as are adepts in phrenology and physiognomy, would, doubtless, wish to have a glimpse of the living man." Buchanan then made his assessment based on a description of Moletsane taken from Thomas Arbousset's *Relation d'un Voyage d'Exploration au Nord-Est de la Colonie du Cap de Bonne-Espérance*, recently translated and reprinted in Cape Town:

> Well, then, there he stands. Black as soot—five feet ten inches high—slender figure—small cranium—narrow forehead—restless, savage eye—melancholy expression—very thoughtful—absent—humbly submissive in trouble.[129]

Like Godlonton and Macartney in Grahamstown, Buchanan was critical of liberals who saw African chiefs in humanitarian terms. Moletsane, the *Natal Witness* warned, was "always ready with a plausible excuse for his crimes, which he adroitly manages to persuade almost every body, blessed with the necessary amount of credulity, is owing to others, he being the victim." And like Godlonton, Buchanan also carefully edited the

Natal Witness to reinforce this connection between African character and frontier violence. The article on the phrenology of Moletsane appeared directly above a "Memoranda of the Movement of Troops." Lest readers be in any doubt about the chief's "savage" tendencies, the *Natal Witness* reported on one of Moletsane's raids "in which 20 persons had lost their lives." Bodies were found "in such a decomposed state as scarcely to be approachable," while "dogs still remain howling piteously near the corpses." The *Natal Witness* concluded that Moletsane was a "pillaging, murdering, and complete bandit chief."[130]

Phrenology flourished in the antiliberal periodical press on the frontier.[131] However, Macartney recognized that in order to really influence colonial policy, he needed to return to the Western Cape, where the decisions were made. In August 1845, Macartney boarded the *Justitia* at Port Elizabeth, spending four days at sea on his way back to Cape Town.[132] Once again, Macartney made good use of the periodical press. He sent advertisements ahead of the ship in July, to be printed in the growing number of newspapers and magazines popping up in Cape Town during the 1840s. This ensured that Macartney's audience was primed for his arrival. The satirical magazine *Sam Sly's African Journal* printed one of his advertisements, explaining that "Dr. Macartney begs leave to inform the Inhabitants of Cape Town of his intention shortly to visit them, and of delivering a few Public Lectures there." As another advertisement announced, these would be "Popular Lectures," and Macartney would be "happy to receive, during his stay, any persons who may wish to consult him, or to have their own or children's Heads examined."[133] Most importantly, these new lectures would describe the place of phrenology in "the Eastern Province—every part of which he has traversed and resided in for the last few years."[134]

Once again, Macartney returned to the Commercial Exchange. But this time he had a very different message for the audience of polite merchants and liberals. Rather than extolling the virtues of improvement and education, Macartney entered into an extended discussion of colonial policy and Xhosa character. According to *Sam Sly's African Journal*, which reported on the lectures, Macartney discussed "Frontier Matters" giving "his opinion and method of dealing with the Kafir question." He set out "the manner of the inhabitants and the native tribes," linking these to "the system of Treaties in general." Macartney even used a blackboard to sketch a map of his proposal to move the "Kafir districts" beyond the Keishkamma River. Finally, Macartney attacked Fairbairn directly, announcing, "Here are the organs of *benevolence* and *caution*. . . . These are the organs that enable him to carry out all those profound and humane projects, with respect to the Kafirs."[135] However, according to Macartney, Fairbairn's phrenological organs were misdirected. The Xhosa were "savages," and the existing treaties were "defective in principle."[136] In effect, the lecture room allowed Macartney to bring what he had been writing for the *Graham's Town Journal* on the eastern frontier back to an audience in Cape Town. In this way, Macartney completed a circuit whereby phrenology traveled between the lecture room, the periodical press, and back again. In the process, phrenology transformed from a liberal science of colonial improvement to a racialized justification of frontier expansion.[137]

SHIPS AND SOCIETY IN COLONIAL INDIA

Cally Coomar Das made the short journey from his house in Calcutta to the docks by the Hooghly River. In February 1848, Combe had promised to send a box full of phrenological books and periodicals from Scotland to India. Das, eager to read the latest issue of the *Phrenological Journal*, was still waiting in December. Even by sail, it shouldn't have taken more than six months. Das feared the worst, admitting it was his "daily expectation" that the vessel had met a "perilous" end. But finally, amid the crowd of ships traveling up and down the Hooghly each day, Combe's package arrived aboard the *Framjee Cowasjee*. Das returned home satisfied and examined the contents of the box: an unbound copy of Combe's *Notes on the United States of North America*, the complete *Transactions of the Phrenological Society*, and the 1847 volume of the *Phrenological Journal*.[138]

Three years earlier, in 1845, Das had founded the Calcutta Phrenological Society.[139] According to the secretary of the society, Nobin Chunder Bose, its members were "all Hindoos by birth," mainly consisting of medical students as well as a few teachers and clerks. (Das himself was a student at the Calcutta Medical College, founded in 1835 to train subassistant surgeons for the East India Company.)[140] These were the *bhadralok*, a new generation of middle-class men earning a living through salaried professions under the colonial state.[141] Each week, Das and his fellow Bengali phrenologists would meet to deliver lectures, read the latest scientific works, and discuss educational and religious reform. Within a few years, the Calcutta Phrenological Society had built up an impressive library. The books and journals sent by Combe sat alongside copies of Spurzheim's *Outlines of Phrenology* and Mackenzie's *Illustrations of Phrenology*. Like many colonial libraries, the collection represented the different ways in which India was connected to the wider world. As well as books imported from Britain, the Calcutta Phrenological Society owned a copy of Fowler's *Matrimony* and Amariah Brigham's *Remarks on the Influence of Mental Cultivation*, both printed in the United States. In total, the Calcutta Phrenological Society's library housed over forty books and eighty issues of periodicals.[142] The Bengali phrenologists, however, weren't just consumers of European and American print culture. They were also authors and publishers. In February 1850, the Calcutta Phrenological Society issued the first number of the *Pamphleteer*, a monthly periodical devoted to "phrenology, sciences in general, literature and arts" (fig. 26). In doing so, the Calcutta Phrenological Society hoped to cultivate a new reading audience of colonized subjects.

Edited by Das, the *Pamphleteer* was just one among a range of new journals that addressed scientific topics published in Calcutta in the middle of the nineteenth century.[143] Through these periodicals, and the voluntary associations to which they were attached, middle-class Bengalis discussed all the latest social and political issues.[144] Many of these debates revolved around questions of nature and religion.[145] The wealthy landowner Debendranath Tagore, for instance, founded the Tattvabodhini Sabha in 1839 in order to promote the compatibility of the Vedas with scientific knowledge.[146] Debendranath believed that "the strata of the earth," "the laws of the body," and even "the subtle laws of the mind" all provided evidence of Brahma's "wonderful and inscrutable power."[147] This understanding of the harmony between nature and Hindu religious beliefs was reflected in the pages of *Tattvabodhini Patrika*. Through this monthly periodical, the Tattvabodhini Sabha countered missionary critiques of Hindu superstition, arguing that the lessons of physics, chemistry, and physiology could all be understood in Vedantic terms.[148] On the other side of the political spectrum, Young Bengal set up the Society for the Acquisition of General Knowledge in 1838. Much more radical than the Tattvabodhini Sabha, Young Bengal attacked both Christian and Hindu beliefs alike, promoting the perfectibility of man in place of religion.[149] Again, these debates played out in the pages of Young Bengal's quarterly scientific journal, *Bidyakalpadrum*.[150] Underpinning all these political, religious, and publishing projects was a nascent understanding of Indian nationalism. Tagore, Young Bengal, and the Calcutta Phrenological Society argued over the status of nature, society, and religion because these issues mattered for the meaning of India as a nation. Throughout the 1850s, Das and the members of the Calcutta Phrenological Society jostled with these radicals, reformers, and conservatives, advancing a phrenological vision of India's place in the world.

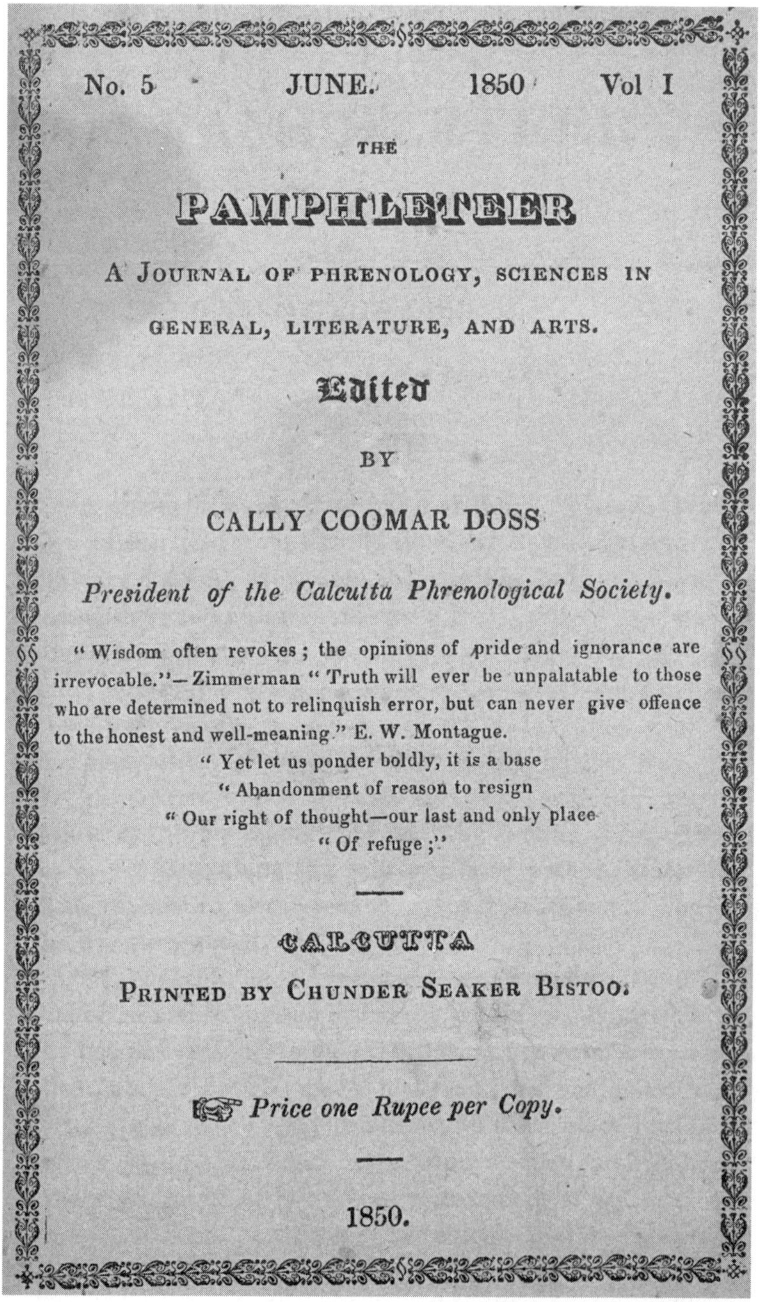

No. 5 JUNE. 1850 Vol I

THE

PAMPHLETEER

A JOURNAL OF PHRENOLOGY, SCIENCES IN GENERAL, LITERATURE, AND ARTS.

Edited

BY

CALLY COOMAR DOSS

President of the Calcutta Phrenological Society.

" Wisdom often revokes ; the opinions of pride and ignorance are irrevocable."—Zimmerman " Truth will ever be unpalatable to those who are determined not to relinquish error, but can never give offence to the honest and well-meaning." E. W. Montague.

" Yet let us ponder boldly, it is a base
" Abandonment of reason to resign
" Our right of thought—our last and only place
" Of refuge ;"

CALCUTTA

PRINTED BY CHUNDER SEAKER BISTOO.

☞ *Price one Rupee per Copy.*

1850.

Fig. 26. The title page of the fifth number of the *Pamphleteer* 1 (1850), The British Library Board.

According to Das, the *Pamphleteer* was a periodical "of a most liberal stamp." The Bengali phrenologists believed that the sciences, particularly the sciences of the mind, had the power to reform colonial society.[151] The opening number explained that readers would be provided with "useful informations calculated to enlighten them on topics connected with the physical wellbeing, and the moral and intellectual improvement of society."[152] The contents of the *Pamphleteer* reflected these concerns, with articles connecting phrenology to issues ranging from education and commerce to caste and religion. Significantly, the Calcutta Phrenological Society linked this liberal political agenda to a growing sense of Indian nationalism, albeit a nationalism centered on Bengal.[153] The *Pamphleteer* referred to its readers as "countrymen" and hoped that phrenology would be directed toward the "education and enlightenment of the native population."[154]

Phrenology itself provided a new language through which to articulate national identity. The second and third numbers of the *Pamphleteer* contained two long articles entitled "Phrenological Development of the Bengalees." These set out the "character of our countrymen," explicitly countering the typical British charges of "Hindoo" inferiority. Crucially, the articles accepted that Bengalis tended to be "less bold, and less enterprizing" than Europeans. But, the *Pamphleteer* argued, this was a result of circumstance rather than "cerebral organization." The "Bengalee head" was in fact characterized by "perseverance, industry, and artistic skill." The intellectual organs in particular were "well developed." However, Indian society needed reform—both internally and externally—in order to release this untapped potential. In the same article on Bengali character, the *Pamphleteer* criticized both the East India Company and traditional Hindu beliefs. India was "oppressed by monopolies, and a fixed revenue system" while at the same time suffering from "the abominable system of castes." The message was politically powerful. Bengalis were "not inferior to Europeans." However, according to the *Pamphleteer*, the emergence of a successful Indian nation depended upon a range of social, political, and religious reforms, all underpinned by a materialist understanding of the mind.[155]

Nationalism did not imply parochialism. With a library of works printed in Britain, the United States, and India, the Bengali phrenologists quickly saw the periodical press as a powerful means of connecting the world.[156] Das marveled at the idea that "a person sitting in a solitary prison in Paris, or London, can within a short period of a few months communicate his thoughts, sentiments, and passions to millions of human beings in Europe, America, and Asia." He concluded that "the press, the publisher, the steamer," provided the "means of communication between distant nations."[157] The Calcutta Phrenological Society therefore imagined reform within India as part of a global history of human improvement. In an article entitled "The Prospect of the Age," the *Pamphleteer* explained that "to whatever region of the globe we direct our eyes . . . we find that improvement is the order of the day." The Bengali phrenologists believed that new understandings of nature, coupled with educational and religious reform, were changing the structure of society, not just in India, but across the globe. The *Pamphleteer* described Indian improvement in relation to political concerns ranging from the French Revolution to the abolition of slavery. The "allied despots," "unholy trade," and "last vestiges of monopolies" were "passing away day by day in all civilized countries." It wouldn't be long before India would take its place "on the first rank among the nations of the world." The *Pamphleteer* concluded with an evocative image, stating "from the banks of the Ganges to that of the Mississipi [*sic*]—from the Indian archepalago [*sic*] to the broad bosom of the Pacific, the march of civilization is onward—the minds of men are on the move."[158]

This attitude, blending together national and global political narratives, structured how the Bengali phrenologists addressed both religious and educational reform in India. In June 1850, the *Pamphleteer* published an article entitled "Education of the People." This article, which discussed Indian educational policy in relation to Britain and the United States, was based on a lecture Das had given at a meeting of the Calcutta Phrenological Society. Throughout, Das made extensive reference to the works of the Scottish American socialist Robert Dale Owen as well as those of the British educational reformer William Johnson Fox.[159] Oral culture, therefore, was not displaced by the growth of a global market for print. Rather, meetings of the Calcutta Phrenological Society acted as a space in which Bengalis discussed European and American texts.[160] These meetings, which took place at Das's home, were not so dissimilar to a traditional *adda*, where Bengalis would gather informally to discuss the latest news and ideas.[161] Now, however, Bengalis brought imported books to the *adda*. Das explained that he was an "enthusiastic admirer" of Fox's *Lectures, Addressed Chiefly to the Working Classes*, published in London in 1846. Quoting Fox directly, Das set out the purpose of "national education," criticizing those who saw schooling as "a contrivance for making good operatives," or worse, "training up proselytes to their church, sect, or creed." Through a series of exchanges between oral and print culture, extracts from Fox's lecture, originally delivered at the Working Men's Association in Holborn, now appeared in the *Pamphleteer* in Calcutta. Similarly, Das quoted directly from Owen's *Tracts on Republican Government and National Education*, complaining "how one class is doomed to toil for bread, and another privileged to wanton in luxurious idleness." Owen's tracts were explicitly "Addressed to the Inhabitants of the United States of America." Yet Das read this socialist critique of educational policy as equally applicable to colonial India.[162]

The use of these American and British sources was a creative act. Das repurposed Fox's and Owen's texts for a new audience of colonized subjects in Calcutta. After describing Owen's arguments, Das asked, "If such be the state of society in one of the most enlightened quarters of the world, how much more are these observations applicable to our own country!" Das complained of the religious and financial motives behind education in India, just as Fox and Owen had done in Britain and the United States. "The missionaries," Das argued, treated education as "an enlightened instrument of conversion," while the East India Company saw education as "a means of raising up a party who will both serve and support its measures." (The Calcutta Medical College was a case in point.) Das insisted that phrenology showed these attitudes to be misguided. He asked his readers rhetorically, "Are not all human beings possessed morally and physically of a similar organization?" Phrenology proved that "every individual of the human species is entitled to receive the benefits of education." And these benefits were not those promoted by "bigotted sectarians [*sic*]" or "rulers whose grand aim was to keep the people always in bondage and poverty." Rather, the "object of education" was "the bettering of the moral and intellectual condition of the people."[163]

When the *Pamphleteer* addressed education, it was also appealing to that sense of nationalism discussed earlier. Das's repeated use of the term "the people" was no accident. A phrenologically inspired system of "national education," as he called it, would "form the man and the citizen." The careful cultivation of the "moral faculties" as well as "physical education" would help to train "the rising generation of this country." Like the print culture that underpinned the activities of the Calcutta Phrenological Society, Das's attitude to education melded the global with the national. Reading Fox and Owen, the Bengali phrenologists were convinced that educational reform was a global movement. But they also saw education as a means of cementing India's place on the world stage, concluding, "If popular industry be guided by popular scientific skill, what enormous wealth may not India produce!"[164] Here, the construction of a "popular" phrenological audience was a product of both national and global political thought.

The Calcutta Phrenological Society's attitude toward education and nationalism was heavily influenced by ongoing religious debates taking place within India and beyond. The Tattvabodhini Sabha had been keen to defend Hindu beliefs following the growth of missionary activity since the 1813 Charter Act.[165] However, as with Christianity, there were significant disagreements among Indians concerning the authority of natural knowledge and sacred texts. Materialism was a sticking point. Following the Vedas, many Hindus understood the natural world to be an illusion (*maya*), one that masked true reality (*brahman*). Yet others, particularly the younger generation of Bengali reformers, believed that Hindus needed to abandon such a dogmatic focus on the Vedas.[166] These debates divided the Tattvabodhini Sabha itself. Akshay Kumar Dutt, who from 1843 edited *Tattvabodhini Patrika*, passionately believed that Brahma was best appreciated through the study of the natural world rather than sacred texts. The world was not an illusion. Rather, universal laws were an expression of Brahma's will.[167] This attitude was reflected in the pages of the *Tattvabodhini Patrika* under Dutt's editorship. Phrenology even played a role. Dutt, although not a member of the Calcutta Phrenological Society, had read *The Constitution of Man*. Influenced by Combe, Dutt then published *Bahya Bastur Sahit Manab Prakritir Sambandha Bicar* in two volumes between 1851 and 1853. This work provided "an analysis of the relationship between human nature and the external world." And although it was printed as a book at the Sanskrit Press, *Bahya Bastur* actually first appeared as an essay in the periodical *Tattvabodhini Patrika*. In the essay and later book, Dutt promoted a materialist understanding of the Hindu world for a vernacular audience.[168] He explained:

> Mr George Combe has written elegantly on this matter in his book entitled *The Constitution of Man*. He has confidently demonstrated that if we obey God's laws we find happiness and if we disobey them we suffer. This book clearly explicates the type of laws by which the Lord protects his kingdom as well as what benefits derive from following those laws and what punishment obtains from transgressing them.[169]

Others were less sure. Tagore was uncomfortable with the materialism promoted by Dutt, declaring in 1850 that "matter and soul are as different as darkness and light." Directly referring to Dutt's reading of Combe, Tagore later reflected on the differences between these two competing approaches to Brahmoism, writing, "I was seeking to know my relations with God, he was seeking to know the relations of man with the outer world." Tagore concluded, "We were poles asunder."[170]

The Calcutta Phrenological Society made its own contribution to these debates. Even more radical than Dutt, the Bengali phrenologists believed that a materialist understanding of the natural world had the power to do away with both Christian and Hindu religious beliefs. Das was the most forthright on this point, having earlier authored *General Reflections on Christianity*, a work he described as a "fearless exposition of the falsehood of the religion." In this, Das criticized both Christian and Hindu deference to sacred texts, declaring, "Instead of spending our time only in looking into the mysterious and uncertain pages of the *Bible* or the *Veda*, let us study the luminous and nevererring pages of nature."[171] This commitment to materialism was bolstered by Das's experience at the Calcutta Medical College. By the late 1830s, students were undertaking human dissections and had access to a large pathological museum.[172] *The Pamphleteer* even reminded its readers that "the professors of the Medical college often say to their pupils that they have no objections to the general principles of Phrenology."[173] In *General Reflections on Christianity*, Das also made explicit the relationship between his critique of religion and a wider program for social and educational reform. He believed that "a thorough change in the religious opinions of man" was required "for the further improvement of his condition." Once again, the Bengali phrenologists imagined religious reform as part of a wider movement. Das wrote that "the melancholy effects of Hindu superstitions in India are as clearly visible as the effects of similar systems in any other parts of the world." Looking outward, Das argued that "in Europe and in America, which are at present the principal seats of the Christian religion, the time of its down-fall is not far distant." Das concluded that he preferred "the everlasting temples of Nature, virtue and knowledge, exchanged for those of any religion now existing on the face of the Globe."[174]

This materialist critique of religion, closer to Young Bengal than the Tattvabodhini Sabha, was reflected in the pages of the *Pamphleteer*. In one striking article entitled "The Earth a Living Body," the Calcutta Phrenological Society even pushed this narrative to include the origins of life. Directly contradicting both Christian and Hindu teaching, the *Pamphleteer* argued, "That the earth was formed out of nothing is perfectly inconceivable." A sort of Bengali *Vestiges of the Natural History of Creation*, the article then described an evolutionary history of the cosmos. It posited that "the germ of the earth, was thrown out like a cell, or huge fluid sphere in its present place, by some of the larger planets of the solar system." This naturalistic account of the origins of the earth was then extended to plants and animals, even mankind. The *Pamphleteer* took its readers through geological evidence, suggesting that "man is a recent inhabitant of the globe." The emergence of new life was not the result of divine intervention. Rather, life was a product of the "chemical agencies" and the "power inherent in the living materials of the globe." The same was true of "mental phenomena," which amounted to nothing more than "the activity of the nervous substance contained within their skulls."[175]

Debates about religion and education in India were closely connected to the growth of the vernacular press. The choice of the Calcutta Phrenological Society to publish the *Pamphleteer* in English is therefore particularly striking. Throughout the 1840s, there was certainly a large market for vernacular works, particularly cheap Bengali books printed to the north of the city in Battala. However, middle-class Bengalis tended to consider Battala books as vulgar, even obscene.[176] Societies like the Tattvabodhini Sabha therefore hoped to sanitize the vernacular press through the production of improving literature. Much like the evangelical publishers in Britain, Dutt's *Tattvabodhini Patrika*, and his Bengali works on physics and geography, were imagined as a wholesome alternative to the lewd chapbooks hawked in Battala.[177] Dutt, like the other members of the Tattvabodhini Sabha, was also insistent that the development of India as a nation relied on the use of vernacular languages. Like many Hindu reformers, he associated English with the activities of Christian missionaries.[178] In contrast, the Bengali phrenologists promoted English, particularly in the context of education. The *Pamphleteer* argued that "instruction should be imparted through the medium of the English in preference to the Vernacular." Drawing again on the importation of books from Britain, the author explained this choice by noting that "the former contains elementary works ready made, in all the branches of literature, science, and art."[179] Beyond this practical concern, it is important to recognize that the *Pamphleteer* was simply not intended as a periodical for the masses. Relatively expensive at 1 rupee an issue, it was the focal point of a small voluntary association of middle-class Bengali medical students. The prestige and respectability associated with English were much more important for the Calcutta Phrenological Society than the prospect of addressing a broader audience.[180]

Nonetheless, these attitudes did not entirely preclude an engagement with the vernacular. While all the articles in the *Pamphleteer* are printed in English, a number of issues feature a small notice in Bengali type on the back cover (fig. 27). This advertisement promotes a Bengali introduction to Combe's and Spurzheim's phrenological works. Titled *Manatatva Sarsangraha*, the book had recently been completed by Radhaballav Das, the Calcutta Phrenological Society's librarian, and printed at the small but respectable Purnochandradoy Press.[181] The Purnochandradoy Press printed other improving works in both Bengali and English, often in translation, including a Bengali edition of *Arabian Nights* and a daily "literary and political journal" entitled *Sungbad Purnochandradoy*.[182] Whether effective in generating sales or not, this advertisement acted as a statement of intent. It provided a means for the Calcutta Phrenological Society to declare its engagement with those among the Indian community, like Young Bengal and the Tattvabodhini Sabha, promoting vernacular audiences for science.

Within this publishing climate, the Bengali phrenologists attempted to advance their own position as moral and intellectual leaders through attempts to reform and sanitize the vernacular press.[183] The *Pamphleteer* even linked the deleterious effects of sensational reading to mental physiology, warning that "the brain which is the organ of mind . . . , if inordi-

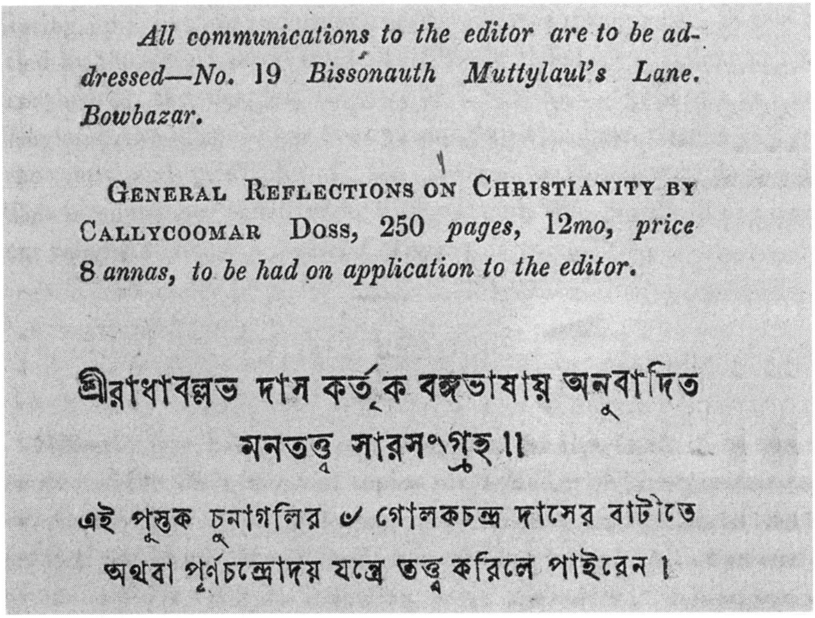

All communications to the editor are to be addressed—No. 19 Bissonauth Muttylaul's Lane. Bowbazar.

GENERAL REFLECTIONS ON CHRISTIANITY BY CALLYCOOMAR DOSS, 250 *pages*, 12*mo*, *price* 8 *annas*, *to be had on application to the editor.*

শ্রীরাধাবল্লভ দাস কর্তৃক বঙ্গভাষায় অনুবাদিত
মনতত্ত্ব সারসংগ্রহ ॥

এই পুস্তক চুনাগলির ৺ গোলকচন্দ্র দাসের বাটীতে
অথবা পূর্ণচন্দ্রোদয় যন্ত্রে তত্ত্ব করিলে পাইবেন ।

Fig. 27. Advertisement for Radhaballav Das, *Manatatva Sarsangraha* (Calcutta: Purnochandradoy Press, 1849), in the *Pamphleteer* 1 (1850), The British Library Board.

nately and incessantly exercized, can produce highly injurious consequences."[184] This intellectual and moral depravity was closely connected to notions of disease and cleanliness. At a meeting of the Bethune Society, Cally Coomar Das later complained of the "effects of the deleterious atmosphere of the native city on the nervous system, and consequently the brain" in response to a paper delivered on "the sanitary position and obligations of the inhabitants of Calcutta."[185] *Manatatva Sarsangraha* was therefore billed as a sober and edifying work, countering the apparent depravity of the vernacular press. In the introduction, Radhaballav Das emphasized the respectable nature of phrenology in India, describing the foundation of the Calcutta Phrenological Society "under the efforts of some learned gentlemen." Radhaballav Das saw his work as part of a much wider translation movement, noting that "this peculiar knowledge has been translated into various languages." Combining notions of global and national improvement once again, he explained that *Manatatva Sarsangraha* had been "compiled from several English phrenology books . . . for the sake of the betterment of the people." The contents reflected this. Following descriptions of different phrenological organs and states of mind, the later sections of *Manatatva Sarsangraha* addressed morality. These sections included a discussion of how phrenology relates to desires, many with religious connotations, such as the desire for meat, alcohol, and wealth.[186]

As the title page announced, *Manatatva Sarsangraha* also contained a "Phrenological Chart" that had been reproduced from Combe's *System of Phrenology* (fig. 28). While the outline of the bust and the placement of the phrenological organs were faithfully reproduced, the image was nonetheless subtly vernacularized. Most likely at the request of Radhaballav Das, the engraver added a small mustache to Combe's previously androgynous image.[187] Once again, this reflected concerns about respectability and improvement. Middle-class Bengali men tended to consider female literature as vulgar, while women themselves were believed to read uncritically. Such attitudes went hand in hand with the rejection of domestic culture and the move toward new voluntary associations such as the Calcutta Phrenological Society itself.[188] The Bengali phrenologists certainly subscribed to this view, deploying phrenology to explain "the difference in mental qualities of man and woman." One member even described how "the anterior region" of the brain, containing the "reflecting faculties," was "narrower and smaller" in women.[189] By gendering phrenology as a masculine science, the Calcutta Phrenological Society hoped to portray *Manatatva Sarsangraha* as edifying, in contrast to the allegedly vulgar literature, particularly literature for women, found in Battala.

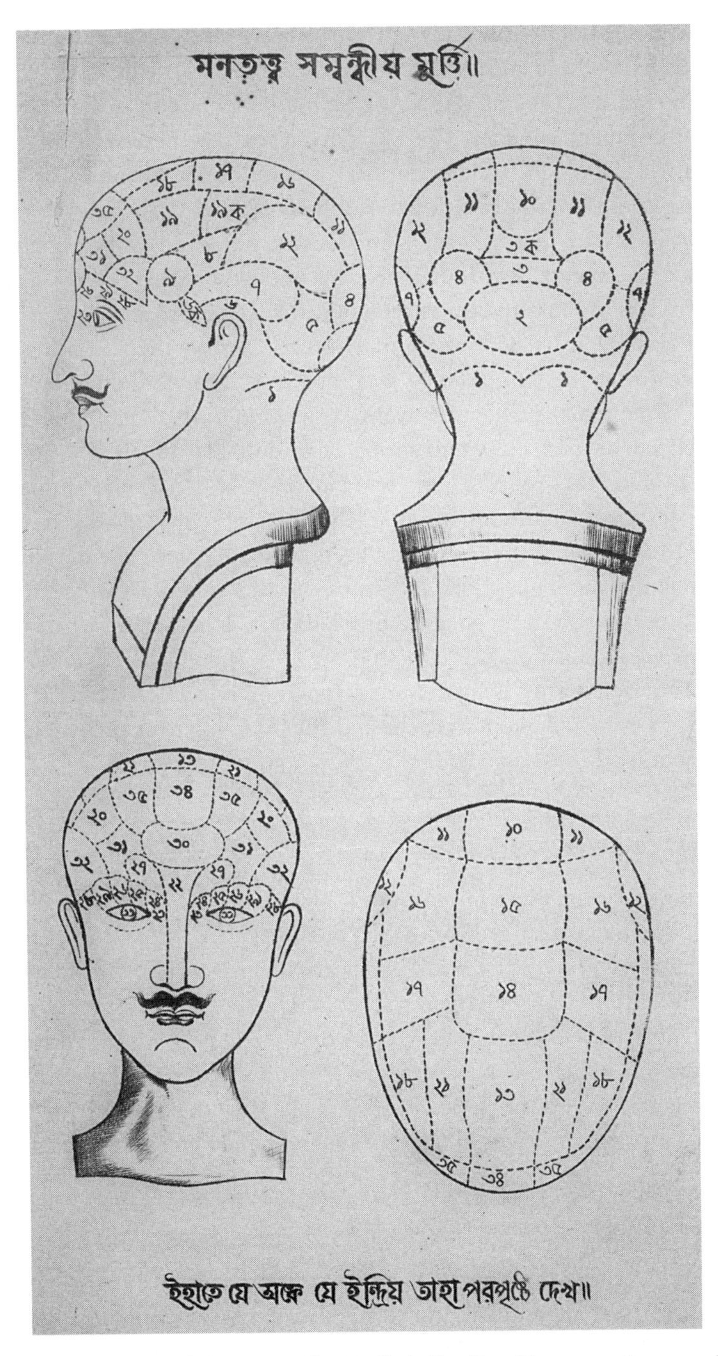

Fig. 28. Phrenological chart printed in Radhaballav Das, *Manatatva Sarsangraha* (Calcutta: Purnochandradoy Press, 1849), The British Library Board.

The Bengali phrenologists imagined works like *Manatatva Sarsangraha* as part of a local movement for the development of vernacular science. In contrast, they imagined the *Pamphleteer* as part of a national and global movement conducted in English. In reality, there were significant inequalities between readers, both within India and beyond. For a start, the *Pamphleteer* actually failed to reach far beyond Calcutta. While phrenology popped up in newspapers and lecture halls in Bombay, there is no evidence to suggest that Indian readers outside of Bengal were aware of the activities of the Calcutta Phrenological Society.[190] And despite the Calcutta Phrenological Society's connection to Combe, there is no evidence to suggest that the *Pamphleteer* was read in Britain or the United States. At the same time, vernacular works like *Manatatva Sarsangraha* failed to reach far beyond the small group of middle-class men who produced them. In part, this reflected the tension between a nationalism centered on Bengal and a nationalism that encompassed the subcontinent as a whole. In the case of the Calcutta Phrenological Society, the development of vernacular audiences was also hampered by a belief in the vulgarity of the press, particularly surrounding Battala. Nonetheless, whether they were successful or not, the *Pamphleteer* and *Manatatva Sarsangraha* do represent attempts by Bengalis to turn colonized subjects into scientific readers. They also demonstrate the significant challenges that such an endeavor entailed.

Traveling between the United States, the Cape Colony, and India, this chapter has demonstrated the variety of ways in which the periodical press fostered new reading audiences for science. These included audiences of settlers, frontiersmen, and colonized subjects. Whether printed on a steam press in New York City or on a hand press in Calcutta, phrenological journals were closely tied to the growth of nationalism. Yet despite this nationalist impulse, phrenologists also imagined themselves as part of a global community in print. The periodical press emerged as a key site in the making of phrenology as both a national and a global scientific movement.

In concluding, it is worth considering how many of the arguments developed in Bernard Lightman's *Victorian Popularizers of Science* might be expanded in the context of a global history. Focusing on middle-class and working-class audiences in Britain, Lightman argues that historians need to pay close attention to the variety of sites of nineteenth-century scientific activity, including zoos, museums, pubs, and churches. Similarly, Lightman emphasizes the need to study publishers, editors, and other middlemen in the creation of scientific readerships.[191] The global history of phrenology uncovered in this chapter builds on both of these themes. The growth of phrenology in the periodical press, and its associated audiences, were closely connected to a range of new sites for science, not just in Britain, but in the United States and the colonies. These sites included centers of imperial trade, such as the Commercial Exchange in Cape Town, as well as frontier lodging houses such as those in Grahamstown. Phrenology was popular on slave plantations as well in the military camps of Union troops. And in the case of the Bengali phrenologists, the sites of science extended to the colonial domestic sphere, with meetings of the Calcutta Phrenological Society taking place at Das's home in Bowbazar. Additionally, this chapter has emphasized the need to consider how the periodical press connected different sites, particularly across national and imperial boundaries. The *American Phrenological Journal* connected readers in the United States to mission stations in India and China. Similarly, the *Pamphleteer* allowed middle-class Bengali readers to learn about lectures delivered to working-class audiences in London. This chapter also expands the range of middlemen responsible for the growth of popular science, whether editors, missionaries, or merchants. Colonial newspaper proprietors, such as Robert Godlonton in Grahamstown, played an active role in connecting phrenology with frontier policy, while missionaries such as the Reverend W. W. Hicks provided American phrenologists with Indian illustrations.

Lightman correctly argues that "popular science" needs to be understood as an actors' category, one that took on a variety of meanings in the nineteenth century. In some cases, "popular" simply meant someone who attracted a large audience, or got paid. In other cases, "popular" implied subversion or a challenge to the elite.[192] The examples presented in this chapter reveal a range of additional meanings of the term "popular." When Das used the term "popular," he was thinking of Bengali nationalism. When Macartney used the term "popular," he meant colonial settlers. And when the Fowlers used the term "popular," they meant a mass audience of Americans, one that they hoped to hold together in the wake of the Civil War. Most significantly, the use of the term "popular" was closely related to an understanding of phrenology as a global science. Reprints of imported articles and illustrations all contributed to a sense that phrenology was spreading across the world. As the case of the Bengali phrenologists demonstrates most clearly, part of what it meant to be popular was to be global—a science stretching "from the banks of the Ganges to that of the Mississipi [*sic*]." Popular science, then, needs to be understood simultaneously as a global, imperial, and national category, both by historians today and from the perspective of historical actors.

CHAPTER 6

Photographs

The phrenological Fowlers never missed a business opportunity. In July 1856, the *American Phrenological Journal* started advertising a new service: "Written Descriptions from Daguerreotypes." For a $4 fee, Lorenzo and Orson Fowler would make a phrenological reading from a photograph. This new technology of reproduction, announced publicly by Louis Daguerre in 1839, allowed the Fowlers to extend phrenology even farther beyond their consultation rooms in New York City. "Many of our friends who reside a considerable distance from us, desire to obtain, for themselves or friends, full Written Descriptions of Character," the *American Phrenological Journal* explained.[1] With studios popping up across the United States, photography provided a solution. The scheme was a huge success. Thirty years later, the Fowlers boasted that "thousands have availed themselves of this method of learning their true character." From parents in California to ladies in Kansas, the Fowlers tapped into new phrenological audiences from the East Coast to the West.[2] Unlike any other technology before it, photography allowed phrenologists to collapse distance. It was, according to the Fowlers, one of the "seven wonders of modern date."[3] Before long, this scheme extended beyond the United States. By the 1890s the Fowlers had received photographs from Canada, the West Indies, Mexico, South Africa, New Zealand, Australia, and Japan.[4] A similar scheme, run by the *Phrenological Magazine* in London, reached an audience across the British Empire. "J.H" from Poona was told he "would make a good explorer," while "G.C.H" from Melbourne was apparently "adapted to farming."[5]

Photography underpinned the first mail-order phrenological service, stretching from the United States to Japan. However, this new visual technology was not without its difficulties. Nor were photographs seen as a uniquely powerful or objective form of evidence.[6] Early photographic techniques such as the daguerreotype and ambrotype needed to be developed on glass. These, the Fowlers complained, were "frequently broken by the postmaster stamping the name of his post-office on the package." With the invention of paper-based printing techniques in the second half of the nineteenth century, these problems were partially resolved. The *American Phrenological Journal* recommended that readers take advantage of salt and albumin prints on paper as they were "not liable to be broken." Even if the photograph arrived intact, it could still be difficult to interpret. "Some likenesses, however, are taken in such a manner that we can not well determine the form and size of all parts of the head," the Fowlers complained. In light of this, the *American Phrenological Journal* printed precise instructions for the composition of suitable photographs. Two photographs were needed, one of the side of the head and one of the front. The hair also had to be smoothed down. Ladies were instructed that "no puffs, braids, or other arrangements of hair or combs" were permitted.[7] Worst of all, studio photographers were in the habit of retouching negatives with a pencil or ink to flatter their subjects. The Fowlers warned against this practice, reminding their readers that "beauty, however desirable, is not the desideratum in a portrait when it does not exist in the type."[8]

Despite these problems, photography proved incredibly popular. It was a powerful medium for extending the basic practices of phrenology into new spaces. This was especially true at the frontiers of the colonial world. Alongside pictures of European settlers and officials, the *American Phrenological Journal* started receiving photographs of colonized people. These often featured in the Fowlers' new "Ethnology" column. In March 1870, American readers were treated to a series of articles, illustrated with photographs, describing the "Aborigines of the Philippine Archipelago" (fig. 29). The photographs themselves had been taken by William Wood, an early pioneer of photography in the Philippines with a studio in Manila.[9] In the Fowlers' *Phrenological Journal*, these photographs appeared as wood engravings, traced from the original prints. Over the course of the articles, readers were invited to look upon the different races of the region, from the "Papuans" and "Negritos" to the "Tagals" and "half-castes." Together, the articles hoped to reveal "the origin of these interesting divisions of the great Malay family."[10] This focus on origins reflected a growing debate over the evolutionary history of man. In the late 1860s, following his return from the East Indies, Alfred Russel Wallace presented the Malay Archipelago as a microcosm of mankind's history. Submitting his early ideas to the Ethnological Society in London, Wallace

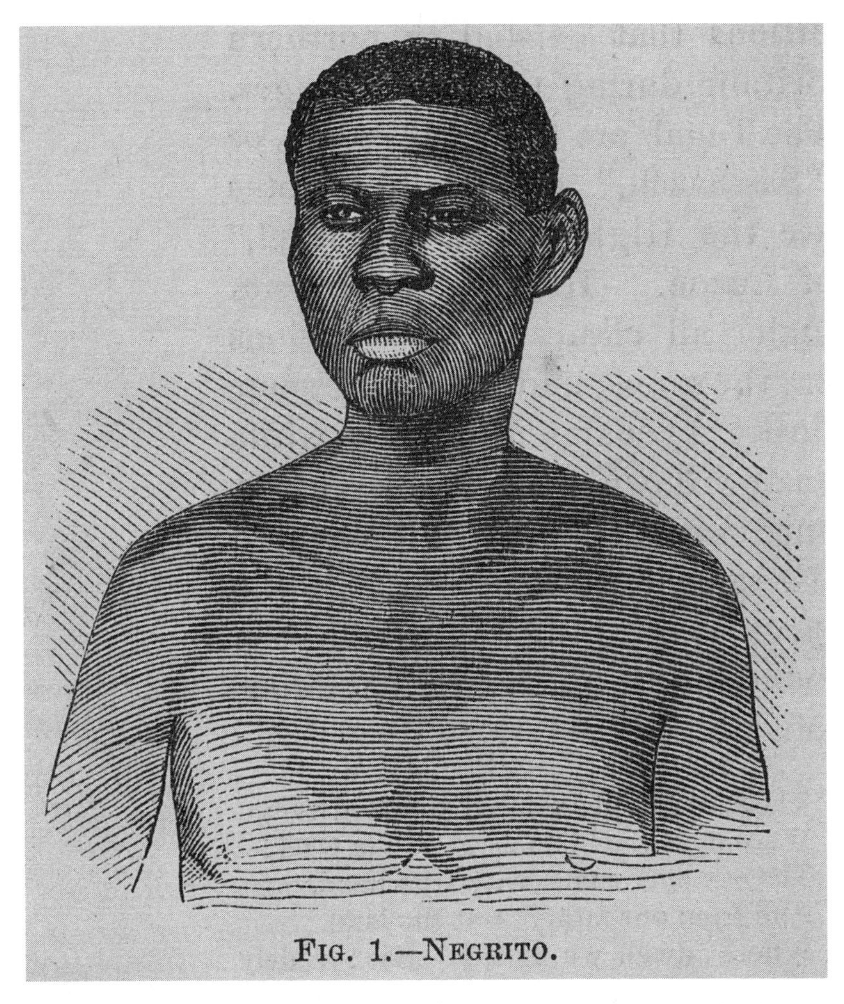

Fig. 29. "A Negrito," in N. W. Beckwith, "Aborigines of the Philippine Archipelago," *Phrenological Journal and Life Illustrated* 50 (1870): 161, Wellcome Collection, CC-BY.

argued that "two very strongly contrasted races are found in this region: the *Malays* who inhabit almost exclusively the western half of the Archipelago, and the *Papuans* whose head-quarters are New Guinea and some of the adjacent islands." Wallace then explained the divergence between these two groups in terms of geological change and geographic distribution. The Malay Archipelago could be divided into two. To the west, the Asian landmass, and to the east, the remains of an ancient Pacific continent. Known as the "Wallace Line," this division of the archipelago apparently accounted for the differences between the Malayans and the Papuans.[11] Now, in the United States, the Fowlers presented phrenology and photography together as part of an evolutionary argument. By looking at the photographs, sent from Manila and copied as wood engravings in New York City, the reader could "judge for himself" the differences between the races of man. The "Papuans" were a "puzzling race," the *Phrenological Journal* argued. They possessed "narrow heads . . . excessive thickness of lips . . . about as ugly and repulsive an exterior as can well be imagined."[12] In contrast, the Malays exhibited "hereditary skill" and "the proverbial refinement of Hindoo art," reflecting an Asian heritage.[13]

Photography allowed phrenologists to connect the material mind to new evolutionary claims about the origins of man. Phrenological photography was also an intensely political activity, reflecting the high imperialism of the late nineteenth century. In 1899, the Fowlers published another article on the Philippines. Things had changed. Beginning in the West Indies and ending in the Pacific, the Spanish-American War of 1898 established the United States as a global imperial power. After over 300 years of Spanish rule, the Philippines were now part of an American empire incorporating Cuba, Guam, and Puerto Rico. The latest phrenological account of the Philippines reflected this. The photographs had been taken by Joseph Steere, professor of zoology at the University of Michigan, for publication in *Scientific American*. Once again, the Fowlers had the photographs copied as wood engravings for their own phrenological purposes. The "origin" of the people of the Philippines was still described as "part of the great problem of the Pacific." But now it was also a political problem. In keeping with efforts to portray American interest in the Pacific as benevolent, the *Phrenological Journal* in New York City offered a positive account of Filipino life. "American enterprise, justice, Christianity, and political administration . . . will work wonders in the islands," the *Phrenological Journal* exclaimed. The photographs depicted the fruits of industry—"women weaving"—alongside members of the Malay elite (fig. 30). "The portraits that we present with this article represents the civilized Indians of the Philippines; they are Malay chiefs of Mindanao," explained the *Phrenological Journal*. These "civilized Indians" were apparently "rapidly increasing in numbers" under American rule. The "Malay Chiefs" in the photograph showed the "superiority of his mental development . . . , a strong perceptive intellect."[14]

Phrenological photography clearly helped promote imperialism back in Europe and America. However, it could also be used to explore the limits of colonial power. In 1885, one of the Fowlers' correspondents in

Fig. 30. "Malay Chiefs, Mindanao," in "American Possessions in the Philippines," *Phrenological Journal and Science of Health* 107 (1899): 178, Wellcome Collection, CC-BY.

Cairo acquired a photograph of Muhammad Ahmad. The photograph appeared alongside an account of "The Mahdi, or Fighting Prophet of the Soudan" in the *Phrenological Journal* in New York City. Having declared himself the Mahdi, or redeemer of Islam, Ahmad launched an anticolonial military campaign against the British in Sudan. "The capture of Khartoum, the stronghold of the Soudan, by the Arabs, and the death of General Gordon, have drawn more attention than ever to the man who leads the fight against the British occupation," explained the *Phrenological Journal*. The article then proceeded to give a phrenological analysis of Ahmad from the photograph. Damningly, the author concluded that "such a cast of expression naturally belongs to the bigot, or the enthusiast, or the monomaniac."[15] The *Phrenological Magazine* in London made a similar use of photography in the wake of the Anglo-Zulu War of 1879. "We have just received a Zulu skull, as well as some photographs of Zulus, brought from South Africa by a gentleman connected with the missions there," explained the *Phrenological Magazine*. These artifacts—both the skull and the photographs—were products of the violent expansion of the British colonial state in southern Africa. The *Phrenological Magazine* blithely informed its readers that "the skull is that of the brother of the chief Moshesh, who was killed in the late war." The photograph, reproduced as a wood engraving, showed the face of "Kreli, a Kaffir Chief." This formed the basis of a detailed phrenological analysis both of racial character and of racial origins. On the history of the Zulus, the *Phrenological Magazine* claimed that "it is evident that they are not Aborigines, but that they have descended upon Southern Africa from some other locality—probably from more northern parts of the same Continent." Then, turning to the photograph of Kreli, the author argued that "his head is one of the most remarkable I have seen in a savage. With the advantages of civilisation and culture he would rank with first-class men for a general intelligence."[16] In London, the *Phrenological Magazine* weaved a complex phrenological argument based on photographic evidence from the Anglo-Zulu War. The phrenologists claimed that the Zulus did not hold aboriginal rights to the lands, while simultaneously suggesting that only British rule would bring civilization.

This chapter follows phrenological photographs as they traveled back and forth across the imperial world. In doing so, it argues that colonial photography needs to be understood as a material practice incorporating both production and reception. Photographs of colonized people were reprinted in periodicals and books, they were compared alongside collections in libraries and museums, and, most importantly, they traveled in the luggage of anthropologists and colonial officials. As a case study, this chapter considers the series of photographs featured in William Elliot Marshall's *A Phrenologist Amongst the Todas*, published in 1873. The nineteen photographs in this book were originally taken in the Nilgiri Hills, in southern India. The negatives were then sent to Longmans and Green in London, along with the manuscript Marshall completed in Faizabad. Together, the photographs and text document the phrenology of the Todas, a pastoral hill tribe living in the Nilgiris. Marshall's book circulated widely. It was read by many of the most influential evolutionary and anthropological thinkers of the late nineteenth century. Edward Burnett Tylor, author of *Primitive Culture* and the first Reader in Anthropology at Oxford University, reviewed Marshall's book in *Nature*. In Paris, *A Phrenologist Amongst the Todas* was read by both Jean Louis Armand de Quatrefages and Paul Topinard, two of the leading French thinkers on the origins of man. And in 1901 the British anthropologist William Halse Rivers Rivers took a copy of the book with him to India. Further copies reached New Zealand, Australia, and the United States.[17]

To date, historians have treated photography in India as relatively disconnected from the wider world. Pioneering work by Christopher Pinney and John Falconer emphasized the connection between photography and colonialism, but failed to address the broader reach of Indian photography. In part this was because both Pinney and Falconer almost entirely ignored the role of reception.[18] As this chapter shows, the history of photography in India needs to be understood, like the history of colonial photography more generally, as part of a global history of material exchange.[19] It was through circulation and reception that photography and phrenology became intertwined with evolutionary thought and colonial power.[20]

This chapter also broadens our understanding of what constitutes an anthropological or phrenological photograph. What is initially striking about the photographs found in *A Phrenologist Amongst the Todas* is their variety. Marshall's book certainly contains photographs that form part of a familiar ethnographic genre: face, profile, and full-length body shots against a screen. But *A Phrenologist Amongst the Todas* also features photographs of the landscape, architecture, and material culture of the Nilgiri Hills. Furthermore, not all the photographs of the Todas are in such a rigid ethnographic format. Many are posed much more like a European studio photograph; some are vignetted; others contain props. Some include artificial backdrops and screens, while others were taken in situ. Previous historians writing about *A Phrenologist Amongst the Todas* have tended to ignore this variety, instead concentrating on the more obviously anthropological photographs.[21] Yet in order to understand Marshall's work, and its reception, *A Phrenologist Amongst the Todas* needs to be considered as a whole. Each of the photographs, whether of a person, a hill, or a hut, played a role in Marshall's argument. The variety of photographic genres incorporated into *A Phrenologist Amongst the Todas* makes a lot more sense once we recognize that at least three different photographers were involved. Marshall took some of the photographs himself, or at least directly supervised their production. However, a number of other photographs featured in *A Phrenologist Amongst the Todas*, including the frontispiece, were taken by Samuel Bourne, the celebrated colonial photographer with studios in Simla and Calcutta. Additional photographs were taken back in Madras by the local firm Nicholas and Curths. The photographs featured in the final published work were therefore produced using different equipment, under different environmental conditions, and with different artistic and scientific genres in mind. This presented various difficulties, but it also allowed Marshall to weave together a detailed visual argument that linked the body, the environment, and material culture to racial history. *A Phrenologist Amongst the Todas* needs to be understood as a compilation of photographs that, together, form an argument about the evolutionary origins of man.

The Todas had long been the subject of intense ethnological debate. Early travelers to the Nilgiris had suggested that the Todas might be the remnants of a postdiluvian migration, perhaps of Jewish ancestry. But opinions quickly diverged. Others argued that the Todas were the descendants of a forgotten Roman colony, or perhaps of Celtic or even Polynesian origin. At the root of many of these debates was the question of whether the Todas were a primitive example of early man, or a degenerate ancestor of a more advanced race.[22] At a time when a number of leading evolutionary thinkers began to address the question of mankind, the Todas proved an ideal case study. In 1870 John Lubbock, president of the new Anthropological Institute, published *Origin of Civilisation and The Primitive Condition of Man*, arguing that anthropologists could trace the evolution of social institutions such as marriage, kinship, and religion from savage to more advanced forms.[23] Tylor made a similar argument in his 1871 work *Primitive Culture*. In this Tylor suggested that "survivals"—seemingly "meaningless customs"—provided "proofs and examples of an older condition of culture out of which a newer has been evolved." In the same year, Charles Darwin finally broke his silence with the publication of *The Descent of Man*.[24] Not everyone was so enthusiastic. In France under the Third Republic, evolutionary accounts of mankind were viewed with suspicion. Darwin's work seemed little more than a return to the bad old days of Lamarckian materialism.[25] In Britain too, many were still skeptical of evolutionary accounts of man. Before his death in 1868, James Hunt had been a leading critic of Darwin, arguing that anthropologists should ignore origins and concentrate solely on "the anatomical aspects of ethnology."[26] This led to a long-running dispute between the Ethnological Society and the Anthropological Society in London, only partially resolved in 1871 with the foundation of the merged Anthropological Institute.[27]

Marshall was certain that phrenology had something to contribute to these debates. In the preface to his book, he argued that phrenology was of "a marked practical value for ethnological purposes." *A Phrenologist Amongst the Todas*, Marshall suggested, could even be read as "a manual for ethnographers." The core argument centered on the evolutionary origins of the Todas. Marshall presented the Todas as living fossils. By studying the Todas phrenologically, the anthropologist could discover "how man lived in days when he had advanced scarce more than one step from the period of his rude simplicity." The Todas were "a surviving sample of some portion of the Turanian race when in its very primitive stage." Clearly influenced by Tylor's *Primitive Culture*, which he had read in India, Marshall concluded that the Todas were representatives of a "prehistoric era," rather than degenerates of a higher race.[28] *A Phrenologist Amongst the Todas* was therefore part of a growing body of phrenological works that began to address human origins.[29]

Discussion of Toda origins played an important role in the growing field of evolutionary anthropology. In India, these debates also had a more direct political purpose. The Todas were at the center of ongoing land disputes. As early as 1829, a group of Todas had submitted a petition to the East India Company claiming "the undisturbed and undisputed possession of the Neelgherries for time beyond the reach of memory." They demanded to be "recognised as the proprietors of the soil." Over the course of the nineteenth century, these land rights were slowly eroded.[30] Ootacamund, at the center of the Nilgiris, had long been a summer retreat for European officials, and, by 1870, it had become the official summer capital of the Madras Presidency. The growth in European presence in the hills led to further encroachment on Toda lands. By the 1860s, large swathes of the forests had been destroyed and replaced with tea, coffee, and cinchona plantations. Ootacamund itself had grown significantly during the second half of the nineteenth century. By the time Marshall arrived in 1870, Ootacamund incorporated a church, a library, botanical gardens, a club, military barracks, a race course, and over 230 houses.[31] The Government of Madras did pay a nominal form of compensation for the loss of territory, known as *gudu*, but insisted that this did not imply Toda ownership of the land.[32]

Ethnological and evolutionary claims played a decisive role in all these disputes. Like many other hill tribes, the Todas were increasingly portrayed as a "dying race." This attitude allowed colonial officials to confine the Todas to reserves known as *munds*, areas of less valuable land far away from the Ootacamund settlement. Additionally, the Todas were routinely described as "primitive." Indeed, Marshall subtitled his book "The Study of a Primitive Tribe in South India." This fed into a broader ideology of improvement in which the colonial state attempted to transform hill tribes like the Todas from pastoralists into cultivators on the new coffee plantations. Finally, colonial officials themselves also debated the exact origins of the Todas. Some suggested that the Todas were not in fact the original inhabitants of the Nilgiri Hills and therefore did not hold any special rights to the land.[33] *A Phrenologist Amongst the Todas* addressed all these questions. Marshall brought together a range of photographic evidence in order to portray the Todas as a "primitive tribe," one at risk of "dying out," and in great need of "improvement."[34] Marshall's book is therefore illustrative of some of the most significant but understudied developments in the later history of phrenology. It shows how phrenologists harnessed photography in order to insert themselves into new debates about the evolutionary origins of man.[35] It also demonstrates how phrenology, like anthropology more generally, found itself increasingly allied with the growth of the colonial state.[36]

PRIMITIVE PHOTOGRAPHY

Marshall spent almost his entire adult life in India. Born into a military family in 1826, he entered the East India Company Seminary at Addiscombe aged sixteen.[37] The curriculum was intense. All cadets were expected to wear military uniform, and lessons covered mathematics, chemistry, engineering, draughtsmanship, Latin, and Hindustani.[38] There was certainly no opportunity to read about phrenology. Marshall was also at Addiscombe too early to learn anything about photography. Both Louis Daguerre and William Talbot's photographic processes were public knowledge in the 1840s. But a course in photography was not introduced as part of the East India Company Seminary's curriculum until 1855.[39] So after two short years at Addiscombe, Marshall successfully completed his military training and immediately entered the Bengal Staff Corps. He arrived in India on 6 April 1845, returning only once to England in 1861 to marry.[40]

On the subcontinent, Marshall's knowledge broadened significantly. It was in India, not in Britain, that he learned about both phrenology and photography. As previous chapters have illustrated, phrenological books were readily available across India during this period. In Madras, Marshall could have easily picked up a copy of Combe's *Elements of Phrenology* or Orson Fowler's *Synopsis of Phrenology*, both of which he later cited in *A Phrenologist Amongst the Todas*.[41] As an officer in the Bengal Staff Corps, Marshall also often found himself in Calcutta, still a hotbed of phrenological activity. Throughout the 1870s and 1880s, both British and Bengali phrenologists continued to operate in the city, promoting their favored brand of mental science alongside mesmerism.[42]

While phrenology was something Marshall picked up of his own accord, photography formed part of his military career. Having demonstrated a certain degree of mathematical prowess at Addiscombe, Marshall was earmarked for further training as an engineer. In the mid-1850s, after recovering from an injury sustained in the First Anglo-Sikh War, Marshall entered the Thomason College of Civil Engineering at Roorkee in the north. Marshall now learned about photography as part of his training to join the Public Works Department.[43] By this time, photography was deemed an essential aspect of survey work, a complement rather than a replacement for drawing. By the end of the 1850s, photographic equipment was routinely issued to military engineers for this very purpose.[44] Later in the century, Thomason College even offered prizes for photography, further encouraging new recruits to take up the practice.[45] Marshall later deployed all these skills, initially developed as a colonial engineer, to turn himself into a colonial anthropologist.[46]

Marshall's training at Roorkee is indicative of the much broader growth in colonial photography during this period. Photography had reached India quickly, with the *Bombay Times* publishing an early account of the daguerreotype process in 1839 just months after its announcement in Paris.[47] By 1860, photographic societies could be found in Calcutta, Bombay, and Madras.[48] The East India Company, and later the Government of India, also saw the potential for photography as a means of cataloguing a colonized land and its people. In 1854 the Court of Directors appointed Captain Thomas Briggs, a military engineer like Marshall, as "Government Photographer," and charged him with compiling a photographic account of the Bombay Presidency. Similarly, in 1867, just before Marshall arrived in the Nilgiri Hills, the Government of India sanctioned the appointment of Captain Edward Lyon as "General Photographer for the Madras Presidency."[49]

Soon enough, cameras, prints, and negatives were all being exchanged both within India and across the British Empire. Exhibitions, organized by local photographic societies, became an almost annual event in many cities. In 1859, the Madras Photographic Society displayed hundreds of prints sent in by both local members and the Photographic Society of Bengal. In 1860, the Madras Photographic Society held another exhibition, this time featuring photographs from beyond India, including photographs taken in Aden, China, Australia, and Egypt. Exhibitions like these allowed colonial photographers to bring together the world in one room. In doing so, they expressed the desire, if not the reality, of an expanding and absolute imperial power. But photographs could also document the limits of empire. The same exhibition in Madras also featured photographs from the Indian Rebellion of 1857. In this case, photographs of the Siege of Delhi, as well as those of the execution of mutineers, served as a powerful reminder of both the need for continued military intervention and the potential fragility of British power in India.[50] No work illustrates this attitude of insecurity better than *The People of India*. Published over eight volumes between 1868 and 1875, this monumental atlas contained over 400 ethnographic photographs of Indian tribes and castes. At the time when Marshall was working on the Todas, a number of volumes had already been published, although the volume on the Madras Presidency did not reach the press until 1875.[51] In the aftermath of the Indian Rebellion of 1857, *The People of India* quickly turned into an exercise in "imperial surveillance." Its publication was organized by the Political and Secret Department, and the photographs were used to document the loyalty of individual tribes and castes. Some groups such as the Jats, *The People of India* suggested, could be trusted and would remain loyal to the British during any future outbreak. Other groups, such as the Pachadas, were viewed with suspicion because of their apparent disloyalty during the events of 1857.[52]

A Phrenologist Amongst the Todas was a product of this world, in which both colonial officials and the colonial state made increasing use of photography. In the preface, Marshall said the work had been undertaken during a "furlough I took in the year 1870 to the Madras sanatorium."[53] This was not strictly true. Colonial archives reveal that Marshall's presence in the Nilgiri Hills was no accident—he was not simply there for rest and recuperation. Rather, Marshall was sent by direct order of the colonial government. His service record states that he was "appointed to the Home Department on an Ethnological duty."[54] It is also clear that Marshall was granted privileged access to the records of the colonial state to aid his study. The secretary to the Government of Madras, Alexander Arbuthnot, allowed Marshall to examine "the records existing in the Revenue Department, on the subject of Toda infanticide." Additionally, Marshall carried out a detailed census of the Todas, for the benefit of the colonial government, at precisely the same time as he undertook his phrenological work.[55] This largely unspoken but close connection between phrenology and the colonial state should be kept in mind when examining the photographs.

For Marshall, photography was particularly useful in documenting the shape of the skull. *A Phrenologist Amongst the Todas* contains four photographs of Toda heads labeled "Male Profile," "Male Full Face," "Female Profile," and "Female Full Face" (fig. 31). These photographs were particularly valued, as they provided the only means of assessing Toda craniology. "As the practice of cremation is the universal method of disposing of the dead . . . a skull without flesh could by no manner of chance be obtained," Marshall explained.[56] In each case the subject was placed against

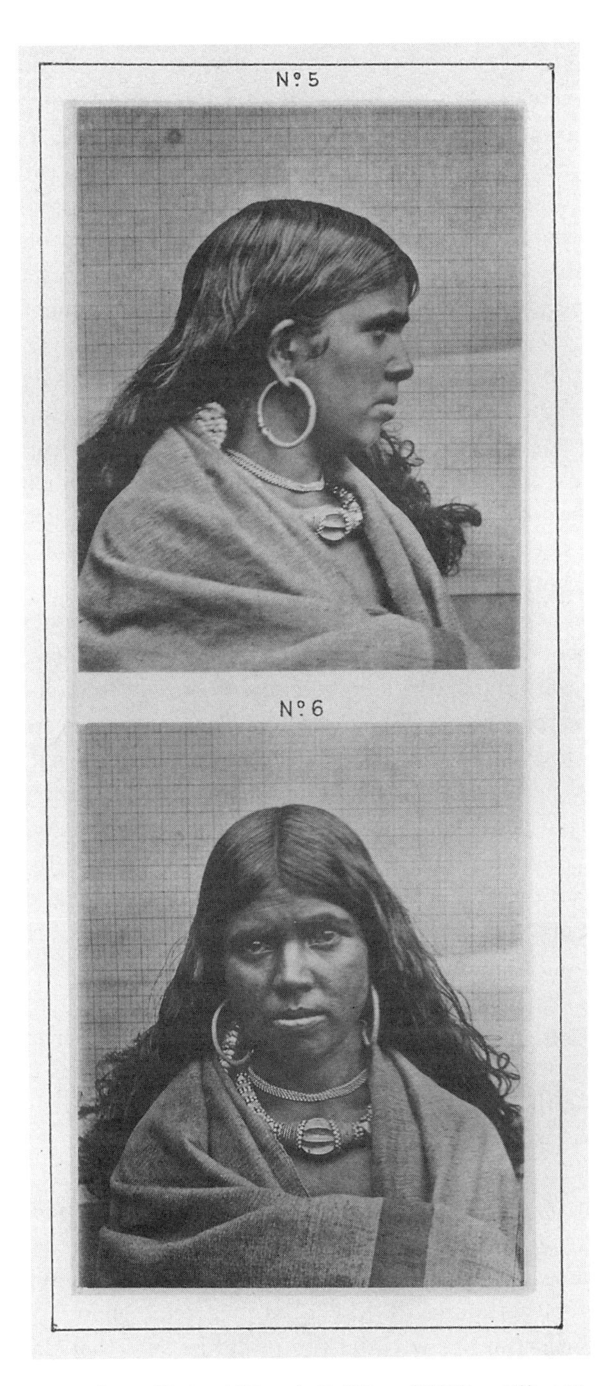

Fig. 31. "Female Profile" and "Female Full Face," William Elliot Marshall, *A Phrenologist Amongst the Todas* (London: Longmans and Green, 1873), nos. 5 and 6. Author's collection.

a checkered screen. This screen was designed to complement the separate phrenological measurements that Marshall undertook with a pair of calipers. The average Toda man measured 5.41 inches between the two organs of Destructiveness on either side of the head. He measured 7.39 inches from the occipital spine at the back of the head to the phrenological organ of Individuality at the front.[57] Together, these measurements and the photographs were advanced in support of Marshall's core argument about Toda phrenology: the Todas were "without individual exception, narrow-long-headed." Significantly, Marshall understood the ratio between the width and length of the head as a direct measure of civilization. In *A Phrenologist Amongst the Todas*, he divided mankind into two basic phrenological categories: "brachycephalic," or wide-headed, and "dolichocephalic," or narrow-headed. The Todas, Marshall argued, were "deficient in every organ at the sides of the skull." This was "prominent physiologic evidence of extreme primitiveness in condition of race."[58]

The photographs were a crucial part of Marshall's argument. The face and profile shots were intended to suggest that the Todas possessed narrow heads. However, Marshall knew he couldn't simply rely on his readers' trust in either phrenology or photography. Phrenology was still routinely parodied in the 1870s, while the Victorian public were well aware that photographs could easily be faked or retouched.[59] So at a time when both phrenology and photography were open to criticism, Marshall suggested that the two sciences could be used to support one another. To begin with, he argued that the photographs backed up his phrenological conclusions. Marshall invited his readers to compare his description of the Todas as dolichocephalic with the photographs:

> Two portraits—in side and full face—of a man and a woman, photographed to scale, are here given, by which some means are afforded of comparing the descriptions which are given in these pages with examples of "real life."[60]

This way the reader could convince themselves of the phrenological account of the narrow-headed Toda. The argument also worked the other way. Phrenology could help bolster the credibility of the photographs. One of the big problems Marshall, like all ethnographic photographers, faced was that of selection. How could the reader be sure that the particular Toda man and woman in the photographs were representative of an entire tribe? The narrow heads they saw might be anomalous rather than typical. Marshall could have picked them to fit his argument. This was clearly something Marshall himself worried about, writing that "the male subject in the picture was selected solely on account of being one of the baldest men of the tribe." (The bald head, Marshall noted, made it easier to trace the outlines of the skull against the screen.)[61] Crucially, Marshall's own phrenological account of the Todas provided a solution to this problem. He argued that primitive races possessed skulls of "extreme simplicity and uniformity." In contrast, "civilised people . . . like the English" showed "an apparently endless variety of combinations of organs and temperaments."[62] According to Marshall, phrenological photography was therefore especially suited to the study of "savage" races. It was the "unvarying type of the Toda head, and the extraordinary uniformity of its general size," that ensured Marshall's photographs were representative. Revealing his attitude to those he considered less civilized than himself, Marshall wrote that the Todas "present scarcely more differences in appearance and character than any one dog does from any other of the same kennel of hounds."[63] Here, Marshall completed the reciprocal connection between phrenology and photography. On the one hand, the photographs corroborated a phrenological account of the Todas as "dolichocephalic." On the other hand, it was that same phrenological description of the Todas as "primitive" that solved the very problem of photographic selection. All individuals of "savage" races were essentially alike.

Marshall was clearly sensitive to the difficulties anthropologists faced in turning photographs into reliable specimens.[64] His use of a divided screen behind the face and profile shots also indicates he was aware of some of the latest advice emanating from London. In 1869 the anatomist and champion of evolution Thomas Huxley sent a circular to all colonial governors requesting a series of photographs of native peoples. For Huxley, anthropometric photography suffered because of a lack of standardization.[65] "Great numbers of Ethnological Photographs already exist, but they lose much of their value from not being taken upon a well-considered and uniform plan," Huxley complained. The circular therefore set out a standardized photographic practice. Huxley suggested that "photographs of the head should be so taken as to give an exact full face, and an exact profile of each head." Photographs of an "average adult male and of an average adult female" were sought. Most importantly, the photographer needed to provide "a scale, divided to inches and tenths."[66] A similar plan was also promoted by John Lamprey at the Ethnological Society that same year. Like Huxley, Lamprey noted that "collectors of photographs illustrative of the races of man, have experienced the greatest difficulty in questions of comparison of measurement of individuals by some common standard." He therefore suggested that all colonial photographers make use of a screen divided "by longitudinal and latitudinal lines into squares of two inches every way."[67]

Marshall seems to have taken this advice seriously. The face and profile photographs featured in *A Phrenologist Amongst the Todas* mirror Huxley's scheme very closely. Still, getting the right shot was easier said than done. Phrenological photography in the colonial field required preparation and skill.[68] Manuals on photography in India, such as F. Fisk Williams's *Guide to the Indian Photographer*, set out the difficulties in detail. The environment itself posed a challenge. Williams explained how the "high temperature has a great influence on the various chemicals used, and, in fact, in many cases destroys their action altogether." Selecting good quality equipment was therefore imperative. Williams recommended a camera manufactured by Thomas Ottewill in London (fig. 32). He found it could "withstand this climate far better than those of any other maker" owing to its construction out of teak.[69] Marshall didn't specify which camera he used, but Ottewill's models were widely available in the presidency capitals. His "Improved Sliding-Body Camera" was available in Calcutta for 220 rupees—expensive, but worth the price.[70] Williams reminded his read-

Fig. 32. Ottewill Folding Box Camera, 1990-5036/7299, Science and Society Picture Library, National Science and Media Museum, Bradford.

ers that "a *cheap* article is perfectly useless." "If the wood is not *perfectly* seasoned, the high temperature to which it is exposed will invariable distory [*sic*] it in a very short period," he explained.[71]

A good photographer knew which lens to use depending on the desired shot. For portraits like those of the Toda face and profile, colonial manuals recommended "Andrew Ross's Portrait Lens," which cost 175 rupees in Calcutta at the time.[72] Using a high-quality lens like this was particularly important for Marshall's purposes, as it helped to reduce spherical aberration. As the Royal Geographical Society's *Hints to Travellers* explained, photographers should opt for a "doublet lens," which "gives perfectly straight lines."[73] A single lens was more likely to introduce aberrations, distorting Marshall's checkered screen. As it turns out, the lens Marshall chose for his face and profile photographs clearly wasn't entirely suitable. All four photographs are distorted, with the ruled lines on the screen appearing curved as they approach the edge of the image. This is characteristic of spherical aberration when taking a photograph using a single lens at an increased focal length. This problem was compounded by the fact that the screen had clearly been damaged in transit, with a large crease visible right across the middle. As a number of reviewers later noted, this rendered the screen almost completely useless as a means of measurement.[74]

The screen was probably the least of Marshall's worries. The valuable camera and lens also had to be carefully packed in order to protect them during the long journey from Madras to the Nilgiri Hills. Most significantly, all the chemicals required to prepare and develop the negative needed to be taken into the field. At this time, Marshall would have been using the collodion wet-plate process developed by Frederick Arthur in 1851. This technique, which largely displaced both the daguerreotype and Talbot's calotype in India by the end of the 1850s, produced sharp images on glass with exposure times over seconds rather than minutes. However, there was one major drawback. The whole process, from preparing the collodion plate to fixing the negative, needed to be performed within around fifteen minutes. It therefore had to be done in the field. First, a solution of collodion (gun cotton) and potassium iodide was poured over a glass plate. Within seconds, this solution would set. The collodion plate then needed to be made light-sensitive. This was achieved by taking the plate into a darkroom and placing it in a bath of silver nitrate. The plate then needed to be set in the camera for exposure while it was still wet. Finally, the exposed plate would be developed and fixed, again in a darkroom, using a combination of chemicals including ferrous sulphate and potassium cyanide. All of Marshall's photographs, including those of the Toda man and woman discussed earlier, would have been developed like this in situ.[75]

A portable darkroom was an absolute necessity. Usually, this was in the form of a simple tent, although various firms promoted specially designed equipment for the colonies.[76] In the 1850s William Rouch advertised a "Portable Dark Operating Chamber," which he suggested would be "preferable to any of the numerous forms of dark tent, all of which are complicated and cumbersome" (fig. 33).[77] On top of this, Marshall would have needed a ready supply of chemicals in advance. Silver nitrate would not have been easy to procure in Ootacamund. In Calcutta and Madras, however, photographers could purchase good-quality chemicals imported from London. If they wanted to save money, locally produced chemicals could also be found in the bazaar.[78]

In total the combined weight of Marshall's camera, lens, plates, chemicals, and darkroom tent would have exceeded 100 pounds.[79] Given that the nearest railway station to Ootacamund was at Mettupalayam, at the foot of the Nilgiri Hills, it is certain that Marshall relied on Indian labor to carry his equipment.[80] Manuals on colonial photography even noted that "in India, porterage is cheap" and should be taken advantage of.[81] Bourne, the photographer responsible for the frontispiece in *A Phrenologist Amongst the Todas*, had an entourage of over fifty people during his expeditions. The very process of taking a wet-plate collodion photograph also typically relied on an assistant, to prepare the plate in the darkroom between shots.[82] All of these technical details are absent in both Marshall's text and in the existing secondary literature on *A Phrenologist Amongst the Todas*. But these details are important. They provide another example of the ways in which photography in India was connected to the wider world. Marshall's photographs certainly circulated widely. Yet, even before publication, the act of making a single negative in the Nilgiri Hills relied on cameras from London, chemicals from Calcutta, and laborers from Madras. What appears as a straightforward photograph of a woman's face is actually an example of global material exchange in action.

HUTS, HILLS, AND HEADS

A Phrenologist Amongst the Todas didn't simply document primitive craniology. It also linked skull configuration to the Indian environment.[83] Like many phrenologists in the late nineteenth century, Marshall was interested in evolution and "natural selection," a term he often used himself. The aim of *A Phrenologist Amongst the Todas* was to uncover the "nature

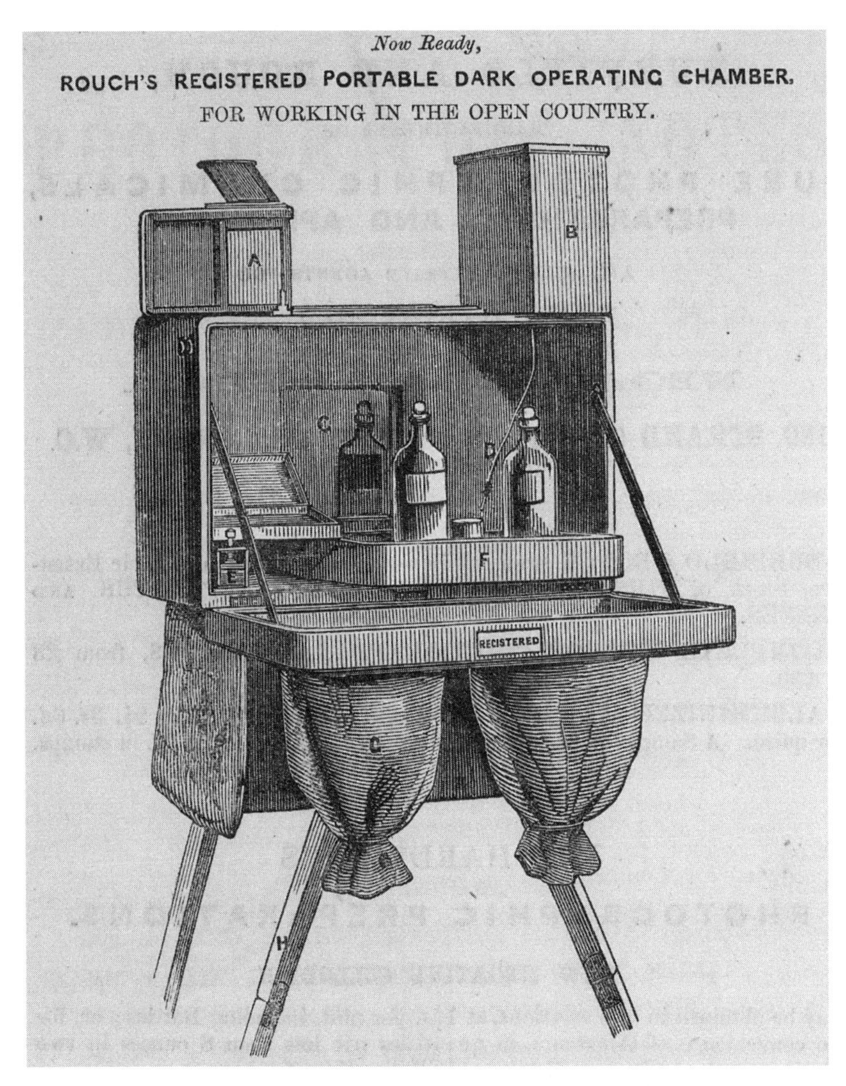

Fig. 33. "Rouch's Registered Portable Dark Operating Chamber," Thomas Sutton, *A Dictionary of Photography* (London: S. Low and Son, 1858). Reproduced by kind permission of the Syndics of Cambridge University Library.

of selection operating on men's characters." Crucially, Marshall presented the Nilgiri Hills as an environment in which natural selection had failed to operate. By "avoiding conflict with nature and man, in the seclusion of the sequestered jungles of warm climates," tribes like the Todas had failed to make any evolutionary progress. The Todas had not been subject to "the vivifying effects of such processes of natural selection as tend to eliminate the weak-minded and the weakly, and produce brachycephalic-headed and broad-shouldered men." Instead, the "Toda cranium" illustrated "what man—as a race—uninfluenced by selection, and living an open-air life, tends to revert to."[84] Here, Marshall made a direct link between phrenology, evolution, and the environment. Marshall's photographs of the man and woman against the checkered screen therefore need to be read, as they were in the 1870s, alongside the accompanying images of the landscape.

Marshall couldn't complete his visual argument alone. For photographs of the Indian environment, he turned to Bourne, an experienced landscape photographer based in the northern hill station of Simla. Bourne arrived in India in 1863 and soon set off to photograph the Himalayas. The images he produced, along with the accompanying narrative published in the *British Journal of Photography*, made him famous. He was celebrated for bringing the remote Himalayas to a wider audience, with cartes de visite of Bourne's photographs circulating both in Britain and in India. Sensing an opportunity for commercial success, Bourne teamed up with another talented photographer, Charles Shepherd, and opened studios in Simla, Calcutta, and Bombay. Over the next seven years, Bourne documented the Indian environment, ranging from Calcutta and Kashmir to Delhi and Lucknow. In 1869, Bourne arrived in the Nilgiri Hills to undertake what would be his final photographic expedition in India.[85]

Unlike Marshall, Bourne was not in the Nilgiri Hills as part of an ethnographic project organized by the colonial state. His photographs were produced for commercial purposes with an international audience in mind. They served two purposes. For those who had spent time in India, Bourne's photographs acted as mementos. Photographs of Ootacamund allowed colonial officers to look back on happy times spent in the hills. These photographs, which were sold in lavish folio albums, acted as advertisements for the hill stations and the British Raj more generally.[86] Back in Britain, Bourne's photographs were popular, as they squeezed India into existing artistic conventions. The "picturesque" landscape of the Nilgiris, with its rolling hills and trees, could then be compared to that of the Lake District, where Bourne had in fact first learned how to use a camera. Bourne himself conceived of photography as part of a global documentary project, writing in 1863 that "there is now scarcely a nook or corner, a glen or valley, or mountain, much less a country, on the face of the globe which the penetrating eye of the camera has not searched."[87]

Marshall wanted to link the Indian environment to phrenology and natural selection. With this in mind, he chose two images from Bourne's commercial collection to include alongside his own photographs in *A Phrenologist Amongst the Todas*. The first, the frontispiece, depicts a group of Toda men and women in front of their homes (fig. 34). The second, taken from a vantage point, depicts the Nilgiri Hills extending into the distance (fig. 35). Both of these images were available to purchase at Bourne and Shepherd's studios. Marshall most likely selected them during one of his summer retreats to Simla with his family.[88] The use of commercial photographs in scientific work was not unusual at the time. Darwin collected photographs produced by the London Stereoscopic Company for his studies of emotion, while anthropologists regularly exchanged commercial postcards taken of living people at international exhibitions.[89] Commercial photography actually solved a number of problems for Marshall. For a start, taking a photograph of the landscape required very different skills and equipment. Colonial manuals recommended "Ross's new Orthographic Petzval Lens" with its wide-angle focal length.[90] Landscape photographers also had to work in the open air. This created various challenges. Williams, in his *Guide to the Indian Photographer*, reminded his readers that "the fearful amount of dust floating in the atmosphere the greater part of the year, renders it difficult, if not impossible, to obtain

Fig. 34. "A Toda Mand," William Elliot Marshall, *A Phrenologist Amongst the Todas* (London: Longmans and Green, 1873), no. 1. Author's collection.

Fig. 35. "The Land He Lives In," William Elliot Marshall, *A Phrenologist Amongst the Todas* (London: Longmans and Green, 1873), no. 10. Author's collection.

pictures free from spots and blemishes."[91] The photographer was recommended "to keep his camera most carefully clean, wiping the lenses very frequently."[92] The camera itself needed to be covered up between shots. "It must not be carelessly exposed to an Indian sun without protection," Williams warned.[93] Mountain photography, Bourne's specialty, was a particular challenge. "For mountainous country it is of great advantage to have a stand with sliding legs," the Royal Geographical Society's *Hints to Travellers* explained. The same manual even recommended carrying "a small pocket level" to check the alignment.[94] Still, even after careful preparation, "the slightest tremor cannot fail to spoil every picture," Williams lamented.[95] Finally, there was the problem of exposure. Taking a photograph of a person was relatively easy. But getting the right balance of exposure between the sky and mountains was an art. "The time of exposure must be estimated according to circumstances, and it requires considerable experience to judge of it accurately," explained *Hints to Travellers*.[96]

Bourne knew what he was doing, and his photographs stood out for their clarity and composition. Still, the use of commercial photography served another purpose. Regardless of technical qualities, no photograph was seen as straightforwardly objective in the nineteenth century.[97] Instead, photographers relied on existing artistic conventions for authority and authenticity. As with paintings, viewers understood the authenticity of an image to be linked to the authority of the person who had produced it. Commercial photography was therefore particularly powerful. Anthropologists could claim independence from the images while simultaneously invoking the authority of a well-known firm or photographer.[98] This served Marshall's purposes well. He was already worried about claims of composing photographs simply to fit his argument. The two images he chose from Bourne's collection didn't suffer from this problem. They were an accepted part of the iconography of the Nilgiri Hills, appearing in expensive albums sold in India and Britain.[99] Bourne's was also a household name.[100] Marshall made sure to credit Bourne in *A Phrenologist Amongst the Todas*. The two images he chose appear above the words "A Photograph by Messrs Bourne & Shepherd, Simla." Bourne's signature, etched onto the negative, is still visible in the bottom right corner of the frontispiece. Even though it would have been easy to remove this during the process of printing, Marshall evidently chose to leave the signature there, hoping it would add further to the authority of the image.

Bourne's landscape photograph—titled "The Land He Lives In. The Nilagiri Plateau"—acted as a pivot in Marshall's argument. It provided the missing link between phrenology and natural selection. The photograph depicts an uninhabited landscape, extending indefinitely into the distance. Marshall explained in the accompanying text that the Nilgiri Hills were an "evergreen surface . . . one continued intermixture of rounded hills, with tracts of rolling prairie." "Picture it an island in a tropical archipelago," Marshall suggested.[101] Bourne's photograph was significant because it provided a visual argument to accompany the one in the text.[102] Marshall knew that many of his readers would never have seen India themselves, much less the Nilgiri Hills. The photograph was intended to convince them that this really was an isolated environment, free from the pressures of competition and selection. Bourne's composition aided in this. Most landscape photographers set the focus at the horizon, ensuring the whole image was sharp. But in this photograph, Bourne decided to keep the foreground in focus, while the mountains in the background are blurred. This added to the impression of distance and isolation, one hill after another continuing on farther than the eye could see. It is important to remember that while Marshall didn't take the photograph, he did play an active role in selecting it for his book. Bourne's studio sold plenty of other prints of the Nilgiri Hills. However, many of these would not have fit Marshall's argument. One of Bourne's photographs of Ootacamund, which appears in the same commercial album as the photographs Marshall selected, shows the hills as they in fact were: occupied and well connected (fig. 36). Houses, government offices, crops, and, most revealingly, a road suggest that the Nilgiris were not in fact as isolated as Marshall wanted his readers to believe. The very fact that Marshall was able to use Bourne's images in his book is itself a testament to the distributed nature of photography in India at this time. *A Phrenologist Amongst the Todas* was a product of a colonial photographic network that extended from Simla in the north to the Nilgiris in the south.

The second image Marshall selected was more unusual, at least for Bourne. It depicted a group of Toda men and women in front of their homes. Bourne rarely took photographs of people, except in order to aid his composition.[103] Yet in this photograph the Todas are most certainly the subject. Indeed, this may have been why Marshall selected it in the first place. This image also served to emphasize the connection between the political goals of the colonial state and Marshall's evolutionary ideas. The Todas were routinely described as "idle." This was chiefly because they did not cultivate the soil. Instead, the Todas sustained themselves

Fig. 36. Samuel Bourne, "26. Panoramic View from a Spur of Dodabet," GBR/0115/Y3022-E, Photographs of India, Royal Commonwealth Society Archives. Reproduced by kind permission of the Syndics of Cambridge University Library.

by raising cattle and trading dairy products. Marshall, like the colonial state more generally, believed the Todas were holding back the development of the Nilgiri Hills: they needed to be turned into cultivators.[104] His evolutionary argument about the lack of selection explained both the problem and the solution. "I will endeavour to account for this persistency in idleness, of primitive and unprogressive races," wrote Marshall. He returned to his previous argument about the environment. The Todas had not been subject to natural selection. They lived in an isolated paradise. "Why should he work? Why should he plough?" Marshall asked rhetorically. The answer was that without competition the Todas had no need to become cultivators. And, according to Marshall, this had held back their evolutionary and phrenological development. "Civilisation owes to labour her first impulses. It gives to nature the opportunity of selection, which the Todas have avoided," he explained. Still, all was not lost. By bringing cash cropping and industry to the Nilgiri Hills, the Todas would be forced to adapt or die. "I see no reason why, if caught young, he should not prove as intelligent and as useful a member of society as the humble Ryot of India," argued Marshall. Together, Bourne's two photographs were intended to illustrate this argument. The first depicted the Nilgiri Hills as a lush environment, isolated from the wider world. The second depicted a group of Todas sitting idly around their camp. The frontispiece, Marshall explained, showed "an old-looking and water-stained log hut, belonging to a primæval tribe of the glacial period." "Their present mode of life precisely suits the constitution of their minds," he concluded.[105] Bourne's photographs of hills and huts were now part of a phrenological argument.

CARBON COPIES

Back in Britain, Longmans and Green began preparing *A Phrenologist Amongst the Todas* for publication. Choosing the right printing technology for the photographs was crucial. Longmans needed to find the right balance between cost and quality, particularly as much of the advertising for Marshall's work centered on the visuals. In the end, the photographs were printed using the carbon process, patented and marketed as "Autotype" in Britain. This process, invented in 1864 by Joseph Swan, produced extremely sharp images that did not degrade over time. This technology differed from the more common albumin process, as the image was formed from a mixture of carbon and gelatin. (In an albumin print, the image is composed of sodium and silver salts.) By the early 1870s, Swan had sold the rights to a large firm, the Autotype Printing and Publishing

Company. Operating from a factory in Ealing, the Autotype Company produced photographic materials and prints on an industrial scale for publishers like Longmans.[106]

The Autotype Company's marketing material emphasized the scientific and artistic merits of the process. Autotype prints were sold as "absolutely permanent," making them particularly suitable for works that might be read in the colonial world. The Autotype Company also emphasized the authority associated with a carbon print. Autotypes could be considered a "fac-similie reproduction" of the original image. They allowed for "the reproduction of an artist's work in Monochrome without the intervention of another hand or eye."[107] Given Marshall's concerns about authority and authenticity, this kind of rhetoric was certainly appealing. In any case, Autotype's marketing seemed to work, and the carbon process was quickly put to a diversity of uses, ranging from printing Julia Margaret Cameron's photographic portraits to the reproduction of Renaissance paintings.[108] In essence, if you could take a photograph of it, you could Autotype it. This versatility suited the visual argument of *A Phrenologist Amongst the Todas*. Marshall's photographs of the people and landscape of the Nilgiris could easily be reproduced from glass negatives, so long as they survived the journey from India to Britain. But it was also possible to print Marshall's sketches and maps using the same technology, just by taking a photograph of the document. This gave the final printed work a coherence in which all of Marshall's distinct media came together in a single format. Pencil sketches, watercolors, and wet-plate photographs could all be printed on the same carbon paper. The Autotype process therefore served as a technique for erasing differences in reproduction, in terms of both practice and location. Whatever their history or content, each of the illustrations in *A Phrenologist Amongst the Todas* appeared with the same words in the bottom right-hand corner: "Autotype, London."

With the finished work printed and bound, Longmans began marketing *A Phrenologist Amongst the Todas* to an audience of professional and learned men in Britain.[109] Pricing the work at a relatively expensive 21s. in cloth, Longmans made sure to promote the quality of the "Permanent Illustrations by the Autotype Process." The publisher also hoped to secure a relatively broad readership in terms of disciplinary backgrounds. Marshall's book, suggested an advertisement in the *Athenaeum*, would be "much appreciated by many classes of readers: phrenologists, ethnographers, and philologists."[110] The timing of publication helped. In Britain, the early 1870s witnessed the simultaneous publication of two major evolutionary works. In 1871 Tylor published *Primitive Culture*, and Darwin published *The Descent of Man*, each advancing subtly different accounts of how natural selection could apply to humans.[111] In this setting, *A Phrenologist Amongst the Todas* was read as a contribution to debates about the evolution of man. The authority of Marshall's work was judged according to different evolutionary standpoints. What's more, Marshall's photographs were co-opted and criticized in ways he did not originally envision.

Tylor was an early and critical reader of *A Phrenologist Amongst the Todas*, penning a review for *Nature* in December 1873. At first glance, Marshall's book had a lot to offer. At this time, Tylor was attempting to define culture in evolutionary terms. Many of the arguments contained within *Primitive Culture* implicitly relied on a materialist understanding of the mind, something that was only gradually coming to be accepted in Britain thanks to the popularity of earlier phrenological works. Tylor talked in terms of "mental evolution" and the "laws of the mind." It was "brain-power" that transformed "primitive man" into "civilized man."[112] With the foundation of the new Anthropological Institute in 1871, Tylor was keen to establish his own brand of cultural evolution.[113] This intellectual climate colored Tylor's reading of *A Phrenologist Amongst the Todas*. "Colonel Marshall calls the Todas a 'primitive tribe,'" explained Tylor. However, Tylor then went on to suggest that "care must be taken . . . to interpret with proper reservation the word 'primitive,' as used in these inquiries." Contradicting Marshall's account, Tylor argued that "the Todas are by no means primitive as representing the earliest known grades of civilisation: they are not savages." According to Tylor, the Todas were not, as Marshall had claimed, an example of "the earliest races of man." Instead, the Todas were "a pastoral tribe much above savagery." The "curious customs" of the Todas, Tylor argued, were not evidence of a primitive condition. Putting his own evolutionary theory to use, Tylor argued that Toda culture could be explained in terms of "the principle of 'survival in culture' . . . , relics of a former condition of the race different from the present."[114]

Tylor didn't want readers confusing Marshall's phrenology with his own work on primitive culture. He was therefore quick to distinguish between what he saw as legitimate anthropological work and dubious phrenological theorizing. "Several points of ethnology will be perceptibly advanced by it," Tylor wrote, "notwithstanding much of the theoretical part of the book which will hardly meet with acceptance."[115] Others were similarly skeptical. The *Examiner*, not as radical as it once was, complained of the "space and importance given to phrenology," while Robert Hardwicke's *Popular Science Review* argued that Marshall had "excluded a good many who simply regard the work as phrenological and therefore absurd."[116] In his review,

Tylor even suggested that "it is to be feared that the title of Col. Marshall's volume may prevent its having all the popularity it deserves."[117] Most publishers followed the reviews, and Longmans was listening. Shortly after the publication of the first edition, Longmans reissued Marshall's book under a new title: *Travels Amongst the Todas*. The text was identical, including the chapter on phrenology. Longmans had simply printed a new title page, hoping to disguise the phrenological content, before binding up the remaining stock. As with *Crania Americana* earlier in the century, books like *A Phrenologist Amongst the Todas* could be reworked, even after publication, to serve different disciplinary ends.

Tylor's criticism also extended to the photographs. He described the illustrations as "beautiful autotypes." But, because of Marshall's phrenological approach, they turned out to be of little use. "It appears from his description, that the Todas are a uniformly long-skulled race," Tylor explained, before going on to complain, "though, among his dimensions, I fail to find anywhere the actual measurements of cranial length and breadth." Tylor was right. *A Phrenologist Amongst the Todas* contained no traditional anatomical measurements. Instead, all the tables were compiled as the distances between different phrenological organs. Tylor complained that Marshall adopted "the now discredited phrenological system of bumps and organs, and tabulates a series of Toda skulls according to their Concentrativeness, Amativeness, Veneration, &c." To make matters worse, the photographs themselves were ambiguous. Tylor could "only guess from the portraits . . . that the proportions of these two diameters may perhaps be something like 100:72 or 75."[118] This was hardly the kind of objective anthropometric measure imagined by Huxley or indeed many historians of colonial photography today. The differences in focal length and composition, as well as the damage to the screen, rendered Marshall's photographs largely useless as a source of numerical data.

There was also a geographical element to Tylor's criticism. "His preface is dated from Faizabad, and he describes himself as 'a solitary Indian, far away from contact with men of science, but fresh from the actual and impressive presence of 'Nature's children.'"[119] Marshall himself had promoted the value of detached fieldwork, and the authority that came with having traveled to the Nilgiri Hills himself. In the preface, Marshall explicitly stated that "I have actually witnessed most of the scenes here described." The photographs were intended as further evidence of this fact, one even featuring Marshall's guide, titled "My Informant."[120] For Marshall, removing himself from the metropolitan elite in London was a measure of authenticity. This did not go down well with Tylor. According to him, Marshall had been reckless in ignoring the disciplinary community in Britain. After having discussed the lack of anthropometric measurements as well as the photographs, Tylor concluded that "this is one of the points which make the reader regret that Col. Marshall did not keep his book waiting till he could bring his opinions under discussion at the Anthropological Institute or the Asiatic Society." This, Tylor argued, "might have led him to modify his views in several ways."[121] When Marshall presented a copy of *A Phrenologist Amongst the Todas* to the Anthropological Institute at the end of 1873, this was simply too little too late.[122]

Not everyone was as critical as Tylor. Alfred Russel Wallace also wrote a review of *A Phrenologist Amongst the Todas*, published anonymously in the *Athenaeum* in November 1873. Wallace was much more sympathetic to Marshall's phrenological leanings, having dabbled in Gall's doctrines as a young man.[123] He studiously avoided this aspect of the book in his review, simply describing Marshall's work as "the best and most exhaustive monograph that has appeared upon the subject." Marshall had done "an important service to anthropology," Wallace explained.[124] And as someone who placed special importance on the environment in shaping evolutionary history, Wallace appreciated Marshall's description of the Nilgiri Hills as an "archipelago." Indeed, it was during his time in the Dutch East Indies that Wallace had proposed a geographical explanation for the differences between Malayan and Papuan peoples.[125] This was an argument that could clearly be applied to the Nilgiri Hills. "There is scarcely another example of a race so completely isolated as the Tudas," Wallace explained enthusiastically in his review. Crucially, unlike Tylor, Wallace did not quibble with regard to the photographic evidence. Marshall's book, Wallace noted, contained "excellent photographs, which convey an accurate idea of their appearance." *A Phrenologist Amongst the Todas*, with its photographs of the hills and huts, was "admirably illustrated."[126]

Wallace wasn't just being polite. Later in the century, he actually returned to these photographs in order to support some of his own evolutionary arguments. In the 1890s, Wallace was editing a collection of essays, published as *Studies Scientific and Social* in 1900. Looking over one of his own articles on the evolution of mankind from the 1870s, Wallace realised that *A Phrenologist Amongst the Todas* provided the visual evidence he had been lacking at the time. His argument rested on the idea that there were "several fragments of primitive Caucasian peoples to be still found in South-Eastern Asia." This fit with Wallace's view that all mankind, including Europeans, shared a common ancestor.[127] In support of this argument, he compared three photographs: the first from Baldwin

Spencer's *The Native Tribes of Central Australia*; the second from Romyn Hitchcock's *The Ainos of Yezo*; and the third from Marshall's *A Phrenologist Amongst the Todas*. Placing these three prints next to one another, Wallace pointed to the similarities. The Ainus "are treated as barbarians and inferiors by their Japanese rulers," Wallace explained.

> Yet it is impossible to avoid a suspicion that they are fully the equals of their masters, when we see the refinement and beauty of their best representatives as shown in the photograph here reproduced (Fig. 81), a type of which is very similar to the man figured, with his wife, in Captain Marshall's *Phrenologist among the Todas*—a tribe in southern India—under the title "Adam and Eve."

This, Wallace concluded, provided evidence that "some of the savage and semi-civilised peoples of southern and eastern Asia really belong to the Caucasian type."[128] Marshall's book had originally been intended as a close study of a single tribe within an isolated geographical region. Back in Britain in the 1890s, those same photographs were redeployed by Wallace as part of a global evolutionary history of mankind, stretching from India across East Asia to Japan, then down the Malay Peninsula and into central Australia. *A Phrenologist Amongst the Todas* was now part of a comparative project.[129] In the reading room of the Anthropological Institute, Wallace could gaze upon the faces of Ainus, Todas, and Aboriginal Australians (figs. 37 and 38). It was therefore through reception that Marshall's phrenological photography found uses within broader evolutionary narratives.

PHRENOLOGY AND THE THIRD REPUBLIC

Darwin was not a popular man in Paris. Although at the end of the eighteenth century, France had been a hotbed of materialist philosophy, by the end of the nineteenth, evolution was considered suspect. By the time of the Franco-Prussian War in 1870, *transformisme*, as the French called it, was typically associated with unwanted radical politics and Lamarckism. The conservative tendencies of the Catholic Church, alongside successive governments wary of revolution, ensured that *On the Origin of Species* was largely ignored. Huxley complained of the "conspiracy of silence" surrounding evolution in France during the 1860s. This was somewhat unfair. French naturalists were certainly sensitive to the political repercussions of advancing transformist theories. But many also simply saw Darwin's

Fig. 37. "Aino Man, Urap," Alfred Russel Wallace, *Studies Scientific and Social* (London: Macmillan and Co., 1900), vol. 1, fig. 81, Wellcome Collection, CC-BY.

ideas as old news. In 1860, the *Revue Européenne* described *On the Origin of Species* as "purely and simply a reproduction of Lamarck's system." Evolution by natural selection did not seem a particularly new or promising theory in late nineteenth-century France. Darwin himself was repeatedly rebuffed by leading French thinkers. Following the publication of the

Fig. 38. "Adam and Eve," William Elliot Marshall, *A Phrenologist Amongst the Todas* (London: Longmans and Green, 1873), no. 16. Author's collection.

first French edition of *On the Origin of Species* in 1862, the physiologist Marie-Jean-Pierre Flourens composed a damning response. In *Examen du Livre de M. Darwin,* Flourens attacked Darwin on his use of metaphorical language and failure to accurately define species. For Flourens, there could be no spontaneous generation as Darwin seemed to suggest: all life was the work of God. Jean Louis Armand de Quatrefages at the Muséum d'Histoire Naturelle was less hostile toward Darwin personally. He had received a presentation copy of the first English edition of *On the Origin of Species* and read it carefully. Still, Quatrefages was skeptical about both the validity and the novelty of the argument. In 1870 he published *Charles Darwin et ses Précurseurs Français: Étude sur le Transformisme,* linking *On the Origin of Species* directly to Lamarck's discredited theories.[130]

Three years later, *A Phrenologist Amongst the Todas* arrived in Paris. It was never translated in its entirety, although Quatrefages did render large extracts into French for his own publications. Others, more or less critical of evolution, also read Marshall's book. These included Paul Topinard, Broca's successor at the Société d'Anthropologie de Paris and cautious supporter of transformism.[131] Following *A Phrenologist Amongst the Todas* to Paris reveals how Marshall's argument was read in a very different political and intellectual environment, one in which Darwin did not command respect. It additionally shows how phrenological photographs traveled beyond the British imperial world. In France, *A Phrenologist Amongst the Todas* was read alongside French anthropological texts, while the photographs were compared to those from across the French Empire.

Quatrefages may not have thought much of Darwin's theories, but this did not preclude an engagement with the history of mankind. In 1856, Quatrefages had been elected to the first chair in anthropology at the Muséum d'Histoire Naturelle.[132] With access to an unparalleled collection of anatomical and fossil specimens, Quatrefages set about studying human difference. As a practicing Catholic, Quatrefages ultimately argued for the unity of mankind and the stability of races over time, publishing his first major work on the subject as *L'Unité de l'Espèce Humaine* in 1861.[133] At this time, it didn't take long for British books to reach Paris, and Quatrefages was able to write a critical review of *A Phrenologist Amongst the Todas* for the *Journal des Savants* in December 1873.[134] This was followed by a much longer discussion of the Todas and Marshall's work in *Hommes Fossiles et Hommes Sauvages,* published in 1884. In both his review and later work, Quatrefages was critical of not only Marshall's phrenology but also his interest in natural selection. With the fall of the July Monarchy in 1848, phrenology had long since lost its institutional status in

Paris. In any case, those who still showed an interest in Gall's doctrines tended to be followers of Broca, Quatrefages's polygenist rival at the Société d'Anthropologie de Paris.[135] Broca's students were also much more sympathetic to transformism, compared with the rest of the French intellectual establishment, particularly those at the Académie des Sciences.[136]

Quatrefages felt that phrenology had impeded Marshall's observational skills, writing, "Unfortunately the zealous colonel is a phrenologist." It was this "absolute faith in a doctrine definitively tried and condemned by science" that had "prevented him from collecting some of the information which would have been the most valuable to us." Like Tylor, Quatrefages complained about the tables of phrenological measurements, asking why Marshall had not included any of "the principal dimensions of the head." But unlike Tylor, Quatrefages extended this criticism to Marshall's use of evolutionary theory. In *Hommes Fossiles et Hommes Sauvages*, Quatrefages summarized Marshall's argument, writing that it was "by the progress of time, and especially selection, that the lateral organs develop and enlarge." Marshall, Quatrefages told his readers, believed that it was "through natural selection, [that] the brachycephalic races had developed." For Quatrefages, "natural selection" could not explain the emergence of different human populations: "Man, *the species*" was "the same in all lands and in all climates." Quatrefages did recognize, as historians should today, that Marshall's evolutionary account would probably be more popular in Britain than in France, writing, "I do not know how this theory of M. Marshall has been received in England." However, in Paris, Quatrefages was able to state confidently that "this theory is in disagreement with the facts."[137]

Combining phrenology and natural selection, Marshall was always going to have a hard time winning over the French intellectual elite of the Third Republic. Yet, as in Britain, a criticism of the text did not prevent French anthropologists from making use of Marshall's photographs. Quatrefages reproduced a number of images from *A Phrenologist Amongst the Todas* in *Hommes Fossiles et Hommes Sauvages* (figs. 39 and 40). Recovering the practices of reproduction pushes the geography of colonial photography beyond British imperial spaces. The illustrations that appeared in *Hommes Fossiles et Hommes Sauvages* included copies of Marshall's frontispiece ("A Toda Mand") and plate 23 ("My Informant"). Given that the final work had over 200 illustrations, Quatrefages chose to reproduce the photographs as cheap wood engravings rather than carbon or albumin prints. A copy of *A Phrenologist Amongst the Todas* was first sent to Édouard Cuyer, professor at the École des Beaux-Arts in Rouen. Cuyer

Fig. 39. Jean Louis Armand de Quatrefages, *Hommes Fossiles et Hommes Sauvages* (Paris: Baillière, 1884), fig. 179. Author's collection.

Fig. 40. Jean Louis Armand de Quatrefages, *Hommes Fossiles et Hommes Sauvages* (Paris: Baillière, 1884), fig. 182. Author's collection.

transferred the two photographs onto wood blocks. Then, one of Cuyer's students, Léon-Louis Chapon, traced the photographs by engraving them into the wood with a burin. The final illustrations have a very different quality from that of those found in Marshall's book. They appear, like the other figures in *Hommes Fossiles et Hommes Sauvages*, as stylized wood engravings, often in-line with the text. And as photographs traveled, they carried traces of the journey. Each of the illustrations features the signatures of Cuyer and Chapon, engraved onto the woodblocks. Marshall was still credited in the text, although, in the process of transit, Bourne's name had been entirely erased. In Paris, far from Simla and the British Raj, Bourne did not carry the same kind of authority. Instead, Quatrefages reminded his readers of the "care and skill" taken in reproducing the photographs by hand.[138] Here, we see once again that there were different routes to visual authority. In Britain, the Autotype Company emphasized the independent nature of photographic reproduction through the carbon process, whereas in Paris, Quatrefages invoked the personal abilities of Cuyer and Chapon alongside their association with the École des Beaux-Arts.

Marshall's phrenology might have been suspect, but Quatrefages argued that the photographs still provided a source of anthropological knowledge. After translating a large section from *A Phrenologist Amongst the Todas* into French, Quatrefages explained that "thanks to photography we can judge the accuracy of the descriptions . . . and complete on some points the details given by M. Marshall."[139] Still, no photograph could be read alone. Authority, Quatrefages implied, was best obtained by comparing different photographs and, if possible, different photographers. Alongside the images from *A Phrenologist Amongst the Todas*, Quatrefages also reproduced a series of photographs and drawings made by the astronomer Jules Janssen. In France, as in Britain, anthropology and astronomy were closely connected.[140] Janssen had traveled to India in 1871 to observe the solar eclipse. After having rejected an offer to join an expedition organized by the British astronomer Norman Lockyer, Janssen managed to convince various French scientific institutions to fund his own voyage. The Académie des Sciences and the Bureau des Longitudes offered 13,600 francs between them, while the French government added another 12,000 francs. Sensing an opportunity, the Muséum d'Histoire Naturelle offered an additional 2,000 francs on the condition that Janssen make a collection of zoological and anthropological specimens when not undertaking astronomical work. Accompanied by his wife, Henriette, Janssen arrived on the Malabar coast in November 1871. Throughout his expedition, Janssen repeatedly moved between French and British colonial spaces, collecting astronomical and anthropological information as he went. At the French outpost of Mahé, Janssen met with the local colonial official, M. Liautaud, who provided two Indian guides. From there, Janssen proceeded to the Nilgiri Hills in order to catch the eclipse due to take place on 12 December. The sighting of the eclipse was a success, but of more interest to Quatrefages was the subsequent anthropological work Janssen undertook with his wife. By the 1870s, photographic equipment was a standard part of an astronomer's equipment. Sensing an opportunity, Janssen turned his camera away from the heavens and back down to earth. While in the Nilgiri Hills he took three photographs of the Todas for the Muséum d'Histoire Naturelle. Janssen also made sketches, some of which were later published alongside an account of the Todas, authored by Henriette, in the periodical *Le Tour du Monde.*[141]

Back in Paris, Quatrefages brought Janssen's astronomical account of the Todas together with Marshall's phrenology. On their own, both Janssen's and Marshall's illustrations were problematic. Janssen's sketches were highly stylized and ornamental, almost an early form of art nouveau (fig. 41). Additionally, the photographs were poorly composed for

FIG. 181. — Femme toda, dessin de M. Janssen, empruntée au *Tour du monde.*

Fig. 41. Jean Louis Armand de Quatrefages, *Hommes Fossiles et Hommes Sauvages* (Paris: Baillière, 1884), fig. 181. Author's collection.

anthropometric purposes. Each featured between four and seven individuals, all adopting different stances. Janssen had also treated his subjects like astronomical phenomena, adopting a long exposure time, which resulted in some blurring on the negative.[142] Finally, Quatrefages complained, "These individuals are all wrapped in the kind of long blanket which . . .

covers the entire body." (Janssen arrived in the Nilgiri Hills during the winter, hence the layers of clothing. This made it difficult to discern the shape of the body and, in the case of the men wearing head coverings, the skull.) Yet despite all these difficulties, Quatrefages found a use for Janssen's illustrations. They were of "a great interest" when read alongside Marshall's descriptions of Toda character.[143] Quatrefages rendered four pages of *A Phrenologist Amongst the Todas* into French. He then inserted references to Janssen's sketches, as in this example translated from Marshall's description of the Toda face:

> Face—Légérement allongée, ovale et d'un contour agreeable sans rien d'étrange ou de heurté (voy. fig. 180 et fig. 181).[144]

In this case, the parenthetical reference "(voy. fig. 180 et fig. 181)" was added by Quatrefages and refers not to Marshall's text, but to the figure numbers in *Hommes Fossiles et Hommes Sauvages*. The figure numbers point the reader to Janssen's Europeanized sketch of a Toda woman, which Quatrefages suggested was of "an agreeable shape." On other occasions, Janssen's sketches even appear on the same page as the translation of Marshall's text, weaving French and British colonial sources even closer together (fig. 42).

At another point, Marshall's and Janssen's photographs were used to corroborate one another. Quatrefages complained that Marshall's frontispiece was not suited to anthropological analysis. It was "too small . . . for the figures to be used seriously." However, Quatrefages went on to explain that, despite being difficult to use on its own, Marshall's photograph could be read in connection with Janssen's work. "Among the men represented in the frontispiece," there were "four individuals who had also posed for the large photographs of M. Janssen." Marshall's frontispiece was of little consequence anthropometrically. Yet the same photograph could still act as a test of authenticity. Readers could be confident that both Janssen's and Marshall's photographs really represented the "Toda type," as the same individuals apparently appeared in both images.[145]

In Paris, Quatrefages brought together French and British colonial photography to produce what he believed was a more accurate picture of Toda character. The Muséum d'Histoire Naturelle was also an ideal site for conducting comparative anthropology. Although Quatrefages disagreed with Wallace and Darwin about natural selection, he too wanted to turn the study of mankind into a global science. At the end of his chapter on the Nilgiri Hills, Quatrefages turned to the "anthropological affinities of

508 LES TODAS.

» *Nez.* — Généralement étroit et d'une hauteur médiocre à la base; les os nasaux larges à leur extrémité inférieure; long; dans les deux sexes, la saillie frontale au-dessus de la racine est souvent fortement accusée. Le nez est quelquefois aquilin, souvent presque aquilin, jamais retroussé (1); légèrement charnu; les narines sont un peu dilatées, mais souvent longues et fines. Le nez acquiert rarement sa perfection avant le milieu de la vie (voy. fig. 177 à 182).

Fig. 177. — Reproduction d'un croquis de M. Janssen.

» *Bouche.* — Un peu charnue; la lèvre supérieure un peu courte, l'inférieure légèrement avancée et pendante. Ce trait est souvent très marqué, surtout dans un âge avancé (voy. fig. 179 et 180) (2).

» *Gencives.* — Généralement de couleur pourpre, mais souvent d'un rouge brillant.

» *Dents.* — Quelquefois courbes et larges, quelquefois

Fig. 42. Jean Louis Armand de Quatrefages's translation of *A Phrenologist Amongst the Todas* appears on the same page as a sketch by the French astronomer Jules Janssen. Jean Louis Armand de Quatrefages, *Hommes Fossiles et Hommes Sauvages* (Paris: Baillière, 1884), 508. Author's collection.

the Todas." He noted that there was great disagreement on this question, setting out the different positions in *Hommes Fossiles et Hommes Sauvages*. Among the range of opinions, from "Jewish" to "Kaffir" ancestry, Quatrefages advanced his own hypothesis. Like Wallace, he believed that the Todas shared many characteristics with the Ainus of Japan. In supporting this conclusion, Quatrefages once again brought together British and French imperial interests. "The museum possesses eight photographs . . . representing many individuals of both sexes, which have been sent by M. Berthelemy, French minister to Japan," explained Quatrefages. From these, he concluded that there were "very real physical resemblances between the Todas and the Aïnos." Far away in Japan, Berthelemy had found the same "horizontal eyes, the prominent nose, the elongated face of the Todas." Quatrefages also suggested that "the colour of the skin appears to be almost exactly the same in the two populations." This, he recognized, was actually quite difficult to judge from both Marshall and Berthelemy's photographs. Different chemicals produced very different effects in the developing process. This was compounded by the fact that the photographs reproduced in *Hommes Fossiles et Hommes Sauvages* were wood engravings rather than albumin prints. With this in mind, Quatrefages admitted that in "the photographs of M. Berthelemy" the Ainus appeared "perfectly white." In this instance, the intervention of the illustrator had compromised the integrity of the image. Quatrefages assured his readers, "This could only be a fantasy of the artist." "All other travellers" found both the Ainus and the Todas to be of "a very dark complexion," he explained.[146]

For added support, Quatrefages then turned to the anatomical collections of the Muséum d'Histoire Naturelle. The museum actually held a series of Ainu skulls, which Quatrefages intended to compare to the Todas. This was made somewhat more difficult by the fact that the Todas cremated their dead, leaving no skulls in any European collection. "The Todas burn their dead with the greatest of care and we can not hope of ever being able to study one of their skulls," Quatrefages explained. Nonetheless, the French anthropologist did suggest that the skulls in the Muséum d'Histoire Naturelle could still be compared with Marshall's face and profile photographs. Here, the practices of craniometry developed by phrenologists earlier in the century found a place alongside late nineteenth-century anthropometric photography. Quatrefages reproduced two photolithographs of Ainu skulls in *Hommes Fossiles et Hommes Sauvages*. Inviting his readers to see for themselves, Quatrefages suggested that "the curve of the head . . . matches exactly that of photograph no. 3 of

M. Marshall." The Toda and Ainu skulls illustrated "much the same character," he concluded.[147]

By the 1880s, *A Phrenologist Amongst the Todas* had become part of a global comparative project. In Paris, a range of different objects, disciplines, and empires came together. Quatrefages's antievolutionary argument invoked photographs and skulls as well as British and French imperial spaces. *Hommes Fossiles et Hommes Sauvages* suggested that this approach, which would have been difficult to undertake outside of the Muséum d'Histoire Naturelle, produced a special kind of authority. On the Todas, Quatrefages claimed that he had "gathered all that is known of this singular race."[148]

RETURN TO THE NATIVES

A Phrenologist Amongst the Todas reached India in December 1873. By this time, readers on the subcontinent were receiving the latest books from Britain by steamship.[149] Just two months after publication in London, the bookseller G. C. Hay and Co. in Calcutta was advertising copies of Marshall's work at 12 rupees.[150] Before long, *A Phrenologist Amongst the Todas* was being sold in Madras, and further advertisements were posted in Allahabad. Recognizing the value of Marshall's book in the south, the bookseller Higginbotham and Co. in Madras marketed *A Phrenologist Amongst the Todas* at an increased price of 16 rupees. The advertisement posted in the *Madras Mail* emphasized the usefulness of the publication for understanding local tribal populations, explaining "the most important part of the investigation turns, beyond doubt, on the practice of infanticide and polyandry among this strange people." *A Phrenologist Amongst the Todas* appeared alongside other "choice stock," including practical manuals on colonial governance, such as Henry Roberts's *Indian Exchange Tables*, as well as the latest works on primitive culture, such as James Fergusson's *Rude Stone Monuments in All Countries*.[151] In Bombay, the *Times of India* welcomed *A Phrenologist Amongst the Todas*, praising Marshall's methodology. Unlike Tylor in Britain, the *Times of India* saw the value in Marshall isolating himself from metropolitan life, writing "he is a 'primitive' man. . . . Discarding tents and the usual paraphernalia of civilization, he lives in the heart of the native villages—conversing all day long, telling stories, listening to the people's songs, and ready to assist at a row or a dance."[152] Marshall played his own part in getting *A Phrenologist Amongst the Todas* to readers in India. He presented a copy to the Asiatic Society of Bengal, and, hoping his depiction of the Todas as a "primitive tribe" would go down well with the colonial bureaucracy, he also sent a copy to the Government of India in Calcutta.[153]

A Phrenologist Amongst the Todas was a book about India. It was also a book for India. Anthropologists used Marshall's work when traveling in the Nilgiri Hills, as did colonial officials and local collectors. Retracing the history of the reception of *A Phrenologist Amongst the Todas* in India reveals how colonial photographs were used in practice, and helps prevent treating the act of taking a photograph as the most important moment in the making of colonial power. On the one hand, *A Phrenologist Amongst the Todas* was incorporated into colonial manuals, where it materially affected the governing of India. On the other hand, the power of colonial photography could be tested. Taking Marshall's book to India provided another opportunity to verify, but also challenge, a phrenological reading of Toda character.

When Henry Grigg, assistant commissioner for Ootacamund, started to compile the *Manual of the Nílagiri District in the Madras Presidency*, he turned to *A Phrenologist Amongst the Todas*. Grigg's mammoth handbook, extending to over 600 pages, addressed everything a local colonial officer might need to know, from climate and population to the revenue and postal departments. A significant portion of the *Manual* was devoted to "Ethnology." Grigg admitted that he had made "constant reference" to *A Phrenologist Amongst the Todas* when writing his manual. In the opening section, Grigg explained that, like Marshall, he was interested in the evolution and origins of the Todas, writing, "Some of these clans would naturally, under favoring circumstances, have progressed, whilst others, under unaltered conditions, would have necessarily remained in their primitive state." Marshall was cited throughout the chapter on the Todas. Grigg told his readers that "Colonel Marshall's researches have led him . . . to look on them as a primitive race still in its infancy." Although the photographs were not reproduced, Grigg used Marshall's visual argument to push Toda origins outside of the Nilgiri Hills. "Their features may best be described as Europeans with Roman noses and bright hazel eyes, good teeth and an abundance of rather coarse but glossy black hair, which is worn in a crop by the men and in long thick ringlets by the women," Grigg explained. He added, paraphrasing Marshall, "Their faces have sometimes a general resemblance to the Jewish type." Given this apparent connection to Jews and Romans, Grigg concluded that "there are not sufficient reasons for considering them to be the earliest inhabitants of the hills." Budding colonial officers were then pointed toward "the chapter on the Revenue History of the District" to help them understand "the action of the Government in regard to the land rights of the Tódas."[154] Four years later, the government of Madras ceased all compensation payments for the loss of tribal land.[155]

In Britain, *A Phrenologist Amongst the Todas* was read as part of a global history of mankind. However, in India, Marshall's work was understood as contributing to a regional project. The *Indian Antiquary*, a monthly periodical printed in Bombay, praised "Colonel W. E. Marshall's investigations into the physical peculiarities, the manners and institutions of the Todas." The work formed "a very welcome addition to our knowledge of the mountain tribes of India." *A Phrenologist Amongst the Todas* appeared alongside other regional accounts of Indian anthropology, such as "Dalton's *Descriptive Ethnology of Bengal*, illustrated by lithographs from photographs" as well as "Sherring's *Tribes and Castes as Represented in Benares*." The *Indian Antiquary* explained that, together with Marshall's account, these works contained a "mass of useful facts for ethnological students." The *Indian Antiquary* also pointed readers toward the "excellent autotype plates," which it claimed "enhanced" the finished work.[156] These carbon prints would have been particularly striking. Photographic studios in 1880s India did not have access to materials produced by the Autotype Company's factory in London. Albumin prints were much more common on the subcontinent, despite the damaging effects of the tropical sun.[157] Additionally, it is worth noting that the *Indian Antiquary* was actually read by members of the local Indian elite. Although we do not know what they thought of Marshall's work, we do know that "Babu Pratab Chundra Ghose" and "Babu Prosanno Chundra Roy" could have read about *A Phrenologist Amongst the Todas* in the *Indian Antiquary*. Pragmalji II, Rao of the princely state of Cutch, was also listed as a subscriber.[158]

Evidence for other Indian readers is sporadic. In 1912 the Bengali legal scholar Bijay Kisor Acharyya turned to Marshall once again. Acharyya, a professor at the University of Calcutta, had been asked to deliver the annual Tagore Law Lectures. His subject was "Codification in British India." This was a very different kind of colonial manual, one reflecting a modest form of Indian nationalism. "Law cannot be properly understood apart from the history and spirit of the nation whose law it is," Acharyya argued. The aim of his lectures was to recover "the legal and constitutional history of India," as it pertained to not only British imperial rule, but also local legal traditions, and especially "the personal laws of the Hindus and Mahomedans." Acharyya hoped that codifying Indian laws would help "the consolidation of different parts of British India." With respect to the Todas, Acharyya did not go into great detail. And although he must have seen them, Acharyya did not comment directly on the photographs. Nonetheless, his reading of Marshall's book was shaped by its earlier reception. The University of Calcutta didn't actually hold a copy of *A Phrenologist Amongst the Todas*. Instead, it was *Travels Amongst the Todas*. Forty years earlier, as noted above, Tylor's criticism in Britain had forced Longmans to bind up the remaining stock with a new title page, hoping to downplay the phrenological content. This suited Acharyya well. Now, in early twentieth-century Calcutta, a Bengali legal scholar could recover "the customs and usages" of the Todas without worrying about Marshall's visual argument.[159] In some cases, phrenological photography could simply be ignored.

Other colonial readers were explicit in their criticism. Helena Blavatsky, founder of the theosophical movement, read *A Phrenologist Amongst the Todas* while making her own tour of the Nilgiri Hills in the 1870s. Blavatsky, an enigmatic critic of materialism and evolution, was preparing her own history of "Oriental psychology" at the time. She traveled widely, entering Tibet and studying Buddhism in the 1860s before heading south and founding the Theosophical Society in Madras in 1875. Throughout her studies, published as *Isis Unveiled* in 1877, Blavatsky rejected the materialism underlying Darwinian theory.[160] Instead, she proposed that the laws of nature were accessible only through occult wisdom, Buddhist meditation, and the practices of "fakirs and jugglers." In *Isis Unveiled*, Blavatsky invoked the authority of her travels. "It is surpassingly strange, that with the thousands of travellers and millions of European residents who have been in India, and have traversed it in every direction, so little is yet known of that country and the lands which surround it," she wrote. Referring to the Nilgiris specifically, she wrote that "fifty years have passed since the discovery; but though since that time towns have been built on these hills and the country has been invaded by Europeans, no more has been learned of the Todas than at the first." Blavatsky valued the close study of local populations. And she took seriously the beliefs and practices of Indian tribes and mystics.[161]

In many ways, this insistence on a close study of Indian life mirrored Marshall's approach. Yet Blavatsky was fiercely critical of *A Phrenologist Amongst the Todas*. In *The Modern Panarion*, published in London and Madras, she wrote that "the phrenological explorer crawled alone by night with infinite pains and—neither saw nor found anything." Bitingly, she concluded that Marshall "saw so much, and discovered so little." Marshall might have spent time among the Todas, but phrenology undermined all his conclusions. "The Todas . . . are not the degenerate remnants of the tribe whose phrenological bumps were measured by Col. Marshall," Blavatsky argued. Next she turned to the photographs. Most strikingly, Blavatsky actually questioned the authority of the images. She wrote that "neither Capt O'Grady, who was born at Madras and was for a time stationed on the Neilgherry hills, nor I, recognized the individuals photographed in Col. Marshall's book as Todas." She continued, "Those we saw wore their dark brown hair very long, and were much fairer than the Badagas, or any other Hindûs, in neither of which particulars do they resemble Col. Marshall's types."[162] For Blavatsky, colonial photography held no special power. Her own personal knowledge, gained over years traveling in India, trumped Marshall's carbon prints.

Blavatsky was not the only one who traveled with a copy of *A Phrenologist Amongst the Todas*. When William Halse Rivers Rivers decided to mount an expedition to South India, he collected all the existing works he could find on the subject. No doubt consulting his colleague Alfred Haddon at the University of Cambridge, who owned a copy of Marshall's book, Rivers decided to pack a copy of *A Phrenologist Amongst the Todas*.[163] This provided a major point of reference during Rivers's time in the Nilgiri Hills between 1901 and 1902. He read it closely, alongside copies of Henry Harkness's *A Description of a Singular Aboriginal Race Inhabiting the Summit of the Neilgherry Hills*, James Breeks's *An Account of the Primitive Tribes and Monuments of the Nilagiris*, and Edgar Thurston's *Anthropology of the Todas and Kotas of the Nilgiri Hills*. These four books, printed in London and Madras, traveled with Rivers as he lived alongside the Todas. They functioned not only as reference works, but also as an opportunity to verify anthropological knowledge. "While I was working I had by me the books or papers of Harkness, Marshall, Breeks, and Thurston . . . and I enquired into most of the details mentioned by them," Rivers explained.[164] From cattle sacrifice to burial rituals, Rivers tried to check all the claims of previous writers on the subject, from Harkness in the 1830s to Thurston in the 1890s. Marshall's photographs proved particularly useful in this respect. Incredibly, Rivers actually showed the photographs to a group of Todas. In doing so, he enacted an anthropological experiment, one designed to test the veracity of Marshall's colonial photography. Worryingly, the Todas did not recognize the individuals they saw:

> When I showed one man the frontispiece in Marshall's book, representing a Toda village and its inhabitants, something unfamiliar in the

arrangement of the scene made the man think that it must have been a picture of the Kamasòdrolam.[165]

Here, Rivers reveals how the Todas challenged Marshall's phrenological photography. The Toda man who looked at the frontispiece in 1901, a photograph originally taken by Bourne over thirty years earlier, did not acknowledge what he saw as a fair representation of either his tribe or his homeland. Instead, the Toda man thought the picture represented "the Kamasòdrolam," a completely separate tribe closely connected with Toda mythology. According to Toda legend, the Kamasòdrolam used to live on the Nilgiri Hills but were driven away by a god known as Kwoten. Rivers wasn't too sure what to make of this story, other than to warn against seeing it as an account of Toda origins.[166] For historians today, however, it does offer a brief impression of Toda attitudes to Marshall's book. In this case, phrenological photography was incorporated into a Toda history of the hills.

A Phrenologist Amongst the Todas reached far beyond the Nilgiri Hills. By following phrenological photography from India to Europe and back again, this chapter has demonstrated how the same photographs could be put to different intellectual and political uses. To date, historians of colonial photography, particularly those working on India, have largely ignored the role of reception. Pinney simply complains that it is "famously difficult" to find evidence of reception, while Falconer briefly notes that *The People of India* had a "disappointing reception" without going into further detail.[167] This has led to an overestimation of the importance of production and the power of photography. Joanna Scherer, in a statement typical of this historiography, writes that "the assumed reality of the photograph invested it with the illusion of 'truth' and gave it much of its power."[168]

When we examine the actual responses of nineteenth-century readers, we get a very different picture. Anthropologists like Tylor in Britain were not impressed with the photographic evidence contained within *A Phrenologist Amongst the Todas*. They found it difficult to extract any accurate anthropometric information, even the dimensions of the head. Quatrefages in France and Blavatsky in India also questioned the veracity of the photographs. Quatrefages found that the frontispiece was too small to work with. Blavatsky went one further and claimed that she did not recognize the men and women in the same photograph as Todas. Even the process of production was not without its difficulties. In the Nilgiri Hills, Marshall and Bourne had to contend with a variety of material challenges. This was an era in which negatives needed to be developed in the field. Chemicals, lenses, and cameras all had to be transported from Madras, Calcutta, and sometimes even London. The weather could turn, and equipment could break. Huxley's scheme to standardize anthropological photography might have been, in his own words, "well-considered," but it was also wishful thinking.

Yet despite all these challenges, phrenological photography was not dismissed outright. Anthropologists and colonial officials developed complex strategies for reading photographs alongside other forms of evidence.[169] Skulls and sketches helped to corroborate questionable photographs. Even the testimony of the Todas was sought. The power of colonial photography stemmed from these techniques of reception, not just production.[170] And it was through the processes of exchange, pushing farther and farther beyond the Nilgiri Hills, that phrenology became linked with evolution and the growth of the colonial state. In Britain and France, *A Phrenologist Amongst the Todas* was incorporated into a global comparative history of race. At the Anthropological Institute in London, Wallace read Marshall's photographs alongside others taken in Japan and Australia, claiming to have identified an ancient race of Caucasians. In Paris, at the Muséum d'Histoire Naturelle, Quatrefages developed his own antievolutionary history of mankind. In this case, phrenological photography was pushed beyond the bounds of the British Empire, incorporating French interests in South Asia and Japan. And finally, in India, the Todas themselves came face-to-face with Marshall's images. It was through reception that phrenological photography gained its power. Yet it was also through reception that those same photographs could be challenged.

EPILOGUE

Electric street lamps flickered into life as motorcars raced down the Nanjing Road. The year was 1919, and the world had changed. Working through the night against the neon glow of the city, the battery of Albright machines owned by the Commercial Press printed over 8,000 sheets per hour.[1] In the 1820s, George Combe's *Constitution of Man* had been painstakingly set in type and printed by hand in Edinburgh. A century later, *Guxiangxue*, the first work on phrenology published in Chinese, raced through electric rotary drums in Shanghai (fig. 43).[2] As previous chapters have suggested, phrenology did not simply fade away during the second half of the nineteenth century. From soldiers fighting in the American Civil War to Indian nationalists seeking independence, phrenology found new audiences and new political uses throughout the nineteenth century. This becomes clear only once we abandon the closed approach to national and imperial contexts that has dominated the history of science for too long. Rather than providing a story of rise and decline, this epilogue examines the sciences of the mind in China in order to reflect on the ways in which the geography of phrenology changed over the course of the century. I then conclude with two broader reflections: the first on the nature of global history, and the second on the chronology of nineteenth-century science.

When the missionary Richard Cobbold arrived in the eastern seaport of Ningbo in the 1850s, he noted that phrenology had "never taken the place which was expected by its advocates" in the East. After all, the Chinese already possessed sophisticated divinatory systems, with woodblocks

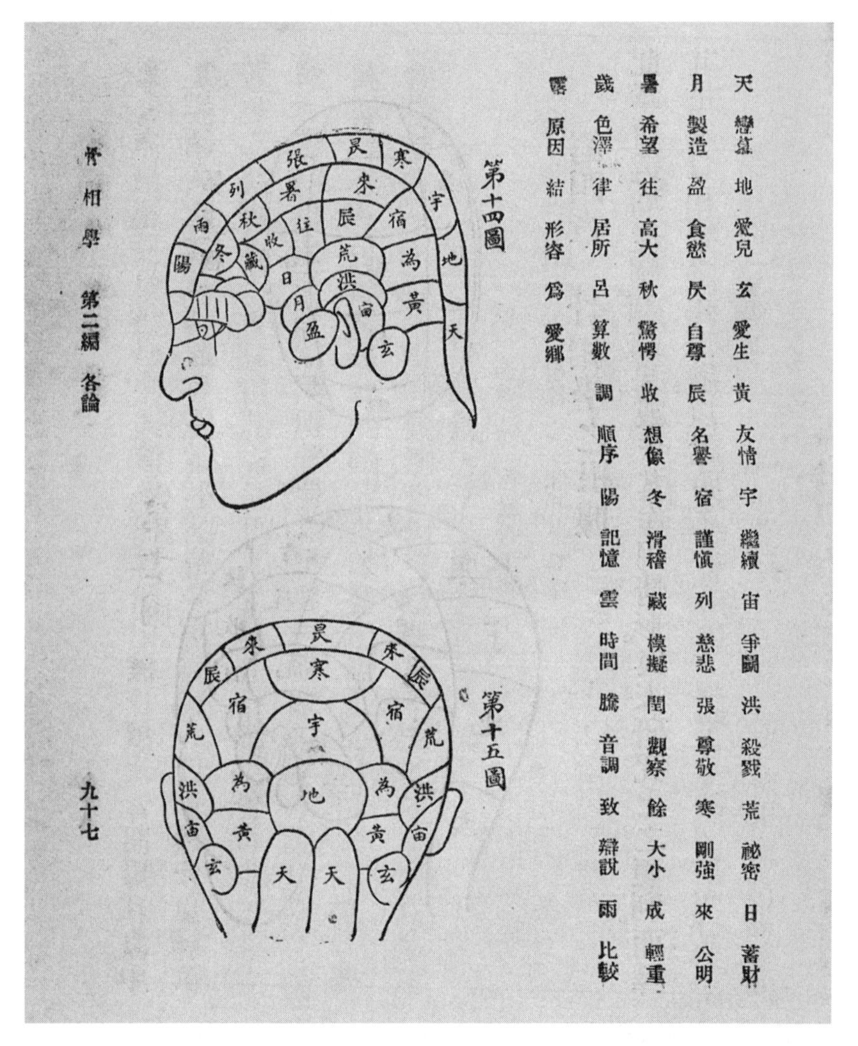

Fig. 43. Phrenological chart, Feng Pingsheng, *Guxiangxue* (Shangha: Commercial Press, 1923), 97, Stanford University Library.

dating from the Ming dynasty illustrating the use of facial features to diagnose illness. These practices developed into sciences of character reading throughout the eighteenth and nineteenth centuries. On the streets of Ningbo, Cobbold recalled seeing what he described as a "physiognomist." The "doctor of the science" occupied a small stall, inviting the "passer-by to make a trial of his skill." On entering, Cobbold found the walls adorned with illustrations of different character types including "the poor, the rich, the noble, and the mean."[3] The American naval officer Robert Burts also remembered seeing "professors of something very like phrenology" when visiting a temple in Canton in the late 1830s. Burts recognized what many did not—studying the shape of the skull was a long-established tradition in China, an "old test of character and genius."[4] For all its global ambitions, phrenology was old news in nineteenth-century East Asia. But then, in the wake of the Revolution of 1911, phrenology suddenly found itself supported by the biggest publisher in the region, with a distribution network stretching from Singapore to Beijing.

The Revolution of 1911 brought an abrupt end to over 2,000 years of dynastic rule. In January 1912, Sun Yat-sen was declared president of the new Republic of China. He was quickly replaced by the military leader Yuan Shikai. Yuan dissolved Parliament and assumed dictatorial powers, plunging Sun's Nationalist Party, the Kuomintang, into chaos. Yuan's death in 1916, along with the outbreak of the First World War, was followed by a decade of conflict between rival warlords. Sun later returned, but the Kuomintang did not manage to secure control of the north of China until 1928. Along with political uncertainty, this was a time of intellectual fervor. The years following the Revolution of 1911 saw growing interest in both nationalism and modernism, as Chinese thinkers looked for a means to strengthen and unify a divided people. This came to a head on 4 May 1919, when over 3,000 students from Peking University marched in protest over the terms of the Treaty of Versailles. They opposed the granting of German concessions to Japan, and worried more generally about the fate of China in the world. This sparked the broader New Culture Movement in which leading intellectuals rejected what they saw as outdated traditions, particularly those related to Confucianism, and promoted a kind of scientific nationalism. Darwin, Spencer, Lenin, and Freud all found a receptive audience in 1920s China.[5] And for the students marching on Tiananmen in 1919, reading the face for signs of character certainly smacked of ancient "superstition."[6] Wang Feng-chieh summed up the ethos of the New Culture Movement when he declared, "The old educational system and old national customs have been destroyed. New education—and education based on science—has begun."[7] This opened up a space in which phrenology could flourish in China, something that always seemed unlikely in the nineteenth century. Phrenology now offered an alternative form of divination, one apparently grounded in modern science rather than Confucianism.

Psychology occupied a prominent place in the Republican vision. In 1921 a group of students in Nanjing founded the Chinese Psychological

Society alongside the journal *Xinli*. The opening issue described psychology as "the most useful science in the world," and over the course of the 1920s *Xinli* published articles on topics ranging from Wilhelm Wundt's thought psychology to Sigmund Freud's theory of dreams.[8] Key political thinkers of the period also believed that psychological research had the potential to transform the nation. Even toward the end of the Qing dynasty, the reformer Liang Qichao had argued that "politics in society is a manifestation of its citizens' psychological states." Consequently, "in order to reform the political system, the key is to reform the citizens' psychological states."[9] By the time of the revolution, this message was firmly entrenched. Cai Yuanpei, one of the leaders of the New Culture Movement, had studied in Leipzig and taken courses under Wundt. In 1917 he approved the foundation of the first psychological laboratory in China at Peking University. By the 1930s, there were twenty-one psychology departments at universities and colleges across China.[10]

Phrenology certainly found a receptive political climate amid a surge of interest in psychological and biological sciences. Self-discipline through understanding of the human mind and body was a common theme, with new periodicals such as *Xing Zazhi* introducing readers to sexology, and Chinese universities offering courses on human and developmental biology. The Commercial Press itself played a major role in cultivating this outlook, publishing early translations of Herbert Spencer's *Study of Sociology* and Thomas Huxley's *Evolution and Ethics*.[11] It also published *A Textbook of Civic Biology* featuring a reproduction of Ernst Haeckel's infamous embryological plates illustrating fetal development. The science editor at the Commercial Press, Zhou Jianren, commissioned *Guxiangxue* as part of this program, advertising it alongside an introduction to anthropology as part of the "Beijing University Series."[12] Like the contemporary works on sex and developmental biology, *Guxiangxue* was pitched as a corrective to those "superstitions" believed to be holding back the new Chinese nation. The pseudonymous author, Feng Pingsheng, blended the old and the new. He acknowledged that "the arts of divination are longstanding in both East and West." However, what China needed was not "mystical divination" but rather "phrenology." *Guxiangxue* would finally allow the reader to understand "man's personality and habits" through "inductive methods." In keeping with the Republican enthusiasm for psychology, Feng Pingsheng went on to argue that a proper understanding of human character would also benefit the Chinese nation. He suggested phrenology could be applied to education, law, and economics and that it would provide a foundation for "social relations and polities as well as . . . laws in order to protect peace and order, to pursue happiness and social progress."[13]

The national political context had certainly undergone a revolutionary change. But accounting for the popularity of phrenology in Republican China also points to the significance of the connections between different imperial powers. In the nineteenth century, phrenology advanced along British and French imperial networks—East India Company ships on their way to Calcutta and the *Astrolabe* voyage to the Pacific. But in early twentieth-century China, phrenology arrived via Germany and Japan. Following defeat in the Sino-Japanese War of 1894–1895, Chinese reformers increasingly looked to Japan as an example of successful modernization. By the 1920s there were over 8,000 Chinese students studying at universities in Japan, while a great number of European scientific works were translated from Japanese editions. These included Ruric Roark's *Psychology in Education* and Joseph Haven's *Mental Philosophy.*[14] *Guxiangxue* was not translated from Japanese, but the author did acknowledge the existing literature on phrenology, including Nagamine Hideki's translation of Combe's *System of Phrenology,* which was published in Tokyo in 1918.[15] The hope was that China could benefit from a scientific practice of character reading as well.

Germany represented another significant connection for Republican China. Throughout the nineteenth century, German industrial and trading firms, such as Siemens and Carlowitz, operated profitable businesses in China. Like other European powers, Germany sought an increasing territorial presence, culminating in the Kiautschou Bay concession of 1898. Under duress, the weakened Qing government agreed to a ninety-nine-year lease on the region surrounding the port of Tsingtao. And while Germany lost Kiautschou Bay along with many of its other colonial territories following the First World War, it nonetheless maintained close trading links with East Asia. China represented a significant export market for military and industrial hardware. This was all the more important given that the Treaty of Versailles had placed serious restrictions on German domestic development.[16] The Weimar Republic provided military advisers in Shanxi Province, and Siemens engineers were hired to work at the Canton military academy.[17] This was accompanied by a growing interest in German biological and psychological literature. The Commercial Press itself hired a number of German-educated Chinese scientists to help write new textbooks, publishing a translation of Haeckel's

Miracle of Life in 1922 that was later advertised alongside *Guxiangxue*.[18] In fact, *Guxiangxue* directly linked phrenology to Haeckel's embryology. His theory of recapitulation held that as an embryo develops, it moves through earlier stages of its own evolutionary history. *Guxiangxue* featured an illustration exploring how this could apply to the development of the brain. The reader is presented with an image of the nervous systems of a number of lower animals, including a fish, a frog, a bird, and a dog (fig. 44). The implication is that if the brain is the organ of the mind, then psychology could also be understood developmentally. In keeping with the literature on sex, *Guxiangxue* therefore suggested that good citizens could understand their psychological strengths and weaknesses through phrenology and improve themselves accordingly. This would then push the Chinese nation up along the evolutionary scale and ensure the new republic won out in the "survival of the fittest," a phrase popularized in East Asia through the Commercial Press's own translation of Spencer's *Study of Sociology*.[19] Germany and Japan both represented imperial nations worthy of emulation in this respect. *Guxiangxue* even featured in the same advertisement as a pamphlet entitled *Reasons for the Wealth and Strength of Germany*.[20]

Phrenology similarly played to emerging debates on the racial origins of the Chinese nation. The first president of the Republic, Sun Yat-sen, had argued that the Chinese people emerged from ancient Babylon. Immediately after the Revolution of 1911, this proved an appealing antidote to dynastic histories grounded in the exceptionalism of the Middle Kingdom. But following the First World War and the rise of the New Culture Movement, political attention once again turned toward forging a strong national identity. In 1921 researchers operating as part of the Geological Survey of China in the north discovered pottery and human remains dating from the Neolithic period. These were presented as evidence of an ancient Chinese people—the Yangshao.[21] A few years later, the Geological Survey made an even more incredible find. Over 500,000 years old, the fossilized remains of the "Peking Man" bolstered the evolutionary case for Chinese national identity.[22] It was also precisely at this time that the idea of "national psychology" gained widespread support in Chinese universities, particularly through translations of Wundt.[23] *Guxiangxue* combined both of these approaches, suggesting a physical basis for a psychology of national character. It featured an image of four skulls, including a "white man," a "black man," an "orangutan," and a "baboon" (fig. 45). Referring to the theory of Petrus Camper, the caption explained that "the angle of the forehead can demonstrate the level of intelligence." There is, however,

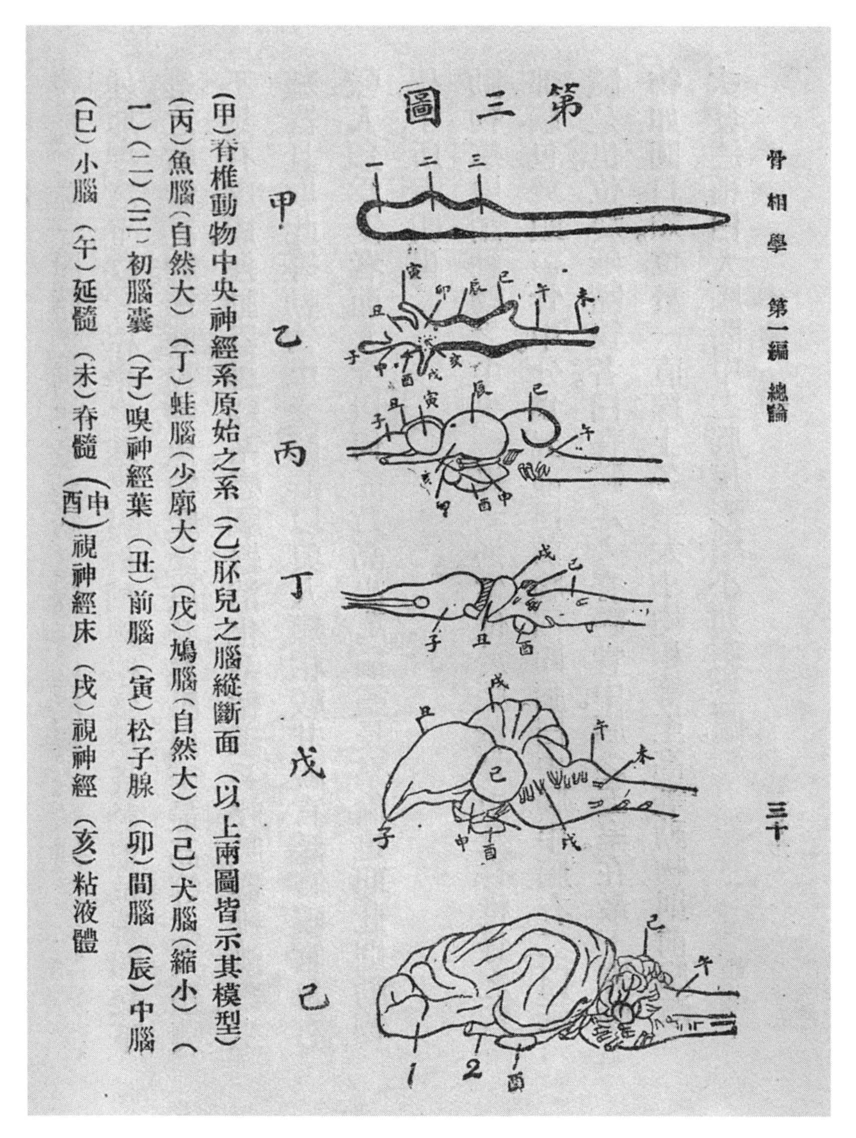

Fig. 44. "Central Nervous System in Vertebrates," Feng Pingsheng, *Guxiangxue* (Shanghai: Commercial Press, 1923), 30, Stanford University Library.

no direct comment on the character of the Chinese people. Nonetheless, the illustration is intended to show how phrenology could be used to determine the mental capacity of different races, and the reader is even invited to find out where they sit on the scale by filling in a table of their own phrenological development.[24]

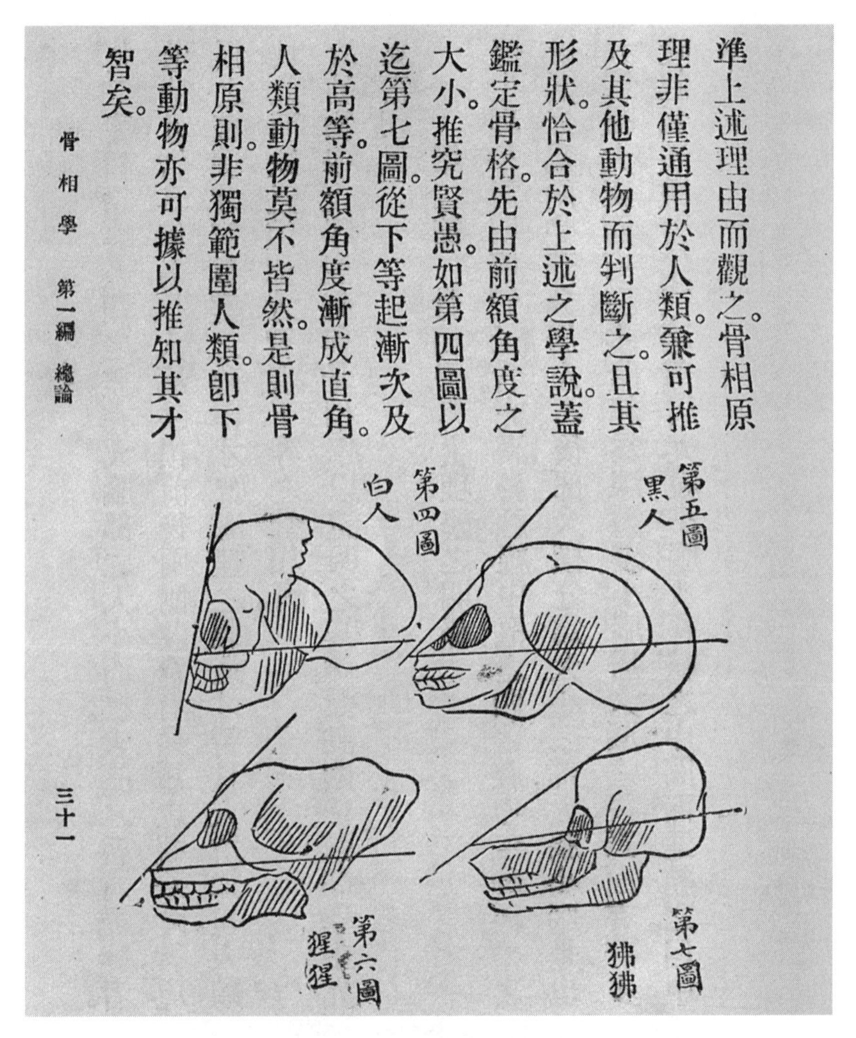

準上述理由而觀之。骨相原理非僅通用於人類。兼可推及其他動物而判斷之。且其形狀。恰合於上述之學說。蓋鑑定骨格。先由前額角度之大小。推究賢愚。如第四圖以迄第七圖。從下等起漸次及於高等。前額角度漸成直角。人類動物莫不皆然。是則骨相原則。非獨範圍人類。即下等動物亦可據以推知其才智矣。

骨相學 第一編 總論 三十一

Fig. 45. "White Man" (*top left*), "Black Man" (*top right*), "Orangutan" (*bottom left*), "Baboon" (*bottom right*), Feng Pingsheng, *Guxiangxue* (Shanghai: Commercial Press, 1923), 31, Stanford University Library.

Guxiangxue ultimately reminds us that phrenology needs to be understood as a global science in two complementary ways. First, in its various uneven material embodiments, phrenology reached an audience of supporters and critics stretching from New York to Shanghai. And second, phrenology was also a science that allowed people to situate themselves within a universal history of human development, something that seemed increasingly relevant in Republican China. *Materials of the Mind* has aimed to understand that relationship: between the material culture of scientific exchange, on the one hand, and the uneasy emergence of universal categories of thought, on the other. Race proved a particularly powerful concept in this respect. From George Combe's lectures in the 1830s to Feng Pingsheng's phrenological charts in the 1920s, phrenology allowed a range of people to both understand and enforce racial difference. Individual chapters have addressed topics ranging from book history and correspondence to popular science and politics, demonstrating how an approach grounded in global history can change our understanding of major themes within the historiography of science. Together, the six chapters reveal a new geography of science. Phrenological objects, from busts of West Indian slaves to photographs of Toda tribesmen, traveled across British and French imperial spaces. And, as the chapters on *Crania Americana* and the *American Phrenological Journal* showed, the history of science needs to recognize the place of American nationalism and imperialism on the world stage. The rise of the United States as an international power was one of the major developments of the nineteenth century, one that had a profound effect on the sciences. Ultimately, in charting this history, I have demonstrated that phrenology cannot be understood solely within national, regional, or imperial contexts. Phrenology was a science forged within a material world that, by the early decades of the twentieth century, stretched across the British, French, American, German, Chinese, and Japanese empires.

This expansive geography was closely tied to phrenology's claim to universality. Phrenologists argued that their theory of the mind worked everywhere, in part by pointing to the geographical reach of exchange. Today, this tension, between local and universal knowledge, is at the heart of the historiography of science.[25] And, given that the sciences make claims to universality, it is important to understand why historians spent so much time studying local contexts. After all, Charles Darwin didn't claim that evolutionary theory worked only in the garden of Down House. Like phrenology, the appeal of Darwin's theory was precisely its ability to explain nature on a planetary scale. Similarly, Robert Boyle's theory of pneumatics would have been a lot less impressive if it held true only for air pumps in London. Again, the power of Boyle's law, as with most sciences, was its claim to universality. For historians of science in the 1970s and 1980s, this was exactly the problem. In order to challenge claims to universal knowledge, historians argued that the sciences were in fact much more situated.[26] Darwin might have claimed his theory worked everywhere, but it nonetheless bore the marks of his life as a gentleman in Victorian Britain.[27] Boyle might have promoted a universal law, but it was exceedingly difficult to replicate his experiments beyond the carefully staged performances at the Royal Society.[28] Since the 1980s, historians of science have extended this argument to a range of other case studies, demonstrating how almost all claims to universal knowledge are the product of a particular time and space. This, however, leaves us with a major unanswered question: if the sciences are so obviously local, and many historical actors were aware of this too, then how did the sciences ever come to be seen as universal?[29] To really understand the nature of science, we therefore need to combine an analysis of science in local contexts with an explanation of how the sciences make claims to universality.[30]

Initially, historians of science pursued two strategies. First, and most commonly, they appealed to practices of replication and disciplining within a local context. Steven Shapin's account of "virtual witnessing" at the Royal Society and Simon Schaffer's history of Victorian astronomical discipline are good examples of this approach.[31] Second, and less often, they appealed to wider networks of circulation and standardization. Schaffer's study of Victorian metrology between laboratories in Cambridge and Berlin is a prime example, as is Joseph O'Connell's study of measures of body mass in the twentieth century.[32] However, even in these cases, the geography of science was still confined to a relatively small set of European and American contexts. There was no sense of how the colonial world, or colonized people, might fit into this history. As this study shows, it isn't just that colonialism and race happened to be part of the making of universal knowledge, they were essential to it.[33] Additionally, these earlier accounts tended to emphasize standardization, particularly in the laboratory context, as the central means through which the sciences could be made to travel.[34] However, once we move beyond the laboratory, to a broader range of natural and human sciences in different sites, the picture becomes more complex. There was an assortment of practices involved in making science move and transforming phrenology into a universal theory of the mind. Notably, phrenologists did not try to discipline observers on a grand scale or create standards of measurement. Nor did they try to reproduce the contexts of experimentation at different sites. And it would be a stretch to describe even a plaster cast as an "immutable mobile."[35] Plaster casts, as phrenologists understood, varied from manufacturer to manufacturer, were liable to break, and were difficult to interpret without the right information. Ultimately, there is no equivalent to the laboratory in the history of phrenology. Instead, phrenologists adopted a different approach. They embraced the connected world in which they inhabited. Phrenologists interpreted a range of material evidence—from skulls and plaster busts to photographs and drawings—and developed complex accounts of how objects, peoples, and cultures from across the globe fit into a universal understanding of human nature.

More fundamentally, my account differs from these earlier histories in what it means for science to be local in the first place. Traditionally, the sciences were treated as something that was first produced locally, and then subsequently moved, and consequently was made universal.[36] In contrast, my account treats the sciences as situated but nonetheless produced through a wider world of exchange from the outset. From the evidence presented in this book, it should be clear that there is no meaningful sense in which phrenology was produced locally in Edinburgh and then moved to places like Cape Town or Calcutta. Instead, phrenology in Edinburgh was fundamentally shaped by a context of production that included skulls arriving from the Arctic, letters arriving from India, and plaster busts arriving from Paris. The same was also true in Calcutta. The arguments made by Bengali phrenologists in the 1840s were a response simultaneously to Hindu reformism in India and to radical socialism back in London. Therefore, historians of science need to recognize that Edinburgh and Calcutta were contexts produced by connection.[37]

These reflections lead to two major conclusions concerning the relationship between the history of science and the emerging field of global history. First, historians of science need to abandon the distinction between the local and the global.[38] The best way to think about this is to consider the analogous debate concerning internal and external histories of science in the 1970s and 1980s. Traditionally, internal histories of science focused on intellectual factors—such as how Darwin interpreted Lyell's geology, or how phrenological controversies could be understood in terms of incommensurable worldviews. In contrast, external histories of science tended to focus on social and cultural factors—such as how Darwin was influenced by his reformist politics, or how disagreements over the nature of the mind were the product of class conflict. Often forgotten today, one of the major contributions made by Shapin and Schaffer in *Leviathan and the Air-Pump* was to break down this distinction. The very boundaries between internal and external factors, they showed, were part of the history of how the sciences came to be seen as authoritative. Early modern natural philosophers like Boyle tried to distinguish between what was part of science and what was not—what was intellectual and what was social—in order to disguise the situated nature of their work. In short, by historicizing the very notions of "internal" and "external," Shapin and Schaffer demonstrated just how problematic such a distinction was for historians of science in the present.[39] I make the same move, but this time with the distinction between "local" and "global." Historical actors themselves reflected on what counted as "local" and what counted as "global," which connections mattered for science, and which didn't. At the same time, those same historical actors were operating within contexts that were produced in part through wider connections and disconnections. Historians of science therefore need to abandon the practice of studying science first locally and then studying how it moves. The sciences are certainly situated, but they are never straightforwardly local.

This brings me to my second major conclusion. If the sciences are never straightforwardly local, then all good history of science must be grounded in global history, at least in the sense advocated in this book. This is of course a strong claim, but one that follows from my earlier discussion. When the sciences were thought of as first local, and then made to move, it was easy to practice a kind of division of intellectual labor. For historians who didn't see what all the fuss was about, it was methodologically acceptable to study science within a bounded local context: botany in eighteenth-century Paris, medicine in nineteenth-century London, and so on. Other historians could then pick up where these studies left off, examining how the sciences moved—say, from London to Calcutta. But once we abandon the distinction between local and global, this starts to look a lot more problematic. In what sense can we study botany in eighteenth-century Paris, when Paris itself was a product of a French Empire incorporating the West Indies and Pacific? In what sense can we study medicine in nineteenth-century London, when London was a space produced in part through connections to Calcutta and New South Wales? In short, if local contexts are always already global contexts—in the sense that they are produced by wider connections and disconnections—then we can't keep practicing the same old history of science. That doesn't mean that every connection is equally relevant, or that all places were equally connected. Judgments still need to be made about what does in fact constitute the context. But for historians working in regions that were so clearly part of a wider world, such as Europe, the Americas, and Asia and Africa, it is hard to sustain a methodology grounded in preexisting assumptions about what counts as local. This is all the more problematic when, as in the case of phrenology, our historical actors were well aware of these connections too. Again, there is an analogy here with the internal-external divide. Once this divide was broken down, it was no longer acceptable to study purely intellectual factors (or for that matter purely social ones). Today, all historians of science are expected to integrate intellectual and social analysis at a fundamental level. What we need now is to draw the same lesson, but this time from the field of global history.

With these arguments in place, the history of science can also contribute to broader theoretical debates within global history at large. At the moment, global histories can be split into two broad categories, typically operating in isolation: histories of connection and histories of ideas.[40] On the one hand, we have the kind of global histories of connection advocated by Sanjay Subrahmanyam.[41] These histories reveal how social, political, and economic events, usually analyzed in local or national terms, were in fact products of a much wider geography. The French Revolution, as Lynn Hunt argues, was just as much a result of colonial unrest in the Atlantic as of social and political instability in Paris.[42] Similarly, Maxine Berg and others have demonstrated how the British Industrial Revolution needs to be understood as a response to increased demand for Chinese and Indian goods.[43] The best of these histories integrate material culture, demonstrating how objects, commodities, and technologies mediate different types of connection and their consequences.[44] On the other hand, we have global intellectual histories.[45] These histories demonstrate how particular ideas—such as "culture" or "liberalism"—were taken up in different contexts across the world. For example, Andrew Sartori demonstrates how the notion of "culture" was reworked by Bengali intellectuals in the nineteenth and early twentieth centuries, while David Armitage has explored the fate of the American Declaration of Independence in contexts ranging from Haiti to Vietnam.[46] More recently, intellectual historians have turned toward the concept of the "global" itself, recognizing it as an actors' category, one that traveled across nations and empires.[47]

Materials of the Mind suggests the need to bring these two approaches together. In many ways, my account of phrenology could be read as a contribution to global intellectual history. It is a history of how a particular idea was reworked as it traveled across the world, and how historical actors reflected upon the universal character of thought. In this sense, it is similar to Vanessa Ogle's *The Global Transformation of Time.* Ogle's work, like my own, is a history of a universal concept that stretches across areas and empires, examining the emergence of a worldwide standard of time around 1900. However, my argument is different from Ogle's in one fundamental way. *The Global Transformation of Time* is by and large a history of ideas.[48] There are allusions to clocks, railways, and telegraph cables, but the material culture of time does not seem to matter much for Ogle's analysis.[49] This reflects the general thrust of global intellectual history today, in which ideas are disembodied, and the only context worth considering is the intellectual context. In this book, I show how the history of an idea—in this case, one concerning the mind itself—was fundamentally shaped by the material culture of exchange. The specific material form of an idea—whether a skull or a photograph—and the specific form of connection—whether an Arctic whaler or an Bengali postal runner—directly affected how historical actors conceptualized the sciences and how they reflected on the world at large. By bringing together global histories of connection with global intellectual history, we can therefore analyze how new ways of thinking were fundamentally shaped by new ways of communicating.

With this in mind, there is a need to integrate the history of science into the study and teaching of global history as a field more generally.[50] As it stands, major surveys of nineteenth-century global history tend to treat the sciences as marginal, occupying no more than a couple of pages in otherwise weighty tomes.[51] Yet for historical actors, the sciences were anything but marginal. Monarchs and artisans, slaves and settlers, Bengalis and Africans, people from across the world and from a variety of backgrounds, all debated the relationship between nature and society in scientific terms. Leading intellectuals, from Rammohun Roy in India to Horace Mann in the United States, discussed phrenology in lectures and letters. Radicals and revolutionaries, such as Gustav Struve in Heidelberg and William Lovett in Britain, read *The Constitution of Man*. And Queen Victoria even invited Combe to make a phrenological assessment of her children, including the future Edward VII.[52]

Core themes in global intellectual and political history would therefore benefit from greater engagement with the history of science.[53] Take nationalism. Although there is ongoing debate, historians broadly argue that modern nationalism emerged in the late nineteenth century. Major world events, ranging from the 1848 Revolutions to the 1857 Indian Rebellion, heralded a wave of nationalist movements both in Europe and across the colonial world. Additionally, historians connect the timing of the emergence of nationalism to a number of key institutions, particularly the periodical press and voluntary associations. Nations were "imagined communities," as famously argued by Benedict Anderson, held together by the common reading experience of the daily newspaper. The "print-capitalism" of the late nineteenth century provided the conditions for the emergence of modern nationalism, whether it be German Unification in

1871 or the Indian National Congress in 1885.[54] This narrative, however, is complicated by the history of nineteenth-century science, particularly the human sciences. Phrenologists, ethnologists, botanists, and zoologists all contributed to the idea of the nation well before the middle of the nineteenth century.[55] As early as the 1820s, the *Phrenological Journal* in Edinburgh featured a long article describing "the cerebral development of nations." The article made an explicit link between individual mental physiology and the nation as a whole. It explained that "a particular form of brain is invariably concomitant of particular dispositions and talents, and that this holds in the case of nations as well as of individuals." Crucially, this was a theory of nationhood that incorporated the entire world. The article went on, arguing that "if we glance over the history of Europe, Asia, Africa, and America, we shall find distinct and permanent features of character which strongly indicate *natural* differences in their mental constitution." The following pages then set out the phrenology of the "French," "Britons," "Germans," "Hindoos," "Turks," Greeks," "Italians," and "Austrians." Similar materialist understandings of national identity were promoted by phrenologists in the United States and the colonial world.[56] And although the periodical press was an important tool for the development of nationalism, including for phrenologists, it was not the only one. In the early nineteenth century, skulls, plaster casts, and lithographs were all used to define national character. The rise of European and colonial nationalism looks like a late nineteenth-century phenomenon only when we ignore the sciences and focus too heavily on traditional political institutions and mass print culture.[57]

Beyond this, phrenological understandings of the world entered into a range of other political debates. Akshay Kumar Dutt and Cally Coomar Das argued over the status of Hinduism and Christianity in Bengali editions of phrenological works. Slaveholders like Charles Caldwell and African Americans like Frederick Douglass deployed phrenology in debates over the relationship between the mind and freedom. Most importantly, the connection between reform and character—those two concepts that underpinned nineteenth-century political thought—was articulated in phrenological terms. The sciences were also one of the principal means through which historical actors imagined the world as a connected and comparative space. Intellectual historians recognize the need to treat the global as an actors' category, but so far this has led to a rather narrow focus on Marx and Hegel.[58] There is also a tendency to consider international consciousness as a late nineteenth- and early twentieth-century phenomenon.[59] As this study reveals, reflections on the connected nature of the world stretch right back into the early nineteenth century, and most likely predate even this. The sciences, and particularly the sciences of the mind, therefore need to be understood as part of an intellectual history with a much broader base. In the nineteenth century, the sciences were a means of doing world history, whether it was the history of man or the history of the earth.[60] In Republican China, phrenology appealed precisely for this reason. Both universal and material, phrenology provided a means for the emerging nation to situate itself in time and space.

Concluding in twentieth-century China also prompts us to consider what global history does to the chronology of science.[61] Individually, the previous six chapters focused on particular objects and geographies. However, taken together, they track the history of phrenology over the course of the nineteenth century. I began with the collection of skulls in the 1810s and 1820s, moved through the circulation of busts, letters, and books in the 1830s and 1840s, and to the rise of periodicals in the 1850s and 1860s, before concluding with the reproduction of photographs in the 1870s and 1880s. In what follows, I discuss how the new geography of science offered by global history also changes our understanding of chronology. What, in other words, was "nineteenth-century" about nineteenth-century science? In answering this question, I pay particular attention to new forms of communication and new understandings of racial character.

The development of cheap print, the penny post, railways, steamships, and telegraphs all signaled an "industrial revolution in communication." The steam press in particular massively increased the scale of scientific publishing in the nineteenth century, bringing works like *The Constitution of Man* to audiences in the hundreds of thousands. For the first time, a gentleman in Cambridge and a working-class reader in Sheffield could read the same text, albeit in different formats and bindings. This significantly altered the meaning of a book like *The Constitution of Man*. Relatively uncontroversial when it was first published in 1828, it was the cheap "People's Edition" published in 1835 that transformed phrenology into a radical science. With railways distributing copies into working-class homes up and down the country, *The Constitution of Man* started to look much more subversive. Conservative clergymen and landowners started to worry that this materialist philosophy of the mind was simply too dangerous to be read by the untutored.[62]

Traditionally, like the Industrial Revolution itself, this is a story told from the perspective of Britain alone. *Materials of the Mind* demonstrates that the communications revolution was in fact part of a transformation that stretched across world, albeit unevenly. These changes in communication then underpinned the growth of new sciences like phrenology. The very same technologies that helped *The Constitution of Man* reach a mass audience in Britain brought phrenology to new readers in Europe, America, Africa, Asia, and the Pacific. Packed aboard steamships, the "People's Edition" found its way into colonial libraries in Calcutta and Cape Town. This too altered the meaning of the text. Colonial governors worried about the spread of materialism among settlers, while phrenologists back in Britain and the United States started to imagine phrenology as part of a universal science, with material minds obeying the same natural laws across the planet. At the same time, when we think of the history of communication, and its relationship to science, it is important to recognize failure as well as success. Looking back at the chronology charted by the previous six chapters, it should be clear that new technologies did not necessarily make communication easier or even quicker, at least not for everyone. Plaster busts proved a cumbersome method of communicating knowledge about the body in the 1830s, yet photography did not solve these problems, instead introducing new challenges in the 1870s. Rather than seeing technology as something that flattened the world, this study reveals how different forms of communication shaped the geography of science.

By concentrating on particular objects, whether plaster busts or lithographic prints, it is possible to be much more specific about both the geography and the chronology of nineteenth-century science. In the 1850s, Chinese engagement with phrenology was limited. The missionary Justus Doolittle had imported a few copies of the *American Phrenological Journal* into Fuzhou, but there were no phrenological books or busts. It wasn't until the early twentieth century, with the Revolution of 1911 and the rise of industrial printing in Asia, that phrenology reached a large audience of Chinese readers, complete with phrenological charts and illustrations of skulls. Similarly, although there was interest in phrenology in the United States during the 1820s, materially this was limited to a few imported busts from Paris alongside pirate editions of *The Constitution of Man*. However, by the 1850s, the Fowlers had transformed American phrenology through the mass production of books and plaster busts, transported by the new railroads opening up across the South and West. In other cases, even rather mundane technologies like the postal service introduced new inequalities between who could and could not communicate on the world stage. African Americans faced racial discrimination whenever they tried to send a letter, while international correspondence proved too expensive for the majority of the world's working population. Similarly, new technologies like the steam press reinforced the status of cities such as London, Edinburgh, and New York as centers of scientific publishing. There were few steam presses in Africa or Asia until much later in the century. The first steam press arrived in the Cape Colony in 1857, while steam printing was not common in India until the 1880s.[63] For much of the century, firms like the Fowlers in New York and the Chambers in Edinburgh exported cheap phrenological books, mirroring the broader trend whereby industrialization turned the colonial world into a market for European and American goods.[64]

Alongside the arrival of new technologies, historians need to recognize the remarkable persistence of older technologies. From the perspective of the world beyond Europe, the most important transformation in communication was not the steam press or the telegraph but rather the introduction of the hand press into Africa, Asia, and the Pacific. Phrenological pamphlets in Cape Town and Calcutta were printed by hand on wooden and iron presses, not by steam. It was only toward the end of the century, as illustrated by *Guxiangxue*, that steam and electric presses really began to contribute to the world of colonial scientific publishing. What's more, there was considerable interaction between old and new technologies. Photography is a good example of this. In the 1880s, the French anthropologist Jean Louis Armand de Quatrefages photographed a set of phrenological busts for his *Histoire Générale des Races Humaines.* The busts had been manufactured much earlier in the century by the phrenologist Pierre Marie Alexandre Dumoutier during his visit to Van Diemen's Land in the 1830s (fig. 46). In this instance, photography generated renewed interest in a set of plaster busts that had been sitting unused for decades in the Muséum d'Histoire Naturelle. These photographs also demonstrate the ways in which technologies could be used to communicate not only across space, but also time. The bust was that of Truganini, described in *Histoire Générale des Races Humaines* as "the last Tasmanian." Truganini was long dead, but the photograph of his bust provided an opportunity to look back to the early nineteenth century from the perspective of the 1880s. Quatrefages read this image as part of recent history, whereby between the 1830s and the 1880s, mankind had witnessed "the total destruction of a race."[65] The image of Truganini that appeared in *Histoire Générale des Races Humaines* represented a combination of media: it was an engraving of a photograph of a plaster bust. It was the work of at least three different artists, geographically and temporally separated.

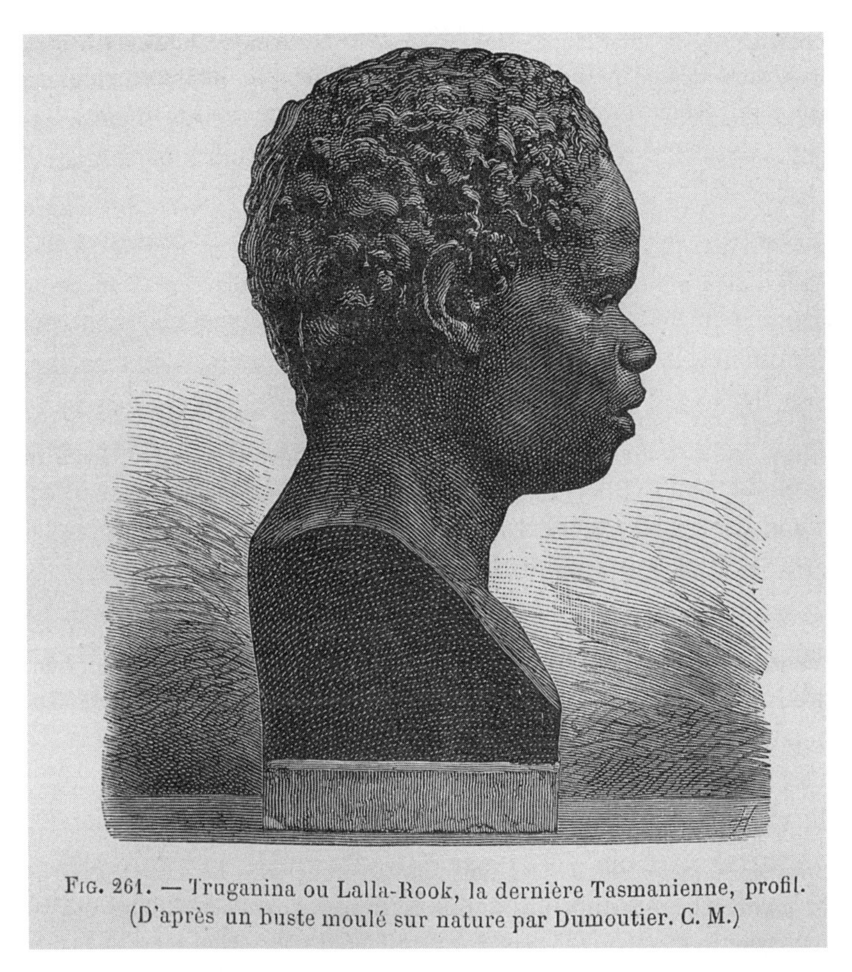

Fig. 261. — Truganina ou Lalla-Rook, la dernière Tasmanienne, profil. (D'après un buste moulé sur nature par Dumoutier. C. M.)

Fig. 46. Engraving of a photograph of a plaster bust, "Truganina or Lalla-Rook, the last Tasmanian," Jean Louis Armand de Quatrefages, *Histoire Générale des Races Humaines* (Paris: A. Hennuyer, 1889), fig. 261. Author's collection.

As the example of Truganini suggests, the history of phrenology is part of the history of race. Traditionally, historians have argued that understandings of race underwent a significant transition over the course of the nineteenth century. Race at the beginning of the nineteenth century is considered to be a broad cultural category, exemplified by the work of ethnologists such as James Cowles Prichard. In contrast, by the end of the nineteenth century, understandings of race are often described in more strictly biological terms. Anthropologists like James Hunt and Robert Knox are seen as having overthrown the broad ethnological approach and replaced it with a hard biological racism.[66] This narrative, however, is seriously complicated by the history of race uncovered across the previous six chapters.[67] Right from the start, with the collection of skulls in the 1810s and 1820s, phrenological understandings of race were clearly grounded in the human body. This material approach to race was more pronounced in the colonial world, where imperial agents were raiding Egyptian tombs and executing Kandyan rebels. Similarly, on the slave plantations in the West Indies and the United States, race was hard to imagine as anything other than material. The collection of skulls and the production of plaster casts were closely linked to the regimes of violence and oppression that underpinned the early nineteenth-century Atlantic world. This partly explains why historians concentrating on Prichard, a man who never left Britain, have overemphasized the earlier ethnological program. Even in the 1830s, phrenologists in Paris and New Orleans held up busts of Africans, Indians, and Pacific Islanders in order to convince audiences that race was ultimately grounded in the brain. The dominance of Prichard's ethnological approach in the early nineteenth century therefore looks less certain, particularly when we examine the world beyond Britain. Equally, however, the second half of the nineteenth century was not entirely dominated by biological theories of race. The place of phrenological photographs in debates concerning evolution shows that, even in the 1870s and 1880s, few anthropologists were willing to take craniological evidence at face value. Marshall's *A Phrenologist Amongst the Todas* featured anthropometric photographs of Indian bodies alongside picturesque landscapes and villages. The appendix even included an analysis of the Toda language. Additionally, leading anthropologists such as Edward Burnett Tylor in Britain and Quatrefages in France considered Marshall's craniological photographs to be highly problematic, instead concentrating on his account of Toda culture and language. Bodily measurements, social habits, and philology all found a place within late nineteenth-century racial science. There was no straightforward transition from a broad cultural understanding of race to a hard biological one.

How then should we characterize nineteenth-century racial science? Rather than focusing on an intellectual change—from culture to biology—the history of phrenology reveals a change in how race was communicated and put to use. The communications revolution discussed earlier had a profound effect on the material culture of racial science. At the beginning of the century, phrenologists largely relied on physical collections of skulls housed in a small number of discrete sites. In 1834, the *Phrenological Jour-*

nal in Edinburgh told its readers that "it is only from the collections of national crania that satisfactory conclusions can be drawn."[68] However, as the century advanced, new technologies provided a means to discuss race without direct access to a collection of skulls. The plaster bust of Eustache Belin allowed supporters and critics of abolition on both sides of the Atlantic to make claims about African character, despite the fact that most of them had never met the man himself. Plaster busts certainly looked impressive during a phrenological lecture. However, it was the rise of new printing technologies that most significantly changed how racial science was communicated. The lithographs featured in Samuel George Morton's *Crania Americana*, as well as cheap woodcut reproductions, brought phrenological understandings of race to a mass audience in Europe and the United States. This process accelerated later in the century. With the introduction of photography and the spread of cheap print in the colonial world, racial sciences like phrenology traveled back to Asia, Africa, and the Americas, where many skulls had originally been collected.

Over the course of the nineteenth century, racial sciences like phrenology were deployed in new political debates ranging from the abolition of slavery to the growth of the East India Company. Strikingly, phrenology often appealed to those on opposite sides of the argument, as illustrated by its popularity among both abolitionists and slaveholders in the United States. At different points, my chapters also took up the history of phrenology from the perspective of colonized and enslaved peoples. Although these voices can be difficult to recover, there is plenty of evidence that phrenology was discussed, promoted, and critiqued by African American, Bengali, and Chinese audiences. Once again, the spread of new communication technologies, particularly cheap print, contributed to the growth of racial science beyond earlier European audiences. By midcentury, enslaved and colonized peoples were attacking phrenology. Yet others, such as Cally Coomar Das in 1850s Calcutta, as well as the editors of the Commercial Press in 1920s China, felt that phrenology did have something to offer. For the first time, racial sciences like phrenology started to be turned against colonialism. Phrenology provided an authoritative language through which to develop a counter narrative. Often, this blended with anticolonial nationalism. This use of racial science extended into the early twentieth century, particularly in India, where phrenology was taken up by a number of figures in the anticolonial movement. Jahangir Bomanji Petit, one of Gandhi's supporters, delivered lectures on phrenology at the Bombay Cercle Littéraire in 1900, while the Bengali artist and musician Jyotirindranath Tagore undertook phrenological examinations of his more famous brother and critic of colonialism, Rabindranath Tagore (fig. 47). Nagendranath Gupta, a member of the Indian National Congress, recalled being examined at the Tagores' residence in Calcutta. Jyotirindranath "proceeded to feel our bumps, jotting down the results of his examination in a note-book." Gupta was impressed, later writing that "his reading of the propensities of our minds by the help of the protuberances on our skulls was exceedingly gratifying."[69] By writing a history of race that extends beyond European thinkers, we can start to move away from the existing intellectual history. Instead, we can better understand the changing ways in which racial sciences traveled and were put to use, whether by American slaveholders or Bengali intellectuals.

Writing in 1898, Alfred Russel Wallace made his own assessment of the nineteenth century. His account of the period is worth considering in further detail, particularly given Wallace's interest in phrenology. In *The*

Fig. 47. Phrenological sketch of Rabindranath Tagore's head, Jyotirindranath Tagore, *Twenty-Five Collotypes* (Hammersmith: Emery Walker Limited, 1914), 22, Victoria and Albert Museum.

Wonderful Century, Wallace sketched "those great material and intellectual achievements which especially distinguish the nineteenth century from any and all of its predecessors." Among the "successes" of the century, Wallace listed "modes of travelling," "labor-saving machinery," and "photography," each of which had transformed phrenology in its own way. He looked back to his boyhood in the Welsh borders before celebrating "the era of railways and of ocean-going steamships," the technologies that had transported him across the world, to South America and the Malay Archipelago. Wallace even pointed readers to the construction of the SS *Great Western*, the steamship that had taken Combe and his phrenological collection to the United States in 1838. Not unlike the history recovered in this book, Wallace argued that a communications revolution had "modified the habits and even the modes of speech of all civilized peoples." For the socialist Wallace, however, the nineteenth century was not all good. Alongside the "successes," Wallace also listed "failures." These included "the plunder of the earth" and the "demon of greed." Wallace lamented the "increase of poverty, of insanity, of suicide, and probably even of crime." He even suggested that "to the historian of the future" these darker aspects of the nineteenth century would "be considered its most striking characteristic." For Wallace, phrenology represented both sides of the century, the wonderful and the not-so-wonderful. He argued that it was now "universally admitted, that the brain is the organ of the mind" and praised phrenology's "practical uses in education, self-discipline, in the reformatory treatment of criminals, and in the remedial treatment of the insane." In a wry remark, Wallace even suggested that Franz Joseph Gall had anticipated Charles Darwin's work on the "expression of emotions." Yet, Wallace also complained of "the neglect of phrenology" as well as the "incredible narrowness and prejudice which prevailed among men of science at the very time they were making such splendid advances in other fields of thought." Nonetheless, Wallace declared confidently that "in the coming century Phrenology will assuredly attain general acceptance." In a utopian vision that would soon be shattered, Wallace concluded that "the coming century will reap the full fruition of that advance."[70]

Wallace was right about the nineteenth century, even if he was wrong about the twentieth. Over the course of the nineteenth century, phrenology transformed how people understood the mind. This revolution in thought was accompanied by a revolution in communication. From Sheffield to New York, from Calcutta to Shanghai, phrenologists exchanged skulls and plaster busts, books and letters, periodicals and photographs.

Wallace's own *Wonderful Century* was even translated into Chinese at around the same time as *Guxiangxue*.[71] Yet, as Wallace suggested, new ideas and new technologies reinforced inequalities, introducing "failures" as well as "successes." Phrenology was a "sign of the times," a global science of the mind for the material age.[72]

FIGURES

Figure 1. Johann Spurzheim, *The Physiognomical System of Drs. Gall and Spurzheim* (London: Baldwin, Cradock and Joy, 1815), pl. 2.

Figure 2. The skull of Alexander Pearce, University of Pennsylvania Museum of Archaeology and Anthropology.

Figure 3. "A Black Lecture on Phrenology," *Follitt's Black Lectures, No. 1* (1846).

Figure 4. Detail from "Bengali," Anatomical Museum collections, University of Edinburgh.

Figure 5. "Singalese Chief of a secluded part of the Interior," John Davy, *An Account of the Interior of Ceylon, and of its Inhabitants* (London: Longman, Hurst, Rees, Orme, and Brown, 1821), pl. 3.

Figure 6. "The death dance of the Upper Nubians," Victoria and Albert Museum.

Figure 7. "An Esquimaux Grave," in George Lyon, *A Brief Narrative of an Unsuccessful Attempt to Reach Repulse Bay* (London: John Murray, 1825).

Figure 8. Plaster cast of Rammohun Roy's head, Science Museum, London.

Figure 9. Plaster cast of the head of Eustache Belin, Warren Anatomical Museum, Harvard University.

Figure 10. Comparing a skull to a bust during a lecture in the United States, George Combe, *Lectures on Phrenology* (London: Simpkin, Marshall and Co., 1839), frontispiece.

Figure 11. Woodcuts of the bust of Eustache Belin, "Character of Eustache," *American Phrenological Journal* 2 (1840).

Figure 12. Frontispiece and title page, Samuel George Morton, *Crania Americana* (Philadelphia: J. Dobson, 1839).

Figure 13. Loose lithographic plates printed to promote Samuel George Morton, *Crania Americana.*

Figure 14. Samuel George Morton, *Crania Americana* (Philadelphia: J. Dobson, 1839), pl. 71.

Figure 15. Detail from pl. 40, Samuel George Morton, *Crania Americana* (Philadelphia: J. Dobson, 1839).

Figure 16. Dedication to James Cowles Prichard in the "foreign edition" of Samuel George Morton, *Crania Americana* (Philadelphia: J. Dobson, 1839).

Figure 17. Woodcut illustrations of "American Indian" and "European" brains, [Combe], "Crania Americana," *American Journal of Science* 38 (1840).

Figure 18. Combe's notes on slavery in the left margin of Charles Caldwell's letter, Caldwell to Combe, 30 August 1839, Combe Papers, National Library of Scotland.

Figure 19. "Am I Not a Woman and a Sister?" glass seal, Collection of the Massachusetts Historical Society, Boston.

Figure 20. Struve to Combe, 9 August 1848, Combe Papers, National Library of Scotland.

Figure 21. Opening of the May 1849 issue of the *American Phrenological Journal,* with advertisements and monthly notices on the inside cover.

Figure 22. Loose printed "Pictorial Prospectus" for the *American Phrenological Journal.*

Figure 23. Sketch of a "Chinese Servant Boy," sent by Justus Doolittle from Fuzhou, *Phrenological Journal and Science of Health* 50 (1870).

Figure 24. Portraits of "Nena Sahib" and "The King of Oude," woodcuts copied from Indian miniatures on ivory, *American Phrenological Journal* 39 (1864).

Figure 25. *Graham's Town Journal,* 10 December 1835, featuring "Government Notice" following the 1834–1835 Xhosa War directly above an advertisement for Henry Macartney's phrenological lecture.

Figure 26. The title page of the fifth number of *The Pamphleteer* 1 (1850).

Figure 27. Advertisement for Radhaballav Das, *Manatatva Sarsangraha* (Calcutta: Purnochandradoy Press, 1849), in *The Pamphleteer* 1 (1850).

Figure 28. Phrenological chart printed in Radhaballav Das, *Manatatva Sarsangraha* (Calcutta: Purnochandradoy Press, 1849).

Figure 29. "A Negrito," in N. W. Beckwith, "Aborigines of the Philippine Archipelago," *Phrenological Journal and Life Illustrated* 50 (1870).

Figure 30. "Malay Chiefs, Mindanao," in "American Possessions in the Philippines," *Phrenological Journal and Science of Health* 107 (1899).

Figure 31. "Female Profile" and "Female Full Face," in William Elliot Marshall, *A Phrenologist Amongst the Todas* (London, 1873), nos. 5 and 6.

Figure 32. Ottewill Folding Box Camera, National Science and Media Museum, Bradford.

Figure 33. "Rouch's Registered Portable Dark Operating Chamber," Thomas Sutton, *A Dictionary of Photography* (London: S. Low and Son, 1858).

Figure 34. "A Toda Mand," William Elliot Marshall, *A Phrenologist Amongst the Todas* (London: Longmans and Green, 1873), no. 1.

Figure 35. "The Land He Lives In," William Elliot Marshall, *A Phrenologist Amongst the Todas* (London: Longmans and Green, 1873), no. 10.

Figure 36. Samuel Bourne, "26. Panoramic View from a Spur of Dodabet," Royal Commonwealth Society Archives, Cambridge University Library.

Figure 37. "Aino Man, Urap," Alfred Russel Wallace, *Studies Scientific and Social* (London: Macmillan and Co., 1900), vol. 1, fig. 81.

Figure 38. "Adam and Eve," William Elliot Marshall, *A Phrenologist Amongst the Todas* (London: Longmans and Green, 1873), no. 16.

Figure 39. Jean Louis Armand de Quatrefages, *Hommes Fossiles et Hommes Sauvages* (Paris: Baillière, 1884), fig. 179.

Figure 40. Jean Louis Armand de Quatrefages, *Hommes Fossiles et Hommes Sauvages* (Paris: Baillière, 1884), fig. 182.

Figure 41. Jean Louis Armand de Quatrefages, *Hommes Fossiles et Hommes Sauvages* (Paris: Baillière, 1884), fig. 181.

Figure 42. Jean Louis Armand de Quatrefages's translation of *A Phrenologist Amongst the Todas* appears on the same page as a sketch by the French astronomer Jules Janssen. Jean Louis Armand de Quatrefages, *Hommes Fossiles et Hommes Sauvages* (Paris: Baillière, 1884), 508.

Figure 43. Phrenological chart, Feng Pingsheng, *Guxiangxue* (Shanghai: Commercial Press, 1923).

Figure 44. "Central Nervous System in Vertebrates," Feng Pingsheng, *Guxiangxue* (Shanghai: Commercial Press, 1923).

Figure 45. "White Man," "Black Man," "Orangutan," "Baboon," Feng Pingsheng, *Guxiangxue* (Shanghai: Commercial Press, 1923).

Figure 46. Engraving of a photograph of a plaster bust, "Truganina or Lalla-Rook, the last Tasmanian," Jean Louis Armand de Quatrefages, *Histoire Générale des Races Humaines* (Paris: A. Hennuyer, 1889), fig. 261.

Figure 47. Phrenological sketch of Rabindranath Tagore's head, Jyotirindranath Tagore, *Twenty-Five Collotypes* (Hammersmith: Emery Walker Limited, 1914).

ACKNOWLEDGMENTS

I always read the acknowledgments page. Now I have the pleasure of writing one. I must begin by adding my name to the long list of students who owe an enormous debt of gratitude to Jim Secord. As many others will attest, Jim is that rare combination of an outstanding scholar and a generous teacher. His immense enthusiasm, knowledge, and attention to detail made me the historian I am today. Janet Browne, Simon Schaffer, and Sujit Sivasundaram also proved incredible mentors, and certainly shaped my intellectual outlook for the better. Many of the arguments contained in this book only became clear following the Exploring Traditions series of workshops, organized by Simon and Sujit, and culminating in a memorable trip to Delhi.

Many people shaped this project through their willingness to offer advice and comment on drafts of various chapters. In this respect I would like to thank Megan Barford, Michael Bravo, Mary Brazelton, Janet Browne, Alix Chartrand, Chris Courtney, Alex Csiszar, Shinjini Das, Christof Dejung, Ann Fabian, Lachlan Fleetwood, James Hall, Hatice Hildiz-Yaman, Sarah Hodges, Shruti Kapila, Jagjeet Lally, David Lambert, Naomi Parkinson, Eòin Phillips, Sadiah Qureshi, Jake Richards, Ricardo Roque, Rohan Deb Roy, Simon Schaffer, Thomas Simpson, Mishka Sinha, Sujit Sivasundaram, Tom Smith, Richard Staley, Kate Stevens, Raphael Uchôa, Aaron Watts, Iain Watts, Paul White, Callie Wilkinson, and James Wilson. It should be clear that global history cannot be written in isolation, particularly when the focus is on a science translated into at least six different languages. I would therefore like to thank Kirie Stromberg for her translation of the Chinese sources, and Sohini Chattopadhyay, Tanima Dey, Ishan Mukherjee, and Rohan Deb Roy for their translation of the Bengali sources. I would also like to thank Raphael Uchôa for pointing me toward the sources on the history of phrenology in Brazil as well as for his help with translations from the Portuguese.

From the outset it was my intention to ground this project in material culture. I am therefore especially grateful to the librarians and museum staff who assisted me during my research, including those at the Asiatic Society of Bengal, Cambridge University Library, Cornell University Library, the National Library of Scotland, the American Philosophical Society, the Houghton Library, the Historical Society of Pennsylvania, the Academy of Natural Sciences of Drexel University, the College of Physicians of Philadelphia, the Firestone Library at Princeton University, the Massachusetts Historical Society, the University of Edinburgh Library, the National Archives of India, the British Library, the Wellcome Library, the Whipple Library, the Warren Anatomical Museum, the National Galleries of Scotland, the Science Museum, the Victoria and Albert Museum, and the University of Edinburgh Anatomy Department. I would also like to thank Ben Garlick for hosting me during a number of trips to Scotland.

Over the course of this research I benefited from the generous financial support of a number of institutions. First and foremost, I would like to thank the Master and Fellows of Trinity College, University of Cambridge, for awarding me the Tarner Studentship to undertake doctoral research in the history of science. I would also like to thank the Master and Fellows of Darwin College, University of Cambridge, for appointing me as the Adrian Research Fellow, and giving me the time and space to write. The Global History and Culture Centre at the University of Warwick provided funds to support image reproduction costs, essential for a book such as this. Additional funding for research in the United States, France, and India was provided by the Rouse Ball Fund at Trinity College, the British Society for the History of Science, the Philadelphia Area Center for the History of Science, and the Countway Library at Harvard University. The Centre for Research in the Arts, Social Sciences, and Humanities at the University of Cambridge also supported the foundation of the Global Science seminar series. Along with my fellow convenors, running these seminars provided a rare opportunity to engage with the global history of science on a regular basis. *Materials of the Mind* would have been a very different history without this support.

At the University of Chicago Press, I would like to extend a special thanks to Karen Darling. Karen is the kind of editor every author hopes for. She took the manuscript and immediately clarified what I was trying to achieve. This, along with exceptionally helpful comments from two anonymous reviewers, helped me push the argument much further and with what I hope is greater precision. I would also like to thank Susannah Engstrom for such excellent editorial support and Marian Rogers for copy-editing the manuscript.

Earlier versions of parts of chapters 3 and 4 appeared as "National Types: The Transatlantic Publication and Reception of *Crania Americana* (1839)," *History of Science* 53 (2015); and as "Phrenology, Correspondence, and the Global Politics of Reform, 1832–1848," *Historical Journal* 60 (2017). I am grateful to Sage Publishing and Cambridge University Press for permission to include this material.

This book is dedicated to my wife, Alice, and my mother, Nancy. Before writing *Materials of the Mind,* I never really understood what people meant when they said that they "couldn't have done it" without someone. Now I do. Thank you.

NOTES

INTRODUCTION

1. George Combe, *Notes on the United States of North America* (Edinburgh: Maclachlan and Stewart, 1841), 1:16.

2. John van Wyhe, "The Authority of Human Nature: The Schädellehre of Franz Joseph Gall," *British Journal for the History of Science* 35 (2002).

3. For the history of *The Constitution of Man* in Britain, see James Secord, *Visions of Science: Books and Readers at the Dawn of the Victorian Age* (Oxford: Oxford University Press, 2014), 173–204; John van Wyhe, *Phrenology and the Origins of Victorian Scientific Naturalism* (Aldershot: Ashgate, 2004), 96–164. *The Constitution of Man* sold well over 300,000 copies by 1900. *On the Origin of Species* managed only 50,000. See also Roger Cooter, *The Cultural Meaning of Popular Science: Phrenology and the Organization of Consent in Nineteenth-Century Britain* (Cambridge: Cambridge University Press, 1984), 120.

4. For Japanese and Bengali translations of Combe's works, see Hideki Nagamine, *Seisogaku Genron* (Tokyo: Senshindo, 1918); Radhaballav Das, *Manatatva Sarsangraha* (Calcutta: Sangbad Purnochandradoy, 1849). Wyhe, *Phrenology*, 217–228, lists editions in European languages including French, German, and Swedish.

5. Stefan Collini, "The Idea of 'Character' in Victorian Political Thought," *Transactions of the Royal Historical Society* 35 (1985).

6. Phrenology's demise in the 1870s and 1880s is a common claim; see John Davies, *Phrenology: Fad and Science* (New Haven, CT: Yale University Press, 1955), 172–174; Nancy Stepan, *The Idea of Race in Science: Great Britain, 1800–1960* (London: Macmillan, 1982), 28. However, it is not clear if this characterization even fully applies to the British context; see Cooter, *The Cultural Meaning*, 16–22.

7. Madeleine Stern, *Heads and Headlines: The Phrenological Fowlers* (Norman: University of Oklahoma Press, 1971), 199–259.

8. Shruti Kapila, "Race Matters: Orientalism and Religion, India and Beyond, c. 1770–1880," *Modern Asian Studies* 41 (2007): 502–511.

9. Steven Shapin, "Phrenological Knowledge and the Social Structure of Early

Nineteenth-Century Edinburgh," *Annals of Science* 32 (1975); Shapin, "The Politics of Observation," in Roy Wallis, ed., *On the Margins of Science: The Social Construction of Rejected Knowledge* (Keele: University of Keele Press, 1979). For the differences between Cooter and Shapin, see Cooter, *The Cultural Meaning*, 9.

10. For Ireland, see Enda Leaney, "Phrenology in Nineteenth-Century Ireland," *New Hibernia Review* 10 (2006); for France, see Marc Renneville, *Le langage des crânes* (Paris: Institut d'édition Sanofi-Synthélabo, 2000).

11. Kapila, "Race Matters"; Andrew Bank, "Of 'Native Skulls' and 'Noble Caucasians': Phrenology in Colonial South Africa," *Journal of Southern African Studies* 22 (1996).

12. Sujit Sivasundaram, "Sciences and the Global: On Methods, Questions, and Theory," *Isis* 101 (2010).

13. George Combe, *The Constitution of Man* (Edinburgh: John Anderson, 1828), 1.

14. My focus on communication is indebted to James Secord, "Knowledge in Transit," *Isis* 95 (2004).

15. On the importance of periodization, see Adam McKeown, "Periodizing Globalization," *History Workshop Journal* 63 (2007). Christophy Bayly, *The Birth of the Modern World, 1780–1914: Global Connections and Comparisons* (Oxford: Blackwell, 2004), 41–48, also highlights this through his discussion of "archaic globalisation."

16. Bayly, *The Birth of the Modern World*, remains the best account of these broader changes.

17. James Secord, *Victorian Sensation: The Extraordinary Publication, Reception, and Secret Authorship of "Vestiges of the Natural History of Creation"* (Chicago: University of Chicago Press, 2000). For readings of *Vestiges* in Bombay and Philadelphia, see "Vestiges of Rain and Wind," *Bombay Times and Journal of Commerce*, 5 February 1845; Mott to Combe, 3 February 1846, MS7281, f. 17, George Combe Papers, National Library of Scotland (henceforth, Combe Papers).

18. Song-Chuan Chen, "An Information War Waged by Merchants and Missionaries at Canton: The Society for the Diffusion of Useful Knowledge in China, 1834–1839," *Modern Asian Studies* 46 (2012).

19. Saul Dubow, *A Commonwealth of Knowledge: Science, Sensibility, and White South Africa, 1820–2000* (Oxford: Oxford University Press, 2006), 38–40; John Gascoigne, *The Enlightenment and the Origins of European Australia* (Cambridge: Cambridge University Press, 2002), 88.

20. Secord, *Victorian Sensation*, 24.

21. Spurzheim to Perier, 18 March 1816, Johann Spurzheim Papers, B MS c22, Countway Library, Harvard University.

22. Wilfred Steel, *The History of the London and North Western Railway* (London: Railway and Travel Monthly, 1914), 1.

23. Iain Watts, "'We want no authors': William Nicholson and the Contested Role of the Scientific Journal in Britain, 1797–1813," *British Journal for the History of Science* 47 (2014).

24. Andrew Carmichael, *A Memoir of the Life and Philosophy of Spurzheim* (Boston: Marsh, Capen and Lyon, 1833), 47–50.

25. For the importance of global history as a form of analysis, see Geoff Eley, "His-

toricizing the Global, Politicizing Capital: Giving the Present a Name," *History Workshop Journal* 63 (2007): 158. For criticism of global history as a description of the past, see Frederick Cooper, *Colonialism in Question: Theory, Knowledge, History* (Berkeley: University of California Press, 2005), 91–112; Sarah Hodges, "Global Menace," *Social History of Medicine* 25 (2012).

26. Hodges, "Global Menace," 720.

27. Cooper, *Colonialism*, 93.

28. Unlike Cooper, Hodges concludes her article more optimistically, arguing it might be possible to write a global history that "turns on analyses of difference and power, rather than sweeping it under a global carpet of sameness and connections." Hodges, "Global Menace," 725.

29. For a classic statement of the need to study science as it travels, see Secord, "Knowledge in Transit." My own approach also owes much to the papers collected in Arjun Appadurai, ed., *The Social Life of Things: Commodities in Cultural Perspective* (Cambridge: Cambridge University Press, 1988). My focus on materiality and practice is indebted to Jim Endersby, *Imperial Nature: Joseph Hooker and the Practices of Victorian Science* (Chicago: University of Chicago Press, 2008). For a critical assessment of the role of circulation in global histories of science, see Kapil Raj, "Introduction: Circulation and Locality in Early Modern Science," *British Journal for the History of Science* 43 (2010); Fa-ti Fan, "The Global Turn in the History of Science," *East Asian Science, Technology and Society* 6 (2012).

30. Cooter, *The Cultural Meaning*, 11, 148–149, and 356; Shapin, "The Politics of Observation," 149–167; Shapin, "Phrenological Knowledge," 223–243.

31. For New South Wales, see David Giustino, *Conquest of Mind: Phrenology and Victorian Social Thought* (London: Croom Helm, 1975), 145–162; for India, see Kim Wagner, "Confessions of a Skull: Phrenology and Colonial Knowledge in Early Nineteenth-Century India," *History Workshop Journal* 69 (2010); for the United States, see Davies, *Phrenology*.

32. Cooter, *The Cultural Meaning*, 8.

33. Shapin, "The Politics of Observation," 149–167.

34. Shapin, "The Politics of Observation," 140; and Cooter, *The Cultural Meaning*, 8.

35. Geoffrey Cantor, "A Critique of Shapin's Social Interpretation of the Edinburgh Phrenology Debate," *Annals of Science* 32 (1975).

36. Shapin, "Phrenological Knowledge," 234.

37. Charles Withers and David Livingstone, "Thinking Geographically about Nineteenth-Century Science," in Charles Withers and David Livingstone, eds., *Geographies of Nineteenth-Century Science* (Chicago: University of Chicago Press, 2011), 2.

38. "Global, adj.," *Oxford English Dictionary Online*, 2009, http://www.oed.com/view/Entry/79019.

39. Arjun Appadurai, "How Histories Make Geographies," *Transcultural Studies* 1 (2010): 11; Fan, "The Global Turn in the History of Science."

40. Buchanan to Capen, 18 June 1839, Nahum Capen Papers, B MS c23, Countway Library, Harvard University.

41. "Phrenology," *Graham's Town Journal*, 10 December, 1835, 1; Das, *Manatatva*.

42. Combe later struggled to manage the copyrights in the United States that were also claimed by the publisher Marsh, Capen and Lyon; see Combe to Howe, 31 January 1841, MS7388, f. 351, Combe Papers.

43. Caldwell to Combe, 3 July 1821, MS7206, f. 51, Combe Papers.

44. Cramer to Morton, 5 August 1848, MS7390, Series I: Correspondence, Samuel George Morton Papers, Historical Society of Pennsylvania, Philadelphia.

45. Bose to Combe, 7 December 1848, MS7289, f. 80, Combe Papers.

46. "Maritime Extracts," *Sailors' Magazine* 9 (1847): 213.

47. Cooper, *Colonialism*, 91.

48. Cooper, *Colonialism*, 91.

49. Appadurai, "How Histories Make Geographies," 7.

50. "On the Progressive Diffusion of Phrenology," *Phrenological Journal* 10 (1836–1837): 352.

51. James Meigs, *Catalogue of Human Crania, in the Collection of the Academy of Natural Sciences of Philadelphia* (Philadelphia: J.B. Lippincott and Co., 1857), 46–47.

52. *Catalogue of the Works in the Library of the Sydney Mechanics' School of Arts* (Sydney: Caxton, 1869), 29.

53. Roy Macleod, "On Visiting the 'Moving Metropolis': Reflections on the Architecture of Imperial Science," in Nathan Reingold and Marc Rothenberg, eds., *Scientific Colonialism* (Washington, DC: Smithsonian Institution Press, 1987); Harm de Blij, *The Power of Place: Geography, Destiny, and Globalization's Rough Landscape* (Oxford: Oxford University Press, 2009), 3–30.

54. "Narrative of the Escape of Eight Convicts from Macquarie Harbour," DL MS 3, State Library of New South Wales, Australia; Morton to Combe, 5 April 1839, MS7251, f. 166, Combe Papers; Dan Sprod, *Alexander Pearce of Macquarie Harbour* (Hobart: Cat and Fiddle Press, 1977).

55. Thomas Bender, "Historians, the Nation, and the Plenitude of Narratives," in Thomas Bender, ed., *Rethinking American History in a Global Age* (Berkeley: University of California Press, 2002).

56. [John Gordon], "The Doctrines of Gall and Spurzheim," *Edinburgh Review* 25 (1815): 227.

57. Bank, "Of 'Native Skulls,'" 391–392.

58. François Magendie, *Précis Élémentaire de Physiologie,* 2nd ed. (Paris: Méquignon-Marvis, 1825), 1:202; David Drummond, *Objections to Phrenology* (Calcutta: Durrumtollah Academy, 1829), vii.

59. Arjun Appadurai, *Modernity at Large: Cultural Dimensions of Globalization* (Minneapolis: University of Minnesota Press. 1996), 33, invites us to write the history of "imagined worlds." Roland Robertson, *Globalization: Social Theory and Global Culture* (London: Sage, 1992), 183, also considers the emergence of a "global consciousness."

60. Hewett Watson, *Statistics of Phrenology* (London: Longman, Rees, Orme, Brown, Green, and Longman, 1836), 69 and 230.

61. "The Prospect of the Age," *The Pamphleteer* 1 (1850): 84; "The Reception of the American Phrenological Journal," *American Phrenological Journal* 1 (1839): 62.

62. Marc Rochette, "Dumont d'Urville's Phrenologist: Dumoutier and the Aesthetics of Races," *Journal of Pacific History* 38 (2003).

63. "Miscellany," *American Phrenological Journal* 1 (1839): 487–488.

64. George Combe, "Answer to Mr Stone's Observations," *Phrenological Journal* 6 (1829–30): 13–14.

65. Paterson to Combe, [May 1823], MS7211, f. 9, Combe Papers.

66. Caldwell to Combe, 24 June 1833, MS7232, f. 44, Combe Papers.

67. George Combe, *Lectures on Phrenology* (London, 1839), 305.

68. Ann Fabian, *The Skull Collectors: Race, Science, and America's Unburied Dead* (Chicago: University of Chicago Press, 2010), 103–112; Ricardo Roque, *Headhunting and Colonialism: Anthropology and the Circulation of Human Skulls in the Portuguese Empire, 1870–1930* (Basingstoke: Palgrave Macmillan, 2010); Paul Turnball, "'Rare Work amongst the Professors': The Capture of Indigenous Skulls within Phrenological Knowledge in Early Colonial Australia," in Barbara Creed and Jeanette Horn, eds., *Body Trade: Captivity, Cannibalism, and Colonialism in the Pacific* (Oxford: Routledge, 2001).

69. Sigrid Schmalzer, *The People's Peking Man: Popular Science and Human Identity in Twentieth-Century China* (Chicago: University of Chicago Press, 2008); Christophe Jaffrelot, "The Idea of the Hindu Race in the Writings of Hindu Nationalist Ideologues in the 1920s and 1930s," in Peter Robb, ed., *The Concept of Race in South Asia* (Oxford: Oxford University Press, 1995).

70. On the need for a global history of race, see Sujit Sivasundaram, "Race, Empire, and Biology before Darwin," in Denis Alexander and Ronald Numbers, eds., *Biology and Ideology from Descartes to Dawkins* (Chicago: University of Chicago Press, 2010). For a twentieth-century perspective on this problem, see Helen Tilley, "Racial Science, Geopolitics, and Empires: Paradoxes of Power," *Isis* 105 (2014).

71. For example, Saul Dubow, *Scientific Racism in Modern South Africa* (Cambridge: Cambridge University Press, 1995); James Belich, "Race," in David Armitage and Alison Bashford, eds., *Pacific Histories: Ocean, Land, People* (Basingstoke: Palgrave Macmillan, 2014); Clare Anderson, *Legible Bodies: Race, Criminality, and Colonialism in South Asia* (Oxford: Berg, 2004); Fabian, *The Skull Collectors.*

72. François-Joseph-Victor Broussais, *Lectures on Phrenology* (London: George Routledge, 1847), 931.

73. George Combe, *A System of Phrenology* (London: John Anderson, 1830), 297; Burgess, "On the Character of the Hindoos," *American Phrenological Journal* 9 (1847): 247–253.

74. For a reflection on the study of race as an actors' category, see Sadiah Qureshi, *Peoples on Parade: Exhibitions, Empire, and Anthropology in Nineteenth-Century Britain* (Chicago: University of Chicago Press, 2011), 9–10 and 271–284.

75. On the need to account for "indigenous notions of race," see Sivasundaram, "Race, Empire, and Biology before Darwin," 130.

76. There is, however, increasing interest in African American science, including racial science, see Britt Rusert, *Fugitive Science: Empiricism and Freedom in Early African American Culture* (New York: New York University Press, 2017).

77. *The Colored American*, 23 September 1837.

78. *Bombay Times*, 13 February 1857.

79. *Barbados Globe*, 27 August 1868.

80. "Phrenological Development of the Bengalees," *The Pamphleteer* 1 (1850): 67–69.

81. For example, Stepan, *The Idea of Race*; Stephen Jay Gould, *The Mismeasure of Man* (New York: Norton, 1981); Peter Robb, "Introduction: The Concept of Race in South Asia," in Peter Robb, ed., *The Concept of Race in South Asia* (Oxford: Oxford University Press, 1995).

82. Sadiah Qureshi, "Displaying Sara Baartman, the 'Hottentot Venus,'" *History of Science* 42 (2004); Samuel Redman, *Bone Rooms: From Scientific Racism to Human Prehistory in Museums* (Cambridge, MA: Harvard University Press, 2016); Robin Bernstein, "Dances with Things: Material Culture and the Performance of Race," *Social Text* 101 (2009). Much of this work is inspired by the earlier collection of essays, George Stocking, ed., *Objects and Others: Essays on Museums and Material Culture* (Madison: University of Wisconsin Press, 1985).

83. Qureshi, "Displaying Sara Baartman," 237.

84. William Mahar, *Behind the Burnt Cork Mask: Early Blackface Minstrelsy and Antebellum American Popular Culture* (Urbana-Champaign: University of Illinois Press, 1999), 70–51.

CHAPTER ONE

1. George Byron, *Voyage of HMS Blonde to the Sandwich Islands* (London: John Murray, 1826), 124–130.

2. Byron, *Voyage of HMS Blonde*, 134. For phrenology in the Pacific, see Marc Rochette, "Dumont d'Urville's Phrenologist: Dumoutier and the Aesthetics of Races," *Journal of Pacific History* 38 (2003).

3. For a complete list, see Descriptive Catalogue of the Museum of the Phrenological Society, Gen 608/5, Records of the Phrenological Society of Edinburgh, Special Collections, Edinburgh University Library (henceforth, Records of the Phrenological Society).

4. For the wider social and cultural significance of death, see Antonius Robben, "Death and Anthropology: An Introduction," in Antonius Robben, ed., *Death, Mourning, and Burial: A Cross-Cultural Reader* (Malden, MA: Blackwell, 2004).

5. Death, mourning, and burial all carried particular social and political meanings back in Britain; see James Curl, *The Victorian Celebration of Death* (Newton Abbot: David and Charles, 1972), 2–7.

6. Ricardo Roque, *Headhunting and Colonialism: Anthropology and the Circulation of Human Skulls in the Portuguese Empire, 1870–1930* (Basingstoke: Palgrave Macmillan, 2010); Kim Wagner, "Confessions of a Skull: Phrenology and Colonial Knowledge in Early Nineteenth-Century India," *History Workshop Journal* 69 (2010); Maya Jasanoff, *Edge of Empire: Conquest and Collecting in the East, 1750–1850* (London: Harper Perennial, 2006), 1–13.

7. Roque, *Headhunting and Colonialism*, 70, rightly argues that skull collecting should be seen as an "entangled system."

8. The use of historical anthropology here is inspired by Marshall Sahlins's account of Captain Cook's death; Marshall Sahlins, *Islands of History* (Chicago: University of Chicago Press, 1985), 104–131.

9. Thomas Metcalf, *Ideologies of the Raj* (Cambridge: Cambridge University Press, 1994), 92–100; Curl, *The Victorian Celebration of Death*, 42–55.

10. Entry dated 2 February 1829, Cash Book, Gen 608/18, Records of the Phrenological Society.

11. "Esquimaux No.8 . . . was found in lat. 69° 21′ 19″ N. and long. 81° 31′ W.," Descriptive Catalogue of the Museum of the Phrenological Society, Gen 608/5, f. 121, Records of the Phrenological Society.

12. Catalogue of Skulls and Casts, Gen 608/4, Records of the Phrenological Society.

13. "Description of an Ashantee Skull," *Phrenological Journal* 4 (1826–1827): 311.

14. "On the Coincidence Between the Natural Talents and Dispositions of Nations, and the Development of their Brains," *Phrenological Journal* 2 (1824–1825): 2.

15. Moffat to Combe, [1825], MS7216, f. 30, Combe Papers. See also "Catalogue of Casts of Skulls of Different Nations," *Phrenological Journal* 6 (1829–1830): 145.

16. This account of the preservation of human remains draws on Christine Quigley, *Skulls and Skeletons: Human Bone Collections and Accumulations* (Jefferson, NC: McFarland, 2001), 9–10.

17. "Woman in Her Social and Domestic Character," *Phrenological Journal* 7 (1831–1832): 412.

18. "Miscellaneous Notices," *Phrenological Journal* 9 (1834–1836): 287.

19. Skulls were not obvious commodities. Rather, the practice of collecting was part of a process of commodification; Igor Kopytoff, "The Cultural Biography of Things: Commodification as Process," in Arjun Appadurai, ed., *The Social Life of Things: Commodities in Cultural Perspective* (Cambridge: Cambridge University Press, 1988), 64–65.

20. On public and private collections in this period, see Samuel Alberti, "Owning and Collecting Natural History Objects in Nineteenth-Century Britain," in Marco Beretta, ed., *From Private to Public: Natural Collections and Museums* (Sagamore Beach, MA: Science History Publications, 2005).

21. "Miscellaneous Notices," *Phrenological Journal* 9 (1834–1836): 287.

22. Authenticity remained a problem well into the late nineteenth and early twentieth century, see Roque, *Headhunting and Colonialism*, 120–124.

23. Descriptive Catalogue, Gen 608/5, f. 15, Records of the Phrenological Society.

24. For the professionalisation of anthropology, see Henrika Kuklick, *The Savage Within: The Social History of British Anthropology, 1885–1945* (Cambridge: Cambridge University Press, 1991), 27–74.

25. On the variety of practices and motivations for collecting, see Michael O'Hanlon, "Introduction," in Michael O'Hanlon and Robert Welsch, eds., *Hunting the Gatherers: Ethnographic Collectors and Agency in Melanesia, 1870s–1930s* (New York: Berghahn Books, 2000), 12.

26. "On the Character and Cerebral Development of the Esquimaux," *Phrenological Journal* 8 (1832–1834): 289.

27. "Third and Final Report of Committee of the Museum," pasted in Descriptive Catalogue, Gen 608/5, Records of the Phrenological Society.

28. Keppetipola's skull was repatriated in 1948 to mark Sri Lankan independence. For an account of this much longer history, see Nira Wickramasinghe, "The Return

of Keppetipola's Cranium: Authenticity in a New Nation," *Economic and Political Weekly* 32 (1997).

29. Henry Marshall, *Ceylon: A General Description of the Island and its Inhabitants, with an Historical Sketch of the Conquest of the Colony by the English* (London: W.H. Allen, 1846), 277; Return of Courts Martial Held in the Kandyan Provinces, CO/54/73, f. 357, The National Archives, London (henceforth, TNA). (I am grateful to James Wilson for pointing me toward this collection.)

30. Marshall, *Ceylon*, 277.

31. George Stronach, "Marshall, Henry (1775–1851)," *Oxford Dictionary of National Biography* (2005), online edition.

32. "On the Character and Cerebral Development of the Inhabitants of Ceylon," *Phrenological Journal* 7 (1831–1832): 634–636. Sri Lankan skulls in the extant collection of the Edinburgh Phrenological Society include XXI.G.5, "Ceylonese," XXI.G.11, "Ceylon," XXI.G.12, "Ceylonese (Kandian)," XXI.G.17, "Ceylon," XXI.H.6, "Vedah (Ceylon)," Anatomy Department, University of Edinburgh.

33. Colvin de Silva, *Ceylon under the British Occupation, 1795–1833* (Colombo: Colombo Apothecaries' Co., 1953), 1:168–200; John Rogers, "Early British Rule and Social Classification in Lanka," *Modern Asian Studies* 38 (2004); Rogers, "Caste as a Social Category and Identity in Colonial Lanka," *Indian Economic and Social History Review* 97 (2006): 61.

34. Stronach, "Marshall, Henry (1775–1851)."

35. John Davy, *An Account of the Interior of Ceylon, and of its Inhabitants* (London: Longman, Hurst, Rees, Orme, and Brown, 1821), 330.

36. Marshall, *Ceylon*, 89, 208, and 249.

37. On the meaning of the tropics, see James Duncan, *In the Shadows of the Tropics: Climate, Race, and Biopower in Nineteenth Century Ceylon* (Aldershot: Ashgate, 2007), particularly 4–17.

38. Davy, *Ceylon*, 331.

39. Marshall, *Ceylon*, 186 and 194.

40. Marshall, *Ceylon*, 259.

41. Marshall, *Ceylon*, 39 and 189.

42. "Inhabitants of Ceylon," 639 and 650.

43. Davy, *Ceylon*, 329–330.

44. Marshall, *Ceylon*, 187–188.

45. Davy, *Ceylon*, 109 and 403.

46. Brownrigg to Bathurst, 9 October 1818, Despatches, No. 306, CO 54/71, f. 386, TNA.

47. Memorandum of Former Insurrectionary Attempts at Ceylon, House of Commons Reports, 1849, CO 882/1, ff. 51–52, TNA.

48. Marshall, *Ceylon*, 187–188.

49. Marshall, *Ceylon*, 200–205.

50. Marshall, *Ceylon*, 198. For the use of war poems in earlier conflicts with the British, see Sujit Sivasundaram, "Ethnicity, Indigeneity, and Migration in the Advent of British Rule to Sri Lanka," *American Historical Review* 115 (2010): 437–439.

51. Marshall, *Ceylon*, 22 and 277.

52. De Silva, *Ceylon under the British Occupation*, 179–181; Marshall, *Ceylon*, 277.

53. "Inhabitants of Ceylon," 647–648.

54. Davy, *Ceylon*, 110.

55. "Inhabitants of Ceylon," 641.

56. Descriptive Catalogue, Gen 608/5, f. 107, Records of the Phrenological Society.

57. Marshall, *Ceylon*, 20.

58. Davy, *Ceylon*, 331.

59. Marshall, *Ceylon*, 186.

60. Descriptive Catalogue, Gen 608/5, f. 107, Records of the Phrenological Society.

61. Vic Gatrell, *The Hanging Tree: Execution and the English People, 1770–1868* (Oxford: Oxford University Press, 1994), 15–16 and 83.

62. Identified as "islanding" in Sujit Sivasundaram, *Islanded: Britain, Sri Lanka, and the Bounds of an Indian Ocean Colony* (Chicago: University of Chicago Press 2013), 3–28.

63. "Inhabitants of Ceylon," 634; Rogers, "Early British Rule and Social Classification in Lanka," 627; Wickramasinghe, "The Return of Keppetipola's Cranium," 85.

64. Sivasundaram, "Ethnicity, Indigeneity, and Migration," 431–436.

65. Return of Courts Martial Held in the Kandyan Provinces, CO/54/73, f. 357, TNA; de Silva, *Ceylon under the British Occupation*, 184.

66. "Inhabitants of Ceylon," 635–650; Davy, *Ceylon*, 109.

67. Marshall, *Ceylon*, 277–279.

68. "Inhabitants of Ceylon," 639.

69. Descriptive Catalogue, Gen 608/5, f. 107, Records of the Phrenological Society.

70. Marshall, *Ceylon*, 278.

71. On the relationship between British and Kandyan attitudes to Buddhism, see Sujit Sivasundaram, "Buddhist Kingship, British Archaeology, and Historical Narratives in Sri Lanka c. 1750–1850," *Past and Present* 197 (2010).

72. Return of Courts Martial Held in the Kandyan Provinces under the Several Proclamations Declaring Martial Law, CO/54/73, f. 357, TNA.

73. Marshall, *Ceylon*, 278.

74. Rita Langer, *Buddhist Rituals of Death and Rebirth: Contemporary Sri Lankan Practice and Its Origins* (London: Routledge, 2007), 11–23. Although in some ways this was a period of transformation, many Buddhist rituals persisted even as they interacted with Christianity; see Anne Blackburn, *Locations of Buddhism: Colonialism and Modernity in Sri Lanka* (Chicago: University of Chicago Press, 2010), xii-xiv.

75. Marshall, *Ceylon*, 278–279.

76. Memorandum of Former Insurrectionary Attempts at Ceylon, House of Commons Reports, 1849, CO 882/1, f. 53, TNA; de Silva, *Ceylon under the British Occupation*, 192–194.

77. Marshall, *Ceylon*, 280.

78. Marshall, *Ceylon*, 278–279.

79. Langer, *Buddhist Rituals*, 22–23.

80. Marshall, *Ceylon*, 279–280.

81. Langer, *Buddhist Rituals*, 15.

82. Marshall, *Ceylon*, 280.

83. Langer, *Buddhist Rituals*, 15–23; Sivasundaram, "Ethnicity, Indigeneity, and Migration."

84. Marshall, *Ceylon*, 279. On the violence of public execution, see Gatrell, *The Hanging Tree*, 6–7.

85. Gatrell, *The Hanging Tree*, 316.

86. Marshall, *Ceylon*, 39 and 138.

87. Davy, *Ceylon*, 500–501.

88. Langer, *Buddhist Rituals*, 70–52 and 140.

89. Gatrell, *The Hanging Tree*, 83–84.

90. Marshall, *Ceylon*, 280. Despite Marshall's claim, I have found no other reference to this practice.

91. Wagner, "Confessions of a Skull," 39.

92. Fabian, *The Skull Collectors*, 15–24.

93. Entering tombs required crawling through confined spaces, sometimes over hundreds of feet, as described in Anthony Sattin, *Lifting the Veil: British Society in Egypt, 1768–1956* (London: J. M. Dent, 1988), 38.

94. This is likely to be the mummified head marked XXVI.A.36, "North Africa," Anatomy Department, University of Edinburgh.

95. "Donations of Mummy Heads, And Irish Skulls," *Phrenological Journal* 6 (1829–1830): 254–256. For Edward Craig, see *The Edinburgh Almanack* (Edinburgh, 1828), 354–355, 359–361, and 406. For Orlando Felix, see "Obituary," *Athenaeum*, 21 April 1860, 544–545; John Ruffle, "Lord Prudhoe and Major Felix, *Hiéroglyphiseurs Décidés*," in Jason Thompson, ed., *Egyptian Encounters* (Cairo: American University in Cairo Press, 2002); Ruffle, "The Journeys of Lord Prudhoe and Major Orlando Felix in Egypt, Nubia, and the Levant, 1826–1829," in Janet Starkey and Paul Starkey, eds., *Travellers in Egypt* (London: I. B. Tauris, 1998); Jason Thompson, *Sir Gardner Wilkinson and His Circle* (Austin: University of Texas Press, 1992), 84.

96. Sattin, *Lifting the Veil*, 26–28; Edward Said, *Orientalism* (London: Routledge and Kegan Paul, 1978), 73–112; Donald Reid, *Whose Pharaohs? Archaeology, Museums, and Egyptian National Identity from Napoleon to World War I* (Berkeley: University of California Press, 2002), 1–36; Stephanie Moser, *Wondrous Curiosities: Ancient Egypt at the British Museum* (Chicago: University of Chicago Press, 2006), 65–67; Efraim Karsh and Inari Karsh, *Empires of the Sand: The Struggle for Mastery in the Middle East, 1789–1923* (Cambridge, MA: Harvard University Press, 1999), 9–17; Jasanoff, *Edge of Empire*, 211–310.

97. Reid, *Whose Pharaohs?* 33–36. The *Description de l'Égypte* is also discussed in Said, *Orientalism*, 85–86.

98. Elliott Colla, *Conflicted Antiquities: Egyptology, Egyptomania, Egyptian Modernity* (Durham, NC: Duke University Press, 2007), 6–11; Moser, *Wondrous Curiosities*, 6–7 and 65–78.

99. David Gange, *Dialogues with the Dead: Egyptology in British Culture and Religion, 1822–1922* (Oxford: Oxford University Press, 2013), 9–27.

100. "Obituary," 544–545; Thompson, *Sir Gardner Wilkinson*, 84.

101. On Mohammad Ali's Egypt, see Afaf al-Sayyid Marsot, *Egypt in the Reign of Muhammad Ali* (Cambridge: Cambridge University Press, 1984).

102. Ruffle, "*Hiéroglyphiseurs Décidés*," 81–82.

103. The separation of "ancient" and "modern" was one aspect of the broader discourse of Orientalism, as detailed in Said, *Orientalism*, particularly 99 and 269–270.

104. "Mummy Heads," 256.

105. "Farther Observations on the Supposed Head of Ramesis II, King of Egypt, and His Queen," *Phrenological Journal* 6 (1829–1830): 524–525.

106. "Mummy Heads," 256.

107. "Head of Ramesis II," 525.

108. "Luke O'Neil and Son," *Transactions of the Phrenological Society* 1 (1820): 6.

109. William Wilde, *Narrative of a Voyage to Madeira, Teneriffe and Along the Shores of the Mediterranean* (Dublin: William Curry and Co., 1840), 2:489–490.

110. Karsh and Karsh, *Empires of the Sand*, 27–41.

111. "Remarkable Discovery of Skulls Near Jerusalem," *Phrenological Journal* 14 (1841): 220–221.

112. Ruffle, "*Hiéroglyphiseurs Décidés*," 81.

113. Ruffle, "*Hiéroglyphiseurs Décidés*," 83.

114. Gange, *Dialogues with the Dead*, 55; Reid, *Whose Pharaohs?* 70; Sattin, *Lifting the Veil*, 30; Robert Richardson, *Travels along the Mediterranean and Parts Adjacent* (London: T. Cadell, 1822), 1:117.

115. Moser, *Wondrous Curiosities*, 67 and 93–95.

116. Ruffle, "The Journeys of Lord Prudhoe and Major Orlando Felix," 75.

117. "Major Felix's Notes," Add. MS 25651, James Burton Papers, Western Manuscripts, British Library.

118. "Mummy Heads," 254–255.

119. "Mummy Heads," 254–255.

120. "Head of Ramesis II," 524–525.

121. Fabian, *The Skull Collectors*, 105–109.

122. "Head of Ramesis II," 525.

123. Ruffle, "The Journeys of Lord Prudhoe and Major Orlando Felix," 79.

124. Reid, *Whose Pharaohs?* 66.

125. Ruffle, "*Hiéroglyphiseurs Décidés*," 81

126. Colla, *Conflicted Antiquities*, 40 and 107.

127. Ruffle, "*Hiéroglyphiseurs Décidés*," 82–86.

128. Mary Louise Pratt, *Imperial Eyes: Travel Writing and Transculturation* (London: Routledge, 1992), 51–52, invites us to consider when and why colonized voices are silenced.

129. Reid, *Whose Pharaohs?* 5; Colla, *Conflicted Antiquities*, 11.

130. Moser, *Wondrous Curiosities*, 95.

131. Ruffle, "The Journeys of Lord Prudhoe and Major Orlando Felix," 78.

132. Ruffle, "*Hiéroglyphiseurs Décidés*," 84; Ruffle, "The Journeys of Lord Prudhoe and Major Orlando Felix," 78.

133. Ruffle, "*Hiéroglyphiseurs Décidés*," 84.

134. "Head of Ramesis II," 526.

135. On "Egyptomania," see James Curl, *Egyptomania: The Egyptian Revival, a Recurring Theme in the History of Taste* (Manchester: Manchester University Press, 1994).

136. Gange, *Dialogues with the Dead*, 80–81.

137. For the relationship between science, showmanship, and Egyptology, see Beverly Rogers, "Unwrapping the Past: Egyptian Mummies on Show," in Joe Kember, John Plunkett, and Jill Sullivan, eds., *Popular Exhibitions, Science, and Showmanship, 1840–1910* (London: Pickering and Chatto, 2012).

138. "Cerebral Development and Inferred Character of Kabooti, an Egyptian Mummy," *Phrenological Journal* 9 (1834–1836): 356–357.

139. "Observations on the Phrenological Standard of Civilization," *Phrenological Journal* 9 (1834–1836).

140. Moser, *Wondrous Curiosities*, 94.

141. Reid, *Whose Pharaohs?* 21.

142. Reid, *Whose Pharaohs?* 52–53; Colla, *Conflicted Antiquities*, 121–126.

143. On the need to recognize how racial science varied, see Sujit Sivasundaram, "Race, Empire, and Biology before Darwin," in Denis Alexander and Ronald Numbers, eds., *Biology and Ideology from Descartes to Dawkins* (Chicago: University of Chicago Press, 2010).

144. William Parry, *Journal of a Second Voyage for the Discovery of a North-West Passage from the Atlantic to the Pacific* (London: John Murray, 1824), 17.

145. *A Supplement to the Appendix of Captain Parry's Voyage for the Discovery of a North-West Passage in the Years 1819–1820* (London: John Murray, 1824), clxxxiii-cxcii and ccxlviii. Human skulls, however, were not listed in this summary of the "Natural History" of the voyage.

146. Lyon to Combe, 29 April 1824, MS7213, f. 97; Ross to Combe, 20 May 1823, MS7211, f. 29, Combe Papers.

147. Skull marked XVI.C.11, "Esquimaux," Anatomy Department, University of Edinburgh.

148. Robert David, *The Arctic in the British Imagination, 1818–1914* (Manchester: Manchester University Press, 2001), 2.

149. Arthur Credland, *The Hull Whaling Trade* (Beverley: Hutton Press, 1995), 11. Whaling surgeons also collected Inuit skulls; see William Ross, *Arctic Whalers, Icy Seas: Narratives of the Davis Strait Whale Fishery* (Toronto: Irwin, 1985), 26.

150. Michael Harbsmeier, "Bodies and Voices from Ultima Thule: Inuit Explorations of the Kablunat from Christian IV to Knud Rasmussen," in Michael Bravo and Sverker Sörlin, eds., *Narrating the Arctic: A Cultural History of Nordic Scientific Practices* (Canton, MA: Science History Publications, 2002).

151. Michael Bravo, "Geographies of Exploration and Improvement: William Scoresby and Arctic Whaling, 1782–1822," *Journal of Historical Geography* (2006): 528; David, *The Arctic in the British Imagination*, xvi-xvii; Trevor Levere, *Science and the Canadian Arctic: A Century of Exploration, 1818–1918* (Cambridge: Cambridge University Press, 2004), 44–45.

152. Michael Bravo, "Ethnological Encounters," in Nick Jardine, James Secord, and Emma Spary, eds., *Cultures of Natural History* (Cambridge: Cambridge University Press, 1994); Bravo, "Ethnographic Navigation and the Geographical Gift," in David Livingstone and Charles Withers, eds., *Geography and Enlightenment* (Chicago: University of Chicago Press, 1996).

153. Parry, *Journal of a Second Voyage*, 390–394.

154. Parry, *Journal of a Second Voyage*, 391; "Cerebral Development of the Esquimaux," 435–436.

155. Parry, *Journal of a Second Voyage*, 391.

156. Parry, *Journal of a Second Voyage*, 392–393.

157. Parry, *Journal of a Second Voyage*, 394–395.

158. "Cerebral Development of the Esquimaux," 432.

159. Sonja Jerkic, "Burials and Bones: A Summary of Burial Patterns and Human Skeletal Research in Newfoundland and Labrador," *Newfoundland Studies* 9 (1993); Barbara Crass, "Gender in Inuit Burial Practices," in Alison Rautman, ed., *Reading the Body: Representations and Remains in the Archaeological Record* (Philadelphia: University of Pennsylvania Press, 2000); and John Bennett and Susan Rowley, *Uqalurait: An Oral History of Nunavut* (Montreal: McGill-Queen's University Press, 2008), 221–225.

160. Curl, *The Victorian Celebration of Death*, 42–55.

161. Parry, *Journal of a Second Voyage*, 396.

162. George Lyon, *The Private Journal of Captain G.F. Lyon During the Recent Voyage of Under Captain Parry* (London: John Murray, 1824), 385 and 398.

163. Parry, *Journal of a Second Voyage*, 397.

164. "Cerebral Development of the Esquimaux," 429–431.

165. These assessments should also be considered in light of ongoing missionary work in the Arctic; see Harbsmeier, "Bodies and Voices from Ultima Thule," 52–64.

166. Parry, *Journal of a Second Voyage*, 280.

167. Peter Whitridge, "Landscapes, Houses, Bodies, Things: 'Place' and the Archaeology of Inuit Imaginaries," *Journal of Archaeological Method and Theory* (2004): 220–226; Bennett and Rowley, *Uqalurait*, 256–257.

168. Parry, *Journal of a Second Voyage*, 394.

169. Parry, *Journal of a Second Voyage*, 285.

170. Frédéric Laugrand and Jarich Oosten, "Canicide and Healing: The Position of the Dog in the Inuit Cultures of the Canadian Arctic," *Anthropos* 97 (2002): 89–75.

171. Bennett and Rowley, *Uqalurait*, 256–257.

172. Robben, "Death and Anthropology," 11, discusses rituals surrounding death as a place to establish kinship ties.

173. Parry, *Journal of a Second Voyage*, 244; Lyon, *The Private Journal*, 200.

174. Dorothy Eber, *Encounters on the Passage: Inuit Meet the Explorers* (Toronto: University of Toronto Press, 2008), 113.

175. These quotations are from an oral history given by Rosie Iqallijuq in 1998; Eber, *Encounters on the Passage*, 21. Further accounts of this creation story are documented in Laugrand and Oosten, "Canicide and Healing," 94–96.

176. On the repatriation of Inuit remains, see Kevin McAleese, "The Reinterment of Thule Inuit Burials and Associated Artifacts," *Études Inuit Studies* 22 (1998). On repatriation more generally, including Aboriginal Australian skulls from the Edinburgh Phrenological Society's collection, see Quigley, *Skulls and Skeletons*, 205–220. However, there are many skulls that still remain to be repatriated, including the Inuit specimens marked XVI.C.1 to XVI.C.11.

177. Susan Pearce, *On Collecting: An Investigation into Collecting in the European Tradition* (London: Routledge, 1994), 28.

178. Wickramasinghe, "The Return of Keppetipola's Cranium."

179. David Lambert's treatment of "captive knowledge" in his history of African geography is exemplary in this respect, as is Michael Bravo's work on Inuit mapping; David Lambert, *Mastering the Niger: James MacQueen's African Geography and the Struggle over Atlantic Slavery* (Chicago: University of Chicago Press, 2013), 88–118; Michael Bravo, *The Accuracy of Ethnoscience: A Study of Inuit Cartography and Cross-Cultural Commensurability* (Manchester: Department of Social Anthropology, University of Manchester, 1996).

180. Parry, *Journal of a Second Voyage*, x. Samuel Alberti, "Objects and the Museum," *Isis* 96 (2005): 563, also argues we must pay attention to the different kinds of objects moving alongside one another.

181. "Notices," *Phrenological Journal* 5 (1828–1829): 324.

182. Alberti, "Objects and the Museum," 565–567.

183. Such as XXI.H.6, "Veddah (Ceylon)," Anatomy Department, University of Edinburgh.

184. George Combe, *Lectures on Phrenology* (London: Simpkin, Marshall and Co., 1839), 144.

185. Samuel Morton, *Crania Americana* (Philadelphia: J. Dobson, 1839), 247–249.

186. Combe to Morton, 28 February 1840, Mss.B.M843, Series I: Correspondence, Samuel George Morton Papers, American Philosophical Society, Philadelphia.

CHAPTER TWO

1. "Execution of the Murderers of Sophia Lewis," *Bendigo Advertiser*, 4 September 1857; *Catalogue of Madame Sohier's Waxworks Exhibition* (Melbourne: Wilson and MacKinnon, 1866), 3, 9, and 10. For a history of phrenology in Australia, see Paul Turnbull, "'Rare Work amongst the Professors': The Capture of Indigenous Skulls within Phrenological Knowledge in Early Colonial Australia," in Barbara Creed and Jeanette Horn, eds., *Body Trade: Captivity, Cannibalism, and Colonialism in the Pacific* (New York: Routledge, 2001).

2. Joseph Davis, *Thesaurus Craniorum* (London: Printed for Subscribers, 1867); *A Catalogue of Phrenological Specimens Belonging to the Boston Phrenological Society* (Boston: John Ford, 1835); Marc Renneville, "Un musée d'anthropologie oublié: Le cabinet phrénologique de Dumoutier," *Bulletins et Mémoires de la Société d'Anthropologie de Paris* 10 (1998).

3. In making this argument I follow Sadiah Qureshi, "Displaying Sara Baartman, the 'Hottentot Venus,'" *History of Science* 42 (2004); Qureshi, "Robert Gordon Latham, Displayed Peoples, and the Natural History of Race, 1854–1866," *Historical Journal* 54 (2011). For an example of intellectual histories, see Nancy Stepan, *The Idea of Race in Science: Great Britain, 1800–1960* (London: Macmillan, 1982).

4. "On the Progressive Diffusion of Phrenology," *Phrenological Journal* 10 (1836–1837): 407–408.

5. Hewett Watson, *Statistics of Phrenology* (London: Longman, Rees, Orme, Brown, Green, and Longman, 1836), 233.

6. Alice Cliff, "Coming Home—Bally's Miniature Phrenological Specimens," *Science Museum Group Journal* 1 (2015).

7. George Combe, *Notes on the United States of North America* (Edinburgh: Maclachlan and Stewart, 1841), 2:238.

8. "Sohier's Waxwork Exhibition," *The Argus*, 5 April 1860.

9. Caldwell to Combe, 3 July 1821, MS7206, f. 51; Combe to Bose, 4 February 1848, MS7391, f. 297; Combe to Otto, 12 May 1825, MS7383, f. 158, Combe Papers.

10. Anthony O'Neil, *Catalogue of Casts of Skulls of Different Nations, selected from the Museum of the Phrenological Society* (Edinburgh, 1834); "Note on Luke O'Neil and Sons Prices," Minute Book of the Edinburgh Phrenological Society, Gen 608/2, Records of the Phrenological Society of Edinburgh.

11. Caldwell to Combe, 3 July 1821, MS7206, f. 51, Combe Papers.

12. Struve to Combe, 28 July 1843, MS7270, f. 78, Combe Papers.

13. Combe to Hunter, 12 October 1839, MS7397, f. 34, Combe Papers.

14. Combe to O. S. Fowler, 30 March 1839, MS7378, f. 126.

15. Peter Nicholson, *The New and Improved Practical Builder* (London: Thomas Kelly, 1837), 2:181–182.

16. "Method of Moulding and Casting Heads," *Annals of Phrenology* 1 (1834): 443–444.

17. *Hunt and Co's Directory and Topography for the Cities of Gloucester & Bristol* (London: B. W. Gardner, 1849), 77; Class: HO107, Piece: 372, Book: 6, Civil Parish: St James, County: Gloucestershire, Enumeration District: 12, Folio: 19, Page: 33, Line: 8, GSU roll: 288782, 1841 England Census, TNA.

18. Timothy Clifford, "The Plaster Shops of the Rococo and Neo-Classical Era in Britain," *Journal of the History of Collections* 4 (1992); Peter Malone, "How the Smiths Made a Living," in Rune Frederiksen and Eckart Marchand, eds., *Plaster Casts: Making, Collecting, and Displaying from Classical Antiquity to the Present* (Berlin: Walter de Gruyter, 2010).

19. Copies of the bust of Rammohun Roy are held at the Science Museum, London (1992–34/32) and the National Galleries of Scotland (PGL 2260).

20. Nicholson, *The New and Improved Practical Builder*, 181–182.

21. Caldwell to Combe, 3 July 1821, MS7206, f. 51, Combe Papers.

22. Combe to Bose, 4 February 1848, MS7391, f. 297, Combe Papers.

23. Douglas Fetherling, *The Gold Crusades: A Social History of Gold Rushes, 1849–1929* (Toronto: University of Toronto Press, 1988), 55–57.

24. *État Présent de la Noblesse Française* (Paris: Librairie Bachelin-Deflorenne, 1866), 1013; Mimi Colligan, "Kreitmayer, Maximilian Ludwig (Max) (1830–1906)," in Christopher Cunneen, Jill Roe, Beverley Kingston, and Stephen Garton, eds., *Australian Dictionary of Biography (Supplement) 1580–1980* (Melbourne: Melbourne University Publishing, 2005).

25. "Catalogue of the Casts of Skulls of Different Nations," *Phrenological Journal* 6 (1829–1830): 144.

26. As James Secord argues, "The cult of the original had not yet taken hold"; James Secord, "Monsters at the Crystal Palace," in Soraya de Chadarevian and Nick Hopwood, eds., *Models: The Third Dimension of Science* (Stanford, CA: Stanford University Press, 2004), 146.

27. Watson, *Statistics*, 136.

28. George Combe, "Remarks on Mr. E. J. Hytch's Communication on the Improvement of Phrenological Busts," *Phrenological Journal* 11 (1838): 110.

29. George Paterson, "On the Phrenology of Hindostan," *Transactions of the Phrenological Society* 1 (1826): 434.

30. "On the Life, Character, Opinions, and Cerebral Development, of Rajah Rammohun Roy," *Phrenological Journal*, 8 (1832–1834): 579; entry dated 18 December 1833, Cash Book, Gen 608/3.18, Records of the Edinburgh Phrenological Society.

31. "Phrenological Developments and Character of the Celebrated Indian Chief and Warrior, Black Hawk," *American Phrenological Journal* 1 (1839). Copies of this bust are held at the New York Historical Society Museum in New York and at the Hauberg Indian Museum in Rock Island, IL.

32. Silas Jones, *Practical Phrenology* (Boston: Russell, Shattuck and Williams, 1836), 327.

33. Sadiah Qureshi also highlights the variety of racial thinking in this period; see Sadiah Qureshi, *Peoples on Parade: Exhibitions, Empire, and Anthropology in Nineteenth-Century Britain* (Chicago: University of Chicago Press, 2011), 146; Qureshi, "Sara Baartman," 234.

34. Spellings vary between "Eustache" and "Eustace" in both English and French sources. I follow the nineteenth-century convention of using Eustache's first name, given that Belin simply identifies his master.

35. Duchesne, "Notice sur le Nègre Eustache," *Journal de la Société Phrénologique de Paris* 3 (1835): 249–250.

36. Alfred Hunt, *Haiti's Influence on Antebellum America: Slumbering Volcano in the Caribbean* (Lafayette: Louisiana State University Press, 1988), 9; Lawrence Jennings, *French Anti-slavery: The Movement for the Abolition of Slavery in France, 1802–1848* (Cambridge: Cambridge University Press, 2000), 25.

37. "Esclave en Marronnage," *Affiches Américaines*, 27 June 1780.

38. Jill Casid, *Sowing Empire: Landscape and Colonization* (Minneapolis: University of Minnesota Press, 2005), 222–223.

39. "Esclave en Marronnage," *Affiches Américaines*, 12 June 1768; "Esclave en Marronnage," *Affiches Américaines*, 12 July 1775; "Esclave en Marronnage," *Affiches Américaines*, 29 June 1774.

40. William B. Cohen, *The French Encounter with Africans: White Response to Blacks, 1530–1880* (Bloomington: Indiana University Press, 1980), 115–117.

41. Jeremy Popkin, *You Are All Free: The Haitian Revolution and the Abolition of Slavery* (Cambridge: Cambridge University Press, 2010), 1–52.

42. Richard Newman, *The Transformation of American Abolitionism: Fighting Slavery in the Early Republic* (Chapel Hill: University of North Carolina Press, 2002), 78,

43. Qureshi, "Sara Baartman," 238, argues that the perceived uniqueness of displayed people did not rely solely on the color of their skin.

44. Émile de la Bédollière, *La Morale en Action Illustrée* (Paris: Alphonse Henriot, 1837), 191–196; François Noël, *Leçons de Philosophie Morale* (Paris: Le Normant, 1834), 81–85; Benjamin Delessert, *Les Bons Exemples* (Paris: G. Kugelmann, 1858), 8–14.

45. Casimir Broussais, "Compte-Rendu des Travaux de la Société Phrénologique de Paris," *Journal de la Société Phrénologique de Paris* 2 (1833): 27 (translation my own).

46. Martin Staum, *Labeling People: French Scholars on Society, Race, and Empire, 1815–1848* (Montreal: McGill-Queen's University Press, 2003), 63.

47. Staum, *Labeling People*, 43.

48. Qureshi, "Sara Baartman," 241–245.

49. Cuvier did believe mankind had a single origin, but argued that different races had separated in the deep past; Colin Kidd, *The Forging of Races: Race and Scripture in the Protestant Atlantic World, 1600–2000* (Cambridge: Cambridge University Press, 2006), 28.

50. Cohen, *The French Encounter*, 197–198.

51. Staum, *Labeling People*, 91–99.

52. Staum, *Labeling People*, 50–51; Martin Staum, "Physiognomy and Phrenology at the Paris Athénée," *Journal of the History of Ideas* 56 (1995): 443; Elizabeth Williams, *The Physical and the Moral: Anthropology, Physiology, and Philosophical Medicine in France, 1750–1850* (Cambridge: Cambridge University Press, 1994), 182; Angus McLaren, "A Prehistory of the Social Sciences: Phrenology in France," *Comparative Studies in Society and History* 23 (1981): 6.

53. Broussais, "Compte-Rendu," 27–28 (translation my own).

54. Erwin Ackerknecht, "P. M. A. Dumoutier et la Collection Phrénologique du Musée de l'Homme," *Bulletins et Mémoires de la Société d'Anthropologie de Paris* 7 (1956): 290–291; Marc Renneville, *Le langage des crânes: Un histoire de la phrénologie* (Paris: Institut d'Édition Sanofi-Synthélabo, 2000), 478–479; Staum, *Labeling People*, 61.

55. Emile Gérards, *Paris souterrain* (Paris: Garnier Frères, 1908), pl. xiv.

56. For a description of this method of casting, see "Method of Moulding and Casting Heads," *Annals of Phrenology* 1 (1834); "Method of Taking Casts," *Phrenological Journal* 14 (1841); "Hints as to the Making of Plaster-Casts," *Phrenological Journal* 18 (1845).

57. See illustration in Jean Fossati, *Manuel Pratique de Phrénologie* (Paris: Baillière, 1845), 257.

58. "Rammohun Roy," 591–601.

59. Jennings, *French Anti-slavery*, 50–51.

60. Staum, *Labeling People*, 59; Renneville, *Le langage des crânes*, 114–115.

61. François-Joseph-Victor Broussais, *Cours de Phrénologie* (Paris: Baillière, 1836), 795 (translation my own).

62. Jennings, *French Anti-slavery*, 22–32.

63. Jennings, *French Anti-slavery*, 218.

64. Broussais, *Cours de Phrénologie*, 796 (translation my own).

65. Broussais, "Compte-Rendu," 27 (translation my own).

66. "Phrenological Journal of Paris," *Phrenological Journal* 9 (1834–1836): 134–135.

67. "Phrenological Journal of Paris," 134–135.

68. James Straton, *Contributions to the Mathematics of Phrenology* (Aberdeen: William Russel, 1845), 21.

69. Straton, *Contributions*, 9–33.

70. "Phrenological Journal of Paris," 134–135.

71. "Intelligence," *Phrenological Journal* 14 (1841): 194–196.

72. Wilson Armistead, *A Tribute for the Negro* (Manchester: W. Irwin, 1848), 539.

73. Frederick Bridges, *The Popular Manual of Phrenology* (London: George Philip and Son, 1860), 43–45.

74. "Phrenological Society of Paris," *Annals of Phrenology* 1 (1834): 405.

75. Combe, *Notes on the United States of North America*, 1:45.

76. David Stack, *Queen Victoria's Skull: George Combe and the Mid-Victorian Mind* (London: Hambledon Continuum, 2008), 125; Donald Scott, "The Popular Lecture and the Creation of a Public in Mid-Nineteenth-Century America," *Journal of American History* 66 (1980): 805.

77. Combe, *Notes on the United States of North America*, 1:59–60.

78. This account builds on recent work reconstructing the practice of phrenological lecturing; John van Wyhe, "The Diffusion of Phrenology through Public Lecturing," in Aileen Fyfe and Bernard Lightman, eds., *Science in the Marketplace: Nineteenth-Century Sites and Experiences* (Chicago: University of Chicago Press, 2007).

79. "Lectures on Phrenology," *New Yorker*, 1 June 1839.

80. Bell to Combe, 11 and 21 December 1838, MS7245, ff. 67–69, Combe Papers.

81. George Combe, *Lectures on Phrenology* (London: Simpkin, Marshall and Co., 1839), vi.

82. "Lectures on Phrenology," *New Yorker*, 1 June 1839.

83. Combe to Bell, 20 September 1822, MS7382, f. 106, Combe Papers.

84. Combe, *Notes on the United States of North America*, 3:378.

85. Combe to Channing, 28 March 1838, MS7395, f. 12, Combe Papers.

86. "Lectures on Phrenology and its Applications," *Southern Literary Messenger* 5 (1839): 810.

87. Betty Fladeland, *Men and Brothers: Anglo-American Antislavery Cooperation* (Urbana-Champaign: University of Illinois Press, 1972); Cohen, *French Encounter*, 194–195.

88. Samuel May, *Some Recollections of the Antislavery Conflict* (Boston: Fields, Osgood and Co., 1869), 316–317.

89. Jennings, *French Anti-slavery*, 29–66.

90. Jennings, *French Anti-slavery*, 171.

91. John Whittier, *The Writings of John Greenleaf Whittier* (London: Macmillan, 1888), 289–290.

92. "Eustace," *Pennsylvania Freeman*, 31 January 1839. The same issue also featured an advertisement for instruction in "Practical Phrenology" from a "Dr Woodward of 325 Race Street, Philadelphia."

93. Seymour Drescher, "British Way, French Way: Opinion Building and Revolution

in the Second French Slave Emancipation," *American Historical Review* 96 (1991): 20; Hunt, *Haiti's Influence*, 140.

94. "Eustace," *Pennsylvania Freeman*, 31 January 1839.

95. "Eustace," *Pennsylvania Freeman*, 31 January 1839.

96. Hunt, *Haiti's Influence*, 95–98.

97. "Eustace," *Pennsylvania Freeman*, 31 January 1839; Charles Tansill, *The United States and Santo Domingo, 1798–1873* (Baltimore: John Hopkins University Press, 1938), 123.

98. Hunt, *Haiti's Influence*, 95–98.

99. Hunt, *Haiti's Influence*, 43–60 and 117–121; Susan Branson and Leslie Patrick, "Étrangers dans un pays étrange: Saint-Domingan Refugees of Color in Philadelphia," in David Geggus, ed., *The Impact of the Haitian Revolution in the Atlantic World* (Columbia: University of South Carolina Press, 2001), 194–196; Paul Lachance, "Repercussions of the Haitian Revolution in Louisiana," in Geggus, *The Impact of the Haitian Revolution*, 210–221.

100. Hunt, *Haiti's Influence*, 147–155.

101. "Eustace," *Pennsylvania Freeman*, 31 January 1839.

102. Hunt, *Haiti's Influence*, 162.

103. "Eustace," *Pennsylvania Freeman*, 31 January 1839.

104. Joseph Vimont, *Traité de Phrénologie Humaine et Comparée* (Paris: Baillière, 1835), 91–93. (This edition features both an English and a French version of the text. I have used the original English translation here.)

105. "Eustace," *Friends' Intelligencer* 14 (1858): 728–729.

106. "The Faithful Slave," *African Repository* 15 (1839): 188.

107. Haiti was also popular because of the reduced costs of transportation ($15 per slave) compared to Liberia ($30 per slave); Hunt, *Haiti's Influence*, 164–170.

108. "The Faithful Slave," *African Repository* 15 (1839): 188.

109. "The Faithful Slave," *Catholic Telegraph*, 30 May 1839; "The Faithful Slave," *Parley's Magazine* 7 (1839): 374.

110. "Character of Eustache," *American Phrenological Journal* 2 (1840): 177; "Eustace," *Pennsylvania Freeman*, 31 January 1839.

111. Unlike the casts presented by anthropologists at the Crystal Palace in the 1850s, phrenological busts were rarely painted or dyed; Qureshi, "Robert Gordon Latham," 147.

112. "Character of Eustache," *American Phrenological Journal* 2 (1840): 177; "Eustace," *Pennsylvania Freeman*, 31 January 1839.

113. Carney Rhinevault and Tatiana Rhinevault, *Hidden History of the Mid-Hudson Valley* (Charleston, SC: History Press, 2011), 90.

114. "The Poughkeepsie Slave Case," *The Liberator*, 12 September 1851. The same abolitionist newspaper also published advertisements for *American Phrenological Journal*, "Prospectus," *The Liberator*, 26 October 1838.

115. Rhinevault and Rhinevault, *Hidden History*, 90.

116. George Combe, *A System of Phrenology* (Edinburgh: Maclachlan and Stewart, 1836), 1:324.

117. "On the Cerebral Development of Rajah Rammohun Roy," *Indian Journal of Medical and Physical Science* 2 (1835).

118. Graham Shaw, "Calcutta: The Birthplace of the Indian Lithographed Book," *Journal of the Printing Historical Society* 27 (1998).

119. Barbara Groseclose, *British Sculpture and the Company Raj* (Newark, DE: University of Delaware Press, 1995), 11; Paterson to Combe, 23 April 1825, MS7216, f. 47; Combe to Bose, 4 February 1848, MS7391, f. 297, Combe Papers.

120. Plaster continued to play an important but distinct role in museum and art collections; Rune Frederiksen and Eckart Marchand, "Introduction," in Rune Frederiksen and Eckart Marchand, eds., *Plaster Casts: Making, Collecting, and Displaying from Classical Antiquity to the Present* (Berlin: Walter de Gruyter, 2010).

CHAPTER THREE

1. Saul Dubow, *A Commonwealth of Knowledge: Science, Sensibility, and White South Africa, 1820–2000* (Oxford: Oxford University Press, 2006), 29–34.

2. Pringle to Combe, 29 January 1820, MS7205, f. 140, Combe Papers.

3. Pringle to Combe, 25 June 1823, MS7211, f. 21, Combe Papers. On the South African Public Library, of which Pringle was briefly the sublibrarian, see Dubow, *A Commonwealth of Knowledge*, 35–36 and 45–51.

4. Wessel de Kock, *A Manner of Speaking: The Origins of the Press in South Africa* (Cape Town: Saayman and Weber, 1982), 2.

5. *A Catalogue of the South African Public Library* (Cape Town: Gazette Office, 1829), 5–8, 26, and 40.

6. *Catalogue of the Calcutta Public Library* (Calcutta: Sanders and Cones, 1846), xvii and 96.

7. *The Catalogue of the Melbourne Public Library for 1861* (Melbourne: Clarson, Shallard and Co., 1861), 100; *A Classified Catalogue of the Mercantile Library of San Francisco* (San Francisco: Mercantile Library Association, 1861), 45.

8. Entry dated 26 July 1842, Diary of Germany, MS7422, f. 2, Combe Papers; *A Catalogue of the Library of the College of St. Margaret and St. Bernard* (London: S. and R. Bentley, 1827), 2:527.

9. *Catalogue of the Works in the Library of the Sydney Mechanics' School of Arts* (Sydney: Caxton Steam Machine Printing Office, 1869), 28–29.

10. *Catalogue of the Mercantile Library in New York* (New York: Baker, Godwin and Company, 1850), 37 and 60.

11. *Library of the Sydney Mechanics' School of Arts*, 34; Bose to Combe, 7 August 1847, MS7283, f. 63, Combe Papers.

12. A notable exception is Aileen Fyfe, *Steam-Powered Knowledge: William Chambers and the Business of Publishing, 1820–1860* (Chicago: University of Chicago Press, 2012), 173–261, which examines scientific publishing as a transatlantic enterprise.

13. James Secord, *Victorian Sensation: The Extraordinary Publication, Reception, and Secret Authorship of "Vestiges of the Natural History of Creation"* (Chicago: University of Chicago Press, 2000). For readings of *Vestiges* in Bombay and Philadelphia,

see "Vestiges of Rain and Wind," *Bombay Times and Journal of Commerce*, 5 February 1845; and Mott to Combe, 3 February 1846, MS7281, f. 17, Combe Papers.

14. John van Wyhe, *Phrenology and the Origins of Victorian Scientific Naturalism* (Aldershot: Ashgate, 2004), 96–164; James Secord, *Visions of Science: Books and Readers at the Dawn of the Victorian Age* (Oxford: Oxford University Press, 2014), 173–204.

15. Secord, *Victorian Sensation*, 54–55.

16. Richard Barnes, Bill Bell, Rimi Chatterjee, Wallace Kirsop, and Michael Winship, "A Place in the World," in David McKitterick, ed., *The Cambridge History of the Book in Britain, 1830–1914* (Cambridge: Cambridge University Press, 2009), 6:600.

17. *The Catalogue of the Melbourne Public Library*, 100; *Catalogue of the Calcutta Public Library*, 96.

18. *Catalogue of the Mercantile Library in New York*, 60.

19. Aileen Fyfe, *Science and Salvation: Evangelical Popular Publishing in Victorian Britain* (Chicago: University of Chicago Press, 2004), 1–15; James Secord, "Progress in Print," in Marina Frasca-Spada and Nick Jardine, eds., *Books and the Sciences in History* (Cambridge: Cambridge University Press, 2000), 369–392.

20. On the "People's Edition" of *The Constitution of Man*, see Secord, *Visions of Science*, 173–204; Wyhe, *Phrenology*, 133–137.

21. The same is true of William and Robert Chambers's "steam-powered" international business; Fyfe, *Steam-Powered Knowledge*, 10–11.

22. *The Catalogue of the Melbourne Public Library*, 100; *Catalogue of the Calcutta Public Library*, 96.

23. Wyhe, *Phrenology*, 225.

24. *Library of the Sydney Mechanics' School of Arts*, 34.

25. Capen to Combe, 13 February 1835, MS7234, f. 94, Combe Papers.

26. Fyfe, *Steam-Powered Knowledge*, 81–87.

27. Capen to Combe, 16 March 1836, MS7237, f. 101, Combe Papers.

28. Combe to Capen, 20 June 1835, Nahum Capen Papers, B MS c23, Countway Library, Harvard University.

29. Capen to Combe, 16 March 1836, MS7237, f. 101, Combe Papers.

30. Combe to Capen, 31 July 1839, MS7396, f. 92, Combe Papers.

31. Capen to Combe, 19 August 1839, MS7249, f. 180, Combe Papers.

32. Combe to Capen, 22 July 1839, MS7396, f. 80; Combe to Capen, 31 July 1839, MS7396, f. 92; Combe to Capen, 22 August 1839, MS7396, f. 98, Combe Papers.

33. Combe to Caldwell, 24 June 1833, MS7385 f. 560; Capen to Combe, 7 June 1835, MS7234, f. 100, Combe Papers.

34. Combe to Struve, 29 November 1845, MS7390, f. 223, Combe Papers.

35. Mariano Cubí y Soler, *La Frenologia i Sus Glorias* (Barcelona: Imprenta Hispana de V. Castaños, 1853); Radhaballav Das, *Manatatva Sarsangraha* (Calcutta: Sangbad Purnochandradoy, 1849). For a list of other European translations, including the Swedish, see Wyhe, *Phrenology*, 217–228.

36. Combe to Caldwell, 28 September 1836, MS7387, f. 140, Combe Papers.

37. Capen to Combe, 7 June 1835, MS7234, f. 100, Combe Papers.

38. For contrasting sciences of reading, see Adrian Johns, "The Physiology of Read-

ing," in Marina Frasca-Spada and Nick Jardine, eds., *Books and the Sciences in History* (Cambridge: Cambridge University Press, 2000).

39. Secord, *Victorian Sensation*, 269–275.

40. Hewett Watson, *Statistics of Phrenology* (London: Longman, Rees, Orme, Brown, Green, and Longman, 1836), 183 and 231.

41. George Combe, *The Constitution of Man* (Edinburgh: Maclachlan and Stewart, 1835), 34.

42. On the role of atlases in disciplinary debates, see Nick Hopwood, "Visual Standards and Disciplinary Change: Normal Plates, Tables, and Stages in Embryology," *History of Science* 43 (2005): 239–303; Lorraine Daston and Peter Galison, *Objectivity* (New York: Zone Books, 2007).

43. Steven Shapin, "The Politics of Observation: Anatomy and Social Interests in the Edinburgh Phrenology Disputes," in Roy Wallis, ed., *On the Margins of Science: The Social Construction of Rejected Knowledge* (Keele: University of Keele Press, 1979), 139–178.

44. Joseph Vimont, *Traité de Phrénologie Humaine et Comparée* (Paris: J. B. Baillière, 1832), pl. 82 (translation my own).

45. *Catalogue of the Library of the Royal College of Surgeons in London* (London: Richard Taylor, 1831), 219.

46. "Review of Dr. Vimont's Work on Comparative Phrenology," *American Phrenological Journal* 2 (1840): 121.

47. Josiah Nott, *Types of Mankind* (Philadelphia: J. B. Lippincott and Co., 1854), 87.

48. Bruce Dain, *A Hideous Monster of the Mind: American Race Theory in the Early Republic* (Cambridge, MA: Harvard University Press, 2002), 197.

49. Ann Fabian, *The Skull Collectors: Race, Science, and America's Unburied Dead* (Chicago: University of Chicago Press, 2010), 1.

50. Barry Joyce, *The Shaping of American Ethnography: The Wilkes Exploring Expedition, 1838–1842* (Lincoln: University of Nebraska Press, 2001), 7.

51. "American Ethnology," *American Review* 9 (1849): 385–398. The term "American School" was coined in Nott, *Types of Mankind*, 87.

52. William Stanton, *The Leopard's Spots: Scientific Attitudes towards Race in America, 1815–59* (Chicago: University of Chicago Press, 1960), 37, implies Morton simply appropriated phrenology for its popularity at the time.

53. Malcolm Chase, *Chartism: A New History* (Manchester: Manchester University Press, 2007), 81–82.

54. *Report of the Ninth Meeting of the British Association for the Advancement of Science held in Birmingham* (London: John Murray, 1840), xxviii.

55. Hannah Augstein, *James Cowles Prichard's Anthropology: Remaking the Science of Man in Early Nineteenth-Century Britain* (Amsterdam: Editions Rodopi, 1999), 25.

56. Jack Morrell and Arnold Thackray, *Gentlemen of Science: Early Years of the British Association for the Advancement of Science* (Oxford: Clarendon Press, 1981), 273–287.

57. A. D. Orange, "The Beginnings of the British Association 1831–1851," in Roy

MacLeod and Peter Collins, eds., *The Parliament of Science: The British Association for the Advancement of Science, 1831–1981* (Northwood: Science Reviews, 1981), 53–59.

58. Morrell and Thackray, *Gentlemen of Science*, xxi and 287–288.

59. Morrell and Thackray, *Gentlemen of Science*, 284.

60. Prichard to Morton, 23 August 1839; Nasmyth to Morton, 18 June 1839, Mss.B.M843, Series I: Correspondence, Samuel George Morton Papers, American Philosophical Society, Philadelphia (henceforth, Morton Papers, APS).

61. Prichard to Morton, 8 October 1839, Morton Papers, APS.

62. Hodgkin to Morton, 12 November 1839, Morton Papers, APS.

63. *The Report of the Eleventh Meeting of the British Association for the Advancement of Science held at Plymouth* (London: John Murray, 1842), 332.

64. *Report of the Ninth Meeting*, xxxvi.

65. *First Report of the Proceedings, Recommendations, and Transactions of the British Association for the Advancement of Science* (York: Thomas Wilson and Sons, 1832), 18.

66. Martin Rudwick, "The Emergence of a Visual Language for Geological Science, 1760–1840," *History of Science* 14 (1976): 149–150.

67. Efram Sera-Shriar, *The Making of British Anthropology, 1813–1871* (London: Pickering and Chatto, 2013), 11; George Stocking, "What's in a Name? The Origins of the Royal Anthropological Institute," *Man* 6 (1971): 371.

68. Augstein, *James Cowles Prichard's Anthropology*, 117.

69. James Cowles Prichard, *Researches into the Physical History of Mankind* (London: John and Arthur Arch 1826), 1:531–544.

70. James Cowles Prichard, *Researches into the Physical History of Mankind* (London: John and Arthur Arch, 1836), 1:275.

71. *The Report of the Eleventh Meeting of the British Association*, 332–333.

72. Prichard, *Researches into the Physical History of Mankind* (1836), 1:281.

73. "The Phrenological Association," *Phrenological Journal* 12 (1839): 29–35.

74. "Phrenology and the British Association," *Phrenological Journal* 12 (1839): 412.

75. James Cowles Prichard, *A Treatise on Insanity and Other Disorders Affecting the Mind* (London: Sherwood, Gilbert and Piper, 1835), 333.

76. Prichard to Combe, 10 October 1836, MS7241, f. 16, Combe Papers.

77. This goal was finally realized in 1851 with the establishment of Section E (Geography and Ethnology); Charles Withers, *Geography and Science in Britain, 1831–1939: A Study of the British Association for the Advancement of Science* (Manchester: Manchester University Press, 2010), 168.

78. "Phrenology and the British Association," 413.

79. Combe to Morton, 7 December 1839, Morton Papers, APS.

80. "Phrenology and the British Association," 413–414.

81. Morton to Combe, 11 October 1838, MS7397, f. 32, Combe Papers.

82. Morton to Combe, undated [1839], MS7251, f. 173, Combe Papers.

83. Morton, *Crania Americana* (Philadelphia: J. Dobson, 1839), 269.

84. "Quarterly List of New Publications," *North American Review* 46 (1838): 268.

85. Shapin, "The Politics of Observation," 139–178.

86. Morton, *Crania Americana*, 275.

87. Morton, *Crania Americana*, iv.

88. Morton, *Crania Americana*, 278.

89. Receipts of Crania Americana, C0199, Princeton University Library (henceforth, Receipts of Crania Americana).

90. Morton to Combe, undated [1839], MS7251, f. 173, Combe Papers.

91. Combe to Morton, 28 October 1838, Morton Papers, APS.

92. Morton to Combe, 1 April 1839, MS7251, f. 169, Combe Papers.

93. Morton, *Crania Americana*, 279.

94. Morton to Combe, undated [1839], MS7251, f. 173, Combe Papers.

95. Duff Hart-Davis, *Audubon's Elephant* (London: Weidenfeld, 2003), 7.

96. Morton to Combe, 8 October 1839, MS7251, f. 177, Combe Papers. This is also confirmed on inspection of the printed text.

97. Morton to Combe, undated [1839], MS7251, f. 173, Combe Papers.

98. Morton, *Crania Americana*, 262.

99. Prichard to Morton, 8 October 1839, Morton Papers, APS.

100. Prichard to Morton, 17 February 1840, Morton Papers, APS.

101. Prichard to Morton, 23 August 1839, Morton Papers, APS.

102. Combe to Morton, 11 October 1839, MS7397, f. 32, Combe Papers.

103. [George Combe], "Crania Americana," *American Journal of Science* 39 (1840): 375.

104. Examples of the "American edition" can be found at the American Philosophical Society, the College of Physicians of Philadelphia, and the Academy of Natural Sciences of Drexel University. Copies of the "foreign edition" can be found at Cambridge University Library, Bristol University Library, and the Linnean Society.

105. Morton to Combe, 24 May 1840, MS7256, f. 50, Combe Papers.

106. Combe to Morton, 31 May 1840, Morton Papers, APS.

107. "Crania Americana," *British Medical and Foreign Review* 10 (1840): 474–475.

108. Chandos Brown, "A Natural History of the Gloucester Sea Serpent: Knowledge, Power, and the Culture of Science in Antebellum America," *American Quarterly* 42 (1990): 431.

109. John Barton and Jennifer Phegley, "Introduction: 'An Age of Sensation . . . across the Atlantic,'" in John Barton, Jennifer Phegley, and Kristin Huston, eds., *Transatlantic Sensations* (Aldershot: Ashgate, 2010), 4–12.

110. Morton to Combe, 24 May 1840, MS7256, f. 50, Combe Papers.

111. James Green, "The Rise of Book Publishing," in Robert Gross and Mary Kelly, eds., *A History of the Book in America: An Extensive Republic, Print Culture, and Society in the New Nation, 1790–1840* (Chapel Hill: University of North Carolina Press, 2010), 125.

112. Jennifer Ambrose and Erika Piola, "The First Fifty Years of Commercial Lithography in Philadelphia," in Erika Piola, ed., *Philadelphia on Stone: Commercial Lithography in Philadelphia, 1828–1878* (University Park: Penn State University Press, 2012), 11.

113. Michael Twyman, "Putting Philadelphia on Stone," in Piola, *Philadelphia on Stone*, 71.

114. Ann Blum, *Picturing Nature: American Nineteenth-Century Zoological Illustration* (Princeton, NJ: Princeton University Press, 1993), 52; Ambrose and Piola, "The First Fifty Years," 3.

115. *Philadelphia Directory* (Philadelphia: A. McElroy, 1839), 47.

116. Georgia Barnhill, "Transformations in Pictorial Printing," in Robert Gross and Mary Kelly, eds., *A History of the Book in America: An Extensive Republic, Print Culture, and Society in the New Nation, 1790–1840* (Chapel Hill: University of North Carolina Press, 2010), 425–426.

117. Blum, *Picturing Nature*, 194.

118. Blum, *Picturing Nature*, 48.

119. Silliman took delivery of his own copy on 17 February 1840; Silliman to Morton, 19 February 1840, Morton Papers, APS.

120. Silliman to Morton, 27 March 1840, Morton Papers, APS.

121. Silliman to Morton, 19 February 1840, Morton Papers, APS.

122. "Crania Americana," *Waldie's Journal of Polite Literature*, 31 January 1840.

123. "Crania Americana," *Christian Examiner* 27 (1840): 251.

124. Combe to Morton, 4 April [1839], Morton Papers, APS.

125. Morton to Combe, 8 October 1839, MS7251, f. 177, Combe Papers.

126. Combe to Morton, 11 October 1839, MS7397, f. 32, Combe Papers.

127. Morton to Combe, 13 December 1839, MS7251, f. 179, Combe Papers.

128. Combe to Morton, 11 October 1839, MS7397, f. 32, Combe Papers.

129. *Awful Calamities; or the Shipwrecks of December 1839* (Boston: J. Howe, 1840), 22–23.

130. Morton to Combe, 8 January 1840, MS7256, f. 41, Combe Papers.

131. Combe to Morton, 7 December 1839, Morton Papers, APS.

132. Combe to Morton, 23 December 1839, Morton Papers, APS.

133. Early copies of the prospectus described the text as "printed on fine paper, in imperial quarto"; "Morton's Crania Americana," *American Medical Intelligencer* 1 (1838): 405.

134. Morton employed J. G. Russell to undertake the majority of the binding; Receipts of Crania Americana.

135. Combe to Morton, 23 December 1839, Morton Papers, APS.

136. Combe to Morton, 11 October 1839, MS7397, f. 32, Combe Papers.

137. Morton still owed money to his lithographers and colorists as late as March 1842; Receipts of Crania Americana.

138. Morton to Combe, 23 July 1840, MS7256, f. 52, Combe Papers.

139. Morton to Combe, 8 January 1840, MS7256, f. 41, Combe Papers.

140. Fyfe, *Steam-Powered Knowledge*, 180.

141. Jonathan Topham, "Science, Print, and Crossing Borders: Importing French Science Books into Britain 1789–1815," in David Livingstone and Charles Withers, eds., *Geographies of Nineteenth-Century Science* (Chicago: University of Chicago Press, 2011), 331, also notes how reviews acted as a means of transnational exchange during this period.

142. George Combe, *Notes on the United States of North America* (Edinburgh: Maclachlan and Stewart, 1841), 2:240.

143. Secord, *Victorian Sensation*, 227, invites us to consider the choice of reviewer.
144. Combe to Morton, 19 February 1840, Morton Papers, APS.
145. Silliman to Combe, undated [1840], MS7257, f. 27, Combe Papers.
146. Silliman to Combe, undated [1840], MS7257, f. 27, Combe Papers.
147. Combe to Morton, 19 February 1840, Morton Papers, APS.
148. Silliman to Combe, undated [1840], MS7257, f. 27, Combe Papers.
149. Combe to Morton, 6 March 1840, Morton Papers, APS.
150. [George Combe], "Crania Americana," 341.
151. [Combe], "Crania Americana," 348.
152. [Combe], "Crania Americana," 354–356.
153. [Combe], "Crania Americana," 354–356.
154. [Combe], "Crania Americana," 358.
155. Augstein, *James Cowles Prichard's Anthropology*, 29.
156. Combe to Morton, 28 February 1840, Morton Papers, APS.
157. Morton, *Crania Americana*, 23.
158. Morton, *Crania Americana*, 97–102.
159. D. Waddell, "International Politics and Latin American Independence," in Leslie Bethell, ed., *The Cambridge History of Latin America: From Independence to 1870* (Cambridge: Cambridge University Press, 1985), 3:209.
160. Charles Walker, *Smoldering Ashes: Cuzco and the Creation of Republican Peru, 1780–1840* (Durham, NC: Duke University Press, 1999), 1 and 140.
161. [Combe], "Crania Americana," 363.
162. "Crania Americana," *North American Review* 50 (1840): 180.
163. John Smith, *Select Discourses on the Functions of the Nervous System in Opposition to Phrenology* (New York: D. Appleton and Co., 1840), 133–134.
164. Morton, *Crania Americana*, 99.
165. [Combe], "Crania Americana," 363.
166. Combe to Morton, 28 February 1840, Morton Papers, APS.
167. Morton to Combe, 3 March 1840, MS7256, f. 45, Combe Papers.
168. Morton to Silliman, 9 March 1840, MS7256, f. 46, Combe Papers.
169. Combe to Morton, 13 March 1840, Morton Papers, APS.
170. [Combe], "Crania Americana," 375.
171. Morton to Combe, 21 March 1840, MS7256, f. 48, Combe Papers.
172. Morton to Combe, 24 May 1840, MS7256, f. 50, Combe Papers.
173. The copy of *Crania Americana* held by the College of Physicians of Philadelphia (ZXc5) contains the canceled sheet signed twenty-five. The text corresponds to Morton's original manuscript, "Crania Americana Text," Series IV, Box 5, Folder 2, Samuel George Morton Papers, Library Company of Philadelphia (henceforth, Morton Papers, LCP).
174. Morton to Combe, 23 July 1840, MS7256, f. 52, Combe Papers.
175. Combe to Morton, 6 March 1840, Morton Papers, APS.
176. This is revealed on close inspection of the *American Phrenological Journal* when compared to the *American Journal of Science*.
177. "Crania Americana," *Christian Examiner* 27 (1840): 252.
178. "Crania Americana," *Eclectic Journal of Medicine* 4 (1840): 99.

179. "Crania Americana," *North American Review* 50 (1840): 173.

180. "Man," *Ladies' Repository* 1 (1841): 72.

181. Jerry Clark, *The Shawnee* (Lexington: University Press of Kentucky, 2007), 25.

182. "Man," *Ladies' Repository* 1 (1841): 73.

183. Certificate of Membership to the Aborigines' Protection Society, 23 January 1839, Morton Papers, APS.

184. James Cowles Prichard, "Crania Americana," *Journal of the Royal Geographical Society* 10 (1841): 552.

185. Prichard to Morton, 24 March 1841, Morton Papers, APS.

186. "Crania Americana," *British Medical and Foreign Review* 10 (1840): 483–485.

187. Combe to Morton, 20 August 1840, Morton Papers, APS.

188. "Crania Americana," *Edinburgh Medical and Surgical Journal* 54 (1840): 443–468; "Crania Americana," *Medico-Chirurgical Review* 37 (1840): 434–464.

189. "Phrenology and Professor Jameson," *Phrenological Journal* 1 (1824): 56.

190. By the 1830s, Edinburgh had emerged as a hub within the transatlantic world of literary exchange, Fyfe, *Steam-Powered Knowledge,* 173–261; Secord, *Victorian Sensation*, 369.

191. "Comparative View of the Skulls of the Various Aboriginal Nations of North and South America," *Edinburgh New Philosophical Journal* 29 (1840): 111–139.

192. Combe to Morton, 18 July 1840, Morton Papers, APS.

193. "Professor Jameson's Illustrations of a Convenient Method for Pruning Reviews that Contain Inconvenient Opinions," *Phrenological Journal* 13 (1840): 312–13.

194. "Comparative View of the Skulls," 111.

195. Fyfe, *Steam-Powered Knowledge,* 173–261.

196. Combe to Morton, 4 November 1840, Morton Papers, APS.

197. Combe to Morton, 4 November 1840, Morton Papers, APS.

198. Kippist to Morton, 6 July 1846, MS7388, Series I, Morton Papers, LCP.

199. *Catalogue of the Glasgow Mechanics' Institution Library* (Glasgow: George Troup, 1848).

200. Flanders to Morton, 3 March 1849, MS7390, f. 39, Series I, Morton Papers, LCP.

201. *Catalogue of the Library of the Mechanics' Institute, of the City of New-York* (New York: A. Baptist Jr, 1844).

202. There are no references to Morton or *Crania Americana* in the major South American newspapers published during the 1830s and 1840s.

203. Linnean Society to Morton, 18 November 1842; Académie Royale des Sciences to Morton, 13 October 1843, Morton Papers, APS.

204. Carlos Costa, *Catalogo Systematico da Bibliotheca da Faculdade de Medicina do Rio de Janeiro* (Rio de Janeiro: Imprensa Nacional, 1892). (I am extremely grateful to Raphael Uchôa for pointing me toward the Brazilian sources as well as for his translations from the Portuguese.)

205. Claudio Simon Hutz, Gustavo Gauer, and William Barbosa Gomes, "Brazil," in David Baker, ed., *The Oxford Handbook of the History of Psychology: Global Perspectives* (Oxford: Oxford University Press, 2012), 36; Erica Windler, "Madame Durocher's Performance: Cross-Dressing, Midwifery, and Authority in Nineteenth-Century Rio de Janeiro, Brazil," in William French and Katherine Bliss, eds., *Gender, Sexuality, and*

Power in Latin America since Independence (Lanham, MD: Rowman and Littlefield, 2007), 52.

206. Ana Maria Jacó Vilela, "History of Psychology in Brazil: A Narrative from Its Teaching," *Psicologia: Ciência e Profissão* 32 (2012): 33; Hutz, Gauer, and Gomes, "Brazil," 34–50.

207. *Catalogo da Exposição Medica Brasileira* (Rio de Janeiro: Typographia Nacional, 1884), 173.

208. Costa, *Catalogo Systematico*, 194.

209. Thomas Skidmore, *Black into White: Race and Nationality in Brazilian Thought* (Durham, NC: Duke University Press, 1993), 24.

210. Skidmore, *Black into White*, 24–60; Nancy Stepan, *"The Hour of Eugenics": Race, Gender, and Nation in Latin America* (Ithaca, NY: Cornell University Press, 1991), 40–45.

211. Skidmore, *Black into White*, 56–60; Juanma Sánchez Arteaga and Charbel Niño El-Hani, "Physical Anthropology and the Description of the 'Savage' in the Brazilian Anthropological Exhibition of 1882," *História, Ciências, Saúde-Manguinhos* 17 (2010).

212. *Revista da Exposição Anthropologica Brazileira* (Rio de Janeiro: Typographia de Pinheiro, 1882), 69–71, 82–83, 91–92 (translation by Raphael Uchôa).

213. *Revista da Exposição Anthropologica Brazileira*, 69–71 (translation by Raphael Uchôa).

214. The strategy seemed to work, as in 1911 João Baptista de Lacerda was invited as the Brazilian representative to the First Universal Races Congress; Lilia Moritz Schwarcz, "Predictions Are Always Deceptive: João Baptista de Lacerda and His White Brazil," *História, Ciências, Saúde-Manguinhos* 18 (2011); Skidmore, *Black into White*, 65. As Nancy Stepan argues, Latin American thinkers did not simply imitate European and American science but reworked it; Stepan, *"The Hour of Eugenics,"* 3.

215. Arteaga and El-Hani, "Physical Anthropology and the Description of the 'Savage,'" 319–414.

216. Daston and Galison, *Objectivity*, 17–54.

217. Arnold Thackray, "R. I. Murchison's *Silurian System* (1839)," *Archives of Natural History* 9 (1978): 61–73; Blum, *Picturing Nature*, 54.

218. Asiatic Society of Bengal to Morton, 18 April 1843, Morton Papers, APS; Martin to Morton, 30 August 1844, MS7388, Series I, Morton Papers, LCP.

219. Mottelettes to Morton, 9 April 1844, MS7388, Series I, Morton Papers, LCP.

220. Claude Blanckaert, "A Jamaican Émigré as the 'Father of Ethnology in France,'" in George Stocking, ed., *Bones, Bodies, and Behavior: Essays in Behavioral Anthropology* (Madison: University of Wisconsin Press, 1988), 18–55.

221. Koeppen to Morton, 27 March 1847, MS7389, Series I, Morton Papers, LCP.

222. Cramer to Morton, 5 August 1848, MS7390, Series I, Morton Papers, LCP.

CHAPTER FOUR

1. "Cheap Postage and Friendly Correspondence," *American Phrenological Journal* 10 (1848): 27.

2. "Excellent Post-Office Suggestion, and the Perfection of Our Post-Office System," *American Phrenological Journal* 11 (1849): 38.

3. Hewett Watson, "Cheap Postage," *Phrenological Journal* 11 (1838): 72.

4. Frederick Bridges, *Phrenology Made Practical and Popularly Explained* (London: Sampson, Low, Son and Co., 1857), 160–161.

5. Combe to Mann, [December 1840], MS7388, f. 278, Combe Papers.

6. Cash Book, Gen. 608/18, Records of the Phrenological Society of Edinburgh.

7. James Straton, "Suggestions for an Interchange of Papers Read Before Phrenological Societies," *Phrenological Journal* 13 (1840): 189.

8. David Armitage and Sanjay Subrahmanyam, eds., *The Age of Revolutions in Global Context, 1760–1840* (Basingstoke: Palgrave, 2010).

9. Marc Renneville, *Le langage des crânes* (Paris: Institut d'Édition Sanofi-Synthélabo, 2000), 20; Charles Colbert, *A Measure of Perfection: Phrenology and the Fine Arts in America* (Chapel Hill: University of North Carolina Press, 1997), xii.

10. Steven Shapin, "Phrenological Knowledge and the Social Structure of Early Nineteenth-Century Edinburgh," *Annals of Science* 32 (1975): 242–243; Enda Leaney, "Phrenology in Nineteenth-Century Ireland," *New Hibernia Review* 10 (2006): 32–42.

11. Andrew Bank, "Of 'Native Skulls' and 'Noble Caucasians': Phrenology in Colonial South Africa," *Journal of Southern African Studies* 22 (1996); Kim Wagner, "Confessions of a Skull: Phrenology and Colonial Knowledge in Early Nineteenth-Century India," *History Workshop Journal* 69 (2010).

12. John van Wyhe, "Was Phrenology a Reform Science? Towards a New Generalization for Phrenology," *History of Science* 42 (2004): 326.

13. Gareth Stedman Jones, *Languages of Class: Studies in English Working Class History, 1832–1982* (Cambridge: Cambridge University Press, 1983), 90–178.

14. Joanna Innes, "'Reform' in English Public Life: the Fortunes of a Word," in Arthur Burns and Joanna Innes, eds., *Rethinking the Age of Reform: Britain, 1780–1850* (Cambridge: Cambridge University Press, 2003), 71–97.

15. Derek Beales, "The Idea of Reform in British Politics, 1829–1850," in Tim Blanning and Peter Wende, eds., *Reform in Great Britain and Germany, 1750–1850* (Oxford: Oxford University Press, 1999), 160–170.

16. Abolition is more often associated with Charles Darwin's evolutionary theories than with phrenology; see Adrian Desmond and James Moore, *Darwin's Sacred Cause: Race, Slavery, and the Quest for Human Origins* (London: Penguin, 2009). Cynthia Hamilton, "'Am I Not a Man and a Brother?' Phrenology and Anti-slavery," *Slavery & Abolition* 29 (2008), does discuss abolition but does not make the connection between phrenology and leading antislavery figures such as Lucretia Mott and Samuel Gridley Howe.

17. Nigel Hall, "The Materiality of Letter Writing: A Nineteenth-Century Perspective," in David Barton and Nigel Hall, eds., *Letter Writing as a Social Practice* (Amsterdam: John Benjamins Publishing Company, 2000).

18. Janet Browne, *Charles Darwin: The Power of Place* (London: Pimlico, 2002), 2:10–13, also suggests the varied uses of correspondence.

19. Thomas Haskell, "Capitalism and the Origins of the Humanitarian Sensibility, Part 1," *American Historical Review* 90 (1985): 353–359.

20. Richard Huzzey, "The Moral Geography of British Anti-slavery Responsibilities," *Transactions of the Royal Historical Society* 22 (2012).

21. Arthur Burns and Joanna Innes, "Introduction," in Arthur Burns and Joanna Innes, eds., *Rethinking the Age of Reform: Britain, 1780–1850* (Cambridge: Cambridge University Press, 2003), 1–3; Anne Secord, "Corresponding Interests: Artisans and Gentlemen in Nineteenth-Century Natural History," *British Journal for the History of Science* 27 (1994): 388–390; Stefan Collini, "The Idea of 'Character' in Victorian Political Thought," *Transactions of the Royal Historical Society* 35 (1985).

22. Paul White, "Lives and Letters: Correspondence and Public Character in the Nineteenth Century," in Rosalind Crone, David Gange, and Katy Jones, eds., *New Perspectives in British Cultural History* (Cambridge: Cambridge Scholars Publishing, 2007), 192–195; Secord, "Corresponding Interests."

23. Peter McCandless, "Mesmerism and Phrenology in Antebellum Charleston: 'Enough of the Marvellous,'" *Journal of Southern History* 58 (1992); Michael O'Brien, *Conjectures of Order: Intellectual Life and the American South, 1810–1860* (Chapel Hill: University of North Carolina Press, 2004), 2:1008–1019.

24. Caldwell to Combe, 1 June 1836, MS7237, f. 94, Combe Papers.

25. Buchanan to Capen, 18 June 1839, Nahum Capen Papers, B MS c23, Countway Library, Harvard University.

26. "Phrenological Facts," *American Phrenological Journal* 7 (1845): 21–23.

27. Orson Fowler, *Phrenology and Physiology Explained and Applied to Education and Self-Improvement* (New York: Orson and Lorenzo Fowler, 1843), 49–50.

28. Charles Caldwell, *The Autobiography of Charles Caldwell* (Philadelphia: Lippincott, Grambo and Co., 1855), 62.

29. Caldwell to Combe, 14 September 1835, MS7234, f. 83, Combe Papers.

30. Andrew Carmichael, *A Memoir of the Life and Philosophy of Spurzheim* (Boston: Marsh, Capen and Lyon, 1833), 27.

31. Caldwell to Combe, 12 August 1837, MS7242, f. 46, Combe Papers; Luke Harlow, "Neither Slavery nor Abolitionism: James M. Pendleton and the Problem of Christian Conservative Antislavery in 1840s Kentucky," *Slavery and Abolition* 27 (2006): 367–368.

32. Caldwell to Combe, 30 August 1839, MS7249, f. 145, Combe Papers.

33. Caldwell to Combe, 12 August 1837, MS7242, f. 46, Combe Papers.

34. Caldwell to Combe, 30 August 1839, MS7249, f. 145, Combe Papers.

35. Caldwell to Combe, 30 August 1839, MS7249, f. 145, Combe Papers.

36. Charles Caldwell, "Phrenology Vindicated," *Annals of Phrenology* 1 (1833).

37. Combe's notes on Caldwell to Combe, 30 August 1839, MS7249, f. 145, Combe Papers.

38. George Combe, *A System of Phrenology* (Edinburgh: Maclachlan and Stewart, 1843), 2:355; George Combe, *Notes on the United States of North America* (Edinburgh: Maclachlan and Stewart, 1841), 2:78.

39. Caldwell to Combe, 25 February 1838, MS7245, f. 97, Combe Papers.

40. Caldwell to Combe, 14 June 1839, MS7249, f. 137, Combe Papers.

41. Huzzey, "The Moral Geography of British Anti-slavery," 111–139.

42. Betty Fladeland, *Men and Brothers: Anglo-American Antislavery Cooperation* (Urbana-Champaign: University of Illinois Press, 1972), 292.

43. Caldwell to Combe, 14 June 1839, MS7249, f. 137, Combe Papers.

44. Caldwell to Combe, 24 December 1838, MS7245, f. 101, Combe Papers.

45. Combe to Channing, 28 March 1838, MS7395, f. 12, Combe Papers.

46. Combe, *Notes on the United States of North America*, 3:333–334.

47. "Colonial Slavery Tested by Phrenology," *Phrenological Journal* 8 (1832–1834): 83.

48. Combe to Chapman, 22 November 1845, MS7390, f. 215, Combe Papers.

49. Combe to Mott, 15 July 1839, MS7396, f. 72, Combe Papers. His view was later confirmed when he met Adams in Washington, DC, in February 1840; Combe, *Notes on the United States of North America*, 2:106–107.

50. Caldwell to Combe, 1 June 1836, MS7237, f. 94, Combe Papers.

51. Combe to Caldwell, 21 November 1834, MS7386, f. 187, Combe Papers.

52. Caldwell to Combe, 25 February 1838, MS7245, f. 97, Combe Papers.

53. Patricia Hollis, "Anti-slavery and British Working Class Radicalism in the Years of Freedom," in Christine Bolt and Seymour Drescher, eds., *Anti-slavery, Religion, and Reform: Essays in Memory of Roger Anstey* (Folkstone: Dawson, 1980), 298–309.

54. Spurzheim to Périer, 7 September 1815, Johann Spurzheim Papers, B MS c22, Countway Library, Harvard University (henceforth, Spurzheim Papers). When he returned to Paris, Spurzheim married Périer and picked up the correspondence, taking it with him on his own American lecture tour of 1832.

55. Spurzheim to Périer, 20 September 1814, Spurzheim Papers. These numbers correspond to the organs of Adhesiveness (4), Self-Esteem (10), Cautiousness (12), Benevolence (13), and Veneration (14) in Johann Spurzheim, *The Physiognomical System of Drs. Gall and Spurzheim* (London: Baldwin, Cradock, and Joy, 1815), 579–580.

56. Spurzheim to Périer, 3 September 1815, Spurzheim Papers.

57. Cynthia Hamilton, "Hercules Subdued: The Visual Rhetoric of the Kneeling Slave," *Slavery and Abolition* 34 (2013): 631–633; Clare Midgley, *Women against Slavery: The British Campaigns, 1780–1870* (London: Routledge, 1995), 97.

58. American letter-writing manuals actually instructed on the appropriate use of wax seals in this period; Lucille Schultz, "Letter-Writing Instruction in 19th-Century Schools in the United States," in David Barton and Nigel Hall, eds., *Letter Writing as a Social Practice* (Amsterdam: John Benjamins Publishing Company, 2000), 117–120.

59. Combe, *Notes on the United States of North America*, 2:49.

60. Mott to Combe, 8 September 1839, MS7251, f. 158; Mott to Combe, 10 April 1840, MS7256, f. 60, Combe Papers.

61. Mott to Combe, 26 April 1847, MS7287, f. 28, Combe Papers.

62. Konstantin Dierks, "The Familiar Letter and Social Refinement in America, 1750–1800," in Barton and Hall, *Letter Writing as a Social Practice*, 31–42.

63. Caldwell to Combe, 7 October 1838, MS7245, f. 99, Combe Papers.

64. Combe to Channing, 23 April 1839, MS7396, f. 52, Combe Papers. Midgley, *Women against Slavery*, 94, argues that abolitionists self-consciously emphasized the value of female "sensibilities" in countering slavery.

65. Beverly Tomek, *Colonization and Its Discontents: Emancipation, Emigration, and Antislavery in Antebellum Pennsylvania* (New York: New York University Press, 2011), xviii.

66. Mott to Combe, 13 June 1839, MS7251, f. 183, Combe Papers.

67. "On the American Scheme of Establishing Colonies of Free Negro Emigrants on the Coast of Africa," *Phrenological Journal* 8 (1832–1834): 145–160.

68. Tomek, *Colonization and Its Discontents*, 5; Combe to Channing, 23 April 1839, MS7396, f. 52, Combe Papers.

69. Tomek, *Colonization and Its Discontents*, 3.

70. "On the American Scheme of Establishing Colonies," 152–160.

71. "On the American Scheme of Establishing Colonies," 159.

72. Fladeland, *Men and Brothers*, 279.

73. Mott to Combe, 8 September 1839, MS7251, f. 52, Combe Papers.

74. Mott to Combe, 13 June 1839, MS7251, f. 183, Combe Papers.

75. Combe to Mott, 15 July 1839, MS7396, f. 72, Combe Papers.

76. Combe to Mott, 15 July 1839, MS7396, f. 72, Combe Papers.

77. Combe, *Notes on the United States of North America*, 3:133–134.

78. Mott to Combe, 8 September 1839, MS7251, f. 185, Combe Papers.

79. Tomek, *Colonization and Its Discontents*, 157.

80. Frederick Douglass, *The Claims of the Negro Ethnologically Considered* (Rochester, NY: Lee, Mann and Co., 1854), 20–35.

81. John Stauffer, "Introduction," in *The Works of James McCune Smith: Black Intellectual and Abolitionist* (Oxford: Oxford University Press, 2007), xiii-xiv; "Phrenology," *The Colored American*, 23 September 1837.

82. James Smith, "The Heads of Colored People," in *The Works of James McCune Smith*, 185–232.

83. James Smith, "Outside Barbarians," in *The Works of James McCune Smith*, 82.

84. Frederick Douglass, "Letter from the Editor," in *The Works of James McCune Smith*, 233–235.

85. Britt Rusert, "The Science of Freedom: Counterarchives of Racial Science on the Antebellum Stage," *African American Review* 45 (2012); Hamilton, "'Am I Not a Man and a Brother?,'" 176–178.

86. Combe, *Notes on the United States of North America*, 2:48.

87. Frederick Douglass, *The Life and Times of Frederick Douglass* (Boston: De Wolfe and Fiske Co., 1882), 299–301.

88. Richard John, *Spreading the News: The American Postal System from Franklin to Morse* (Cambridge, MA: Harvard University Press, 1995), 140–164.

89. Combe's notes on Caldwell to Combe, 30 August 1839, MS7249, f. 145, Combe Papers.

90. Merval Hoare, *Norfolk Island: An Outline of Its History, 1774–1977* (Brisbane: University of Queensland Press, 1969), 4.

91. Timings based on study of postmarks, particularly Maconochie to Combe, 16 June 1841, MS7261, f. 44, Combe Papers.

92. Robert Hughes, *The Fatal Shore: A History of the Transportation of Convicts to Australia, 1787–1868* (London: Collins Harvill, 1986), 499–502.

93. Maconochie to Combe, 16 June 1841, MS7261, f. 44, Combe Papers.

94. Maconochie to Combe, 27 September 1834, MS7233, f. 28, Combe Papers.

95. Maconochie to Combe, 16 June 1841, MS7261, f. 44, Combe Papers.

96. Hughes, *The Fatal Shore*, 463.

97. Hoare, *Norfolk Island*, 36.

98. Maconochie to Combe, 16 June 1841, MS7261, f. 44, Combe Papers.

99. Hughes, *The Fatal Shore*, 500–501.

100. Maconochie to Combe, 16 June 1841, MS7261, f. 44, Combe Papers.

101. Maconochie to Gipps, 25 February 1840, in Frederick Watson, ed., *Historical Records of Australia* (Sydney: Library Committee of the Commonwealth Parliament, 1914), 1:535.

102. Hughes, *The Fatal Shore*, 501.

103. Dario Melossi and Massimo Parvani, *The Prison and the Factory: Origins of the Penitentiary System* (London: Macmillan, 1981).

104. Maconochie to Combe, 16 June 1841, MS7261, f. 44, Combe Papers.

105. Maconochie to Combe, 16 June 1841, MS7261 f. 44; Maconochie to Combe, 12 November 1844, MS7273, f.44, Combe Papers.

106. Hughes, *The Fatal Shore*, 409.

107. Hughes, *The Fatal Shore*, 503–513.

108. Maconochie to Combe, 16 June 1841, MS7261 f. 44; Maconochie to Combe, 12 November 1844, MS7273, f. 44, Combe Papers.

109. Combe to Maconochie, 31 October 1844, MS7388, f. 782, Combe Papers.

110. Combe, *Notes on the United States of North America*, 2:2 and 326; George Combe, "Mr Combe on the Institutions of Germany," *Phrenological Journal* 10 (1836–1837): 698.

111. Wagner, "Confessions of a Skull," 38–45.

112. George Combe, "Norfolk Island—Reform in Convict Treatment," *Phrenological Journal* 15 (1842): 23.

113. George Combe, "Penal Colonies," *Phrenological Journal* 18 (1845): 101–122.

114. Michael Meranze, *Laboratories of Virtue: Punishment, Revolution, and Authority in Philadelphia, 1760–1835* (Chapel Hill: University of North Carolina Press, 1996), 1.

115. Meranze, *Laboratories of Virtue*, 295–296.

116. Combe, *Notes on the United States of North America*, 2:13–15.

117. David Rothman, *The Discovery of the Asylum: Social Order and Disorder in the New Republic* (Boston: Little Brown, 1971), 142.

118. Howe to Combe, [undated 1846], MS7275, f. 110, Combe Papers.

119. Meranze, *Laboratories of Virtue*, 294.

120. Mackenzie to Combe, 29 November 1839, MS7251, f. 130, Combe Papers.

121. Hurlbut to Combe, 27 February 1846, MS7280, f. 79, Combe Papers.

122. "A Phrenological Fact," *American Phrenological Journal* 5 (1843): 48.

123. Combe to Howe, 23 February 1846, MS7390, f. 321, Combe Papers.

124. Hurlbut to Combe, 7 October 1846, MS7280, f. 80; Mann to Combe, 28 February 1845, MS7276, f. 111, Combe Papers.

125. Michael Hindus, *Prison and Plantation: Crime, Justice, and Authority in Massachusetts and South Carolina, 1767–1878* (Chapel Hill: University of North Carolina Press, 1980), 220.

126. Mann to Combe, 1 October 1840, MS7256, f. 27e, Combe Papers.

127. Mittermaier to Combe, 23 July 1842, MS7265, f. 52, Combe Papers.

128. German Lecture Notes, 1842, MS7443, f. 11, Combe Papers.

129. Mittermaier to Combe, 23 July 1842, MS7265, f. 52, Combe Papers.

130. "On the Application of Phrenology to Criminal Legislation and Prison Discipline," *Phrenological Journal*, 16 (1843): 2–15.

131. Combe to Maconochie, 23 August 1846, MS7390, f. 496, Combe Papers.

132. Kapil Raj, *Relocating Modern Science: Circulation and the Construction of Knowledge in South Asia and Europe, 1650–1900* (New Delhi: Permanent Black, 2006), 159–179.

133. Paterson to Combe, 20 July 1825, MS7216, f. 46, Combe Papers; Paterson to Bell, [1825], MSS2/0232-01, College of Physicians of Philadelphia.

134. Assistant-Surgeons' Papers, IOR/L/MIL/9/370, f. 170, British Library.

135. Paterson to Combe, 10 May 1823, MS7211, f. 7, Combe Papers.

136. Timings based on study of postmarks, particularly Paterson to Combe, 23 April 1825, MS7216, f. 47, Combe Papers.

137. Paterson to Combe, 20 July 1825, MS7216, f. 46, Combe Papers.

138. Paterson to Combe, 23 April 1825, MS7216, f. 47, Combe Papers; Paterson to Bell, [1825], MSS2/0232-01, College of Physicians of Philadelphia; *Calcutta Annual Register and Directory* (Calcutta: Scott and Company, 1831), 313.

139. Paterson to Combe, 20 July 1825, MS7216, f. 46, Combe Papers; Paterson to Bell, [1825], MSS2/0232-01, College of Physicians of Philadelphia.

140. Paterson to Combe, 23 April 1825, MS7216, f. 47, Combe Papers.

141. Paterson to Bell, [1825], MSS2/0232-01, College of Physicians of Philadelphia.

142. Paterson to Bell, [1825], MSS2/0232-01, College of Physicians of Philadelphia.

143. Paterson to Wilson, 12 February 1828, AR/L-441; Paterson to Wilson, 16 February 1828, AR/L-442; Paterson to Wilson, 31 March 1828, AR/L-447; Paterson to Wilson, 13 May 1828, AR/L-452, Asiatic Society of Bengal, Kolkata.

144. Raj, *Relocating Modern Science*, 178; Savithri Preetha Nair, "'Bungallee House Set on Fire by Galvanism': Natural and Experimental Philosophy as Public Science in a Colonial Metropolis (1794–1806)," in Bernard Lightman, Gordon McOuat, and Larry Stewart, eds., *The Circulation of Knowledge between Britain, India, and China: The Early-Modern World to the Twentieth Century* (Leiden: Brill, 2013), 52.

145. Paterson to Bell, [1825], MSS2/0232-01, College of Physicians of Philadelphia.

146. Poromesh Acharya, "Education in Old Calcutta," in Sukanta Chaudhuri, ed., *Calcutta: The Living City* (Oxford: Oxford University Press, 1990), 1:86–88.

147. Paterson to Bell, [1825], MSS2/0232-01, College of Physicians of Philadelphia.

148. Paterson to Combe, 6 June 1823, MS7211, f. 11, Combe Papers.

149. Assistant-Surgeons' Papers, IOR/L/MIL/9/370, f. 174, British Library.

150. Paterson to Combe, 20 July 1825, MS7216, f. 46, Combe Papers.

151. George Paterson, "The Phrenology of Hindostan," *Transactions of the Phrenological Society* 1 (1824): 436.

152. Sujit Sivasundaram, "'A Christian Benares': Orientalism, Science and the Serampore Mission of Bengal," *Indian Economic and Social History Review* 44 (2007): 115.

153. Heather Streets, *Martial Races: The Military, Race, and Masculinity in British Imperial Culture, 1857–1914* (Manchester: Manchester University Press, 2004), 162–163.

154. Paterson to Combe, 23 April 1825, MS7216, f. 47, Combe Papers.

155. Paterson, "The Phrenology of Hindostan," 431.

156. Paterson to Bell, [1825], MSS2/0232-01, College of Physicians of Philadelphia.

157. Sivasundaram, "'A Christian Benares,'" 133; Paterson, "The Phrenology of Hindostan," 445.

158. Paterson to Bell, [1825], MSS2/0232-01, College of Physicians of Philadelphia.

159. David Drummond, *Objections to Phrenology: Being the Substance of a Series of Papers Communicated to the Calcutta Phrenological Society* (Calcutta: Printed for the Author, 1829), 62, 89, and 113,

160. Beatson to Combe, 28 January 1827, MS7219, f. 3, Combe Papers.

161. Paterson to Combe, 10 May 1823, MS7211, f. 7, Combe Papers.

162. Drummond, *Objections,* 89; Paterson to Bell, [1825], MSS2/0232-01, College of Physicians of Philadelphia.

163. Combe to Beatson, 23 May 1826, MS7383, f. 317, Combe Papers.

164. Paterson to Bell, [1825], MSS2/0232-01, College of Physicians of Philadelphia.

165. Paterson to Combe, 23 April 1825, MS7216, f. 47, Combe Papers.

166. Combe to Rait, [1852], MS7392, f. 579, Combe Papers.

167. Helen Ludlow, "The Government Teacher as Mediator of a 'Superior' Education in Colesberg, 1849–1858," *Historia* 57 (2012): 150.

168. *Infant Education* (Edinburgh: William and Robert Chambers, 1837), 110.

169. Advertisement in the *Athenaeum*, 27 November 1847.

170. Ludlow, "The Government Teacher," 148; *Rudiments of Knowledge* (Edinburgh: William and Robert Chambers, 1838), 76.

171. Stephen Tomlinson, *Head Masters: Phrenology, Secular Education, and Nineteenth-Century Social Thought* (Tuscaloosa: University of Alabama Press, 2005), x.

172. Combe, *Notes on United States of North America*, 1:64–65.

173. Combe to Mann, 25 April 1839, MS7396, f. 55, Combe Papers.

174. Combe to Mann, 29 December 1844, MS7390, f. 16, Combe Papers.

175. Combe to Mann, 5 July 1844, MS7398, f. 26, Combe Papers.

176. Combe to Mann, [December 1840], MS7388, f. 278, Combe Papers.

177. Tomlinson, *Head Masters*, ix-x.

178. Howe to Combe, 5 March 1839, MS7251, f. 49, Combe Papers; Combe, *Notes on the United States of North America*, 1:61.

179. Howe to Mann, 19 March 1838, 1137, MS Am 2119, Howe Family Papers, Houghton Library, Harvard University.

180. Mann to Combe, 22 July 1853, MS7335, f. 17, Combe Papers.

181. Scott Sanders, "Antioch College: Establishing the Faith," in James Hodges, James O'Donnell, and John Oliver, eds., *Cradles of Conscience: Ohio's Independent Colleges and Universities* (Kent, OH: Kent State University Press, 2003), 13; Tomlinson, *Head Masters*, 291.

182. Howe to Combe, 29 September 1840, MS7255, f. 101, Combe Papers.

183. Tomlinson, *Head Masters*, 241.

184. England and Scotland had very different educational systems at this time. However, neither had a national system; Robert Anderson, *Education and the Scottish People, 1750–1918* (Oxford: Clarendon Press, 1995), 1–47; Andy Green, *Education and*

State Formation: The Rise of Education Systems in England, France, and the USA (London: Palgrave, 1992), 6–11.

185. Combe to Mann, 31 December 1840, MS7388, f. 322, Combe Papers.

186. Howe to Combe, 30 May 1842, MS7264, f. 145, Combe Papers.

187. Combe to Mann, 28 December 1842, MS7388, f. 600, Combe Papers.

188. Combe to Mann, 19 July 1839, MS7396, f. 77, Combe Papers.

189. Mann to Combe, 28 July 1842, MS7265, f. 28a, Combe Papers.

190. Carl Kaestle and Maris Vinovskis, *Education and Social Change in Nineteenth-Century Massachusetts* (Cambridge: Cambridge University Press, 1980), 210.

191. Kaestle and Vinovskis, *Education and Social Change,* 190 and 231.

192. Mann to Combe, 1 October 1840, MS7256, f. 26e, Combe Papers.

193. Combe to Mann, 15 November 1841, MS7398, f. 11, Combe Papers.

194. Mann to Combe, 13 October 1841, MS7261, f. 56, Combe Papers.

195. Tomlinson, *Head Masters,* 248.

196. Combe to Howe, 24 March 1839, MS7396, f.42, Combe Papers.

197. Mann to Combe, 1 December 1844, MS7273, f.61 and Combe to Howe, 24 March 1839, MS7396, f.42, Combe Papers.

198. Tomlinson, *Head Masters,* 248.

199. Mann to Combe, 25 March 1839, MS7251, f. 136, Combe Papers.

200. Combe to Mann, 30 April 1841, MS7388, f. 511, Combe Papers.

201. [George Combe], "Education in America," *Edinburgh Review* 73 (1841): 492.

202. Maconochie to Combe, 31 August 1844, MS7273, f. 30, Combe Papers.

203. Combe, "Penal Colonies," 114.

204. Combe to Mann, 29 December 1844, MS7390, f. 16, Combe Papers.

205. Combe to Mott, 28 October 1847, MS7391, f. 139, Combe Papers.

206. Mott to Combe, 28 May 1850, MS7310, f. 27, Combe Papers.

207. "Miscellany," *American Phrenological Journal* 5 (1843): 288.

208. Charles Caldwell, "New Views of Penitentiary Discipline and Moral Education and Reformation of Criminals," *Phrenological Journal* 7 (1831–1832): 387–388.

209. Hindus, *Prison and Plantation,* 178 and 236–237.

210. It wasn't until the 1820s that abolitionists came to associate their work with the term "reform"; David Turley, *The Culture of English Antislavery, 1780–1860* (London: Routledge, 2003), 183.

211. On the development of "reform" as a political ideology, see Burns and Innes, "Introduction."

212. Struve to Combe, 3 July 1848, MS7297, f. 115, Combe Papers.

213. Struve to Combe, 28 August 1846, MS7282, f. 103, Combe Papers.

214. Struve to Combe, 9 August 1848, MS7297, f. 117, Combe Papers.

215. Peter Wende, "1848: Reform or Revolution in Germany and Great Britain," in Tim Blanning and Peter Wende, eds., *Reform in Great Britain and Germany, 1750–1850* (Oxford: Oxford University Press, 1999); Innes, "'Reform' in English Public Life," 86–88.

216. Combe to Struve, 31 July 1848, MS7391, f. 496, Combe Papers.

217. Struve to Combe, 30 August 1849, MS7304, f. 15, Combe Papers.

CHAPTER FIVE

1. *Barbados Globe*, 27 August 1868; *Barbados Agricultural Reporter*, 25 August 1868; *The Times*, 2 September 1868 and 24 October 1868.

2. *Daily Chronicle*, 16 June 1888, 12 July 1888, and 18 December 1890.

3. Aileen Fyfe and Bernard Lightman, "Science in the Marketplace: An Introduction," in Aileen Fyfe and Bernard Lightman, eds., *Science in the Marketplace: Nineteenth-Century Sites and Experiences* (Chicago: University of Chicago Press, 2007).

4. For the development of scientific audiences in the British national context, see the papers collected in Geoffrey Cantor, Gowan Dawson, Graeme Gooday, Richard Noakes, Sally Shuttleworth, and Jonathan Topham, eds., *Science in the Nineteenth-Century Periodical: Reading the Magazine of Nature* (Cambridge: Cambridge University Press, 2004); Geoffrey Cantor and Sally Shuttleworth, "Introduction," in Geoffrey Cantor and Sally Shuttleworth, eds., *Science Serialized: Representation of the Sciences in Nineteenth-Century Periodicals* (Cambridge, MA: MIT Press, 2004); Jonathan Topham, "Scientific Publishing and the Reading of Science in Nineteenth-Century Britain: A Historiographical Survey and Guide to Sources," *Studies in History and Philosophy of Science Part A* 31 (2000). As discussed in chapter 3, an important exception to these nationally oriented studies is Aileen Fyfe, *Steam-Powered Knowledge: William Chambers and the Business of Publishing, 1820–1860* (Chicago: University of Chicago Press, 2012), particularly 173–261.

5. As with much of the original work on scientific audiences, my own account is conceptually indebted to Jon Klancher, *The Making of English Reading Audiences, 1790–1832* (Madison: University of Wisconsin Press, 1987).

6. Like others, I am making use of and extending Benedict Anderson, *Imagined Communities: Reflections on the Origin and Spread of Nationalism* (London: Verso, 2006), 39–48.

7. *Daily Chronicle*, 18 December 1890.

8. "Phrenology," *The Australian*, 12 December 1827; "Philophrenologist," *The Australian*, 9 December 1834. On the politics of this paper, see Victoria Goff, 'Convicts and Clerics: Their Roles in the Infancy of the Press in Sydney, 1803–1840," *Media History* 4 (1998): 105–107; Anna Johnston, *The Paper War: Morality, Print Culture, and Power in Colonial New South Wales* (Perth: University of Western Australia Press, 2011), 108.

9. Title page of *Annals of Phrenology* 1 (1833). For the role of periodicals in gathering together foreign articles, see Jonathan Topham, "Anthologizing the Book of Nature: The Circulation of Knowledge and the Origins of the Scientific Journal in Late Georgian Britain," in Bernard Lightman, Gordon McOuat, and Larry Stewart, eds., *The Circulation of Knowledge between Britain, India, and China: The Early-Modern World to the Twentieth Century* (Leiden: Brill, 2013).

10. "Edinburgh Phrenological Journal," *Annals of Phrenology* 1 (1833): 86–87.

11. "Notices," *Phrenological Journal* 7 (1831–1832): 96; "Miscellaneous Notices," *Phrenological Journal* 10 (1836–1837): 510; 'Intelligence," *Phrenological Journal* 12 (1839): 302.

12. "Address to the Subscribers to the Phrenological Journal," *Phrenological Journal* 6 (1829–1830): 153.

13. "Notices," *Annals of Phrenology* 1 (1833): 272.

14. "Notices," *Annals of Phrenology* 2 (1835): 132; "Notices," *Annals of Phrenology* 2 (1835): 260.

15. "Liste des Membres," *Journal de la Société Phrénologique de Paris* 1 (1832): 28.

16. Christine Haynes, *Lost Illusions: The Politics of Publishing in Nineteenth-Century France* (Cambridge, MA: Harvard University Press, 2010), 52–53 and 103.

17. Advertisement in *Annals of Phrenology* 1 (1833). For Nahum Capen's role in the publication of the *Annals of Phrenology*, see Capen to Combe, 13 February 1835, MS7234, f. 94, Combe Papers; Brigham to Capen, 17 May 1833, Nahum Capen Papers, B MS c23, Countway Library, Harvard University.

18. Kirsten McKenzie, "'Franklins of the Cape': *The South African Commercial Advertiser* and the Creation of a Colonial Public Sphere, 1824–1854," *Kronos* 25 (1998): 93; Herbert Heaton, "The Early Tasmanian Press, and Its Struggle for Freedom," *Papers and Proceedings of the Royal Society of Tasmania* 42 (1916): 3–10.

19. Saul Dubow, *A Commonwealth of Knowledge: Science, Sensibility, and White South Africa, 1820–2000* (Oxford: Oxford University Press, 2006), 34–36; Wessel de Kock, *A Manner of Speaking: The Origins of the Press in South Africa* (Cape Town: Saayman and Weber, 1982), 2; Sandy Blair, "'The convict press': Edward Smith Hall and *The Sydney Monitor*," in Denis Cryle, ed., *Disreputable Profession: Journalists and Journalism in Colonial Australia* (Rockhampton: Central Queensland University Press, 1997), 25–27.

20. "Journal de la Société Phrénologique de Paris," *Phrenological Journal* 9 (1834–1836): 131.

21. Struve to Combe, 12 March 1845, MS7277, f. 87; Combe to Otto, 28 March 1832, MS7385, f. 267, Combe Papers.

22. "Notices," *Annals of Phrenology* 1 (1833): 271.

23. Mann to Combe, 28 February 1843, MS7269, f. 91, Combe Papers; "Miscellany," *American Phrenological Journal* 2 (1839): 384.

24. "Notices," *Phrenological Journal* 3 (1825–1826): 325–326.

25. Fyfe, *Steam-Powered Knowledge*, 1–26; Iain Watts, "'We want no authors': William Nicholson and the Contested Role of the Scientific Journal in Britain, 1797–1813," *British Journal for the History of Science* 47 (2014); both emphasize the close connection between scientific publishing and business concerns.

26. Madeleine Stern, *Heads and Headlines: The Phrenological Fowlers* (Norman: University of Oklahoma Press, 1971), 7–20, 26, and 220.

27. Donald Scott, "The Popular Lecture and the Creation of a Public in Mid-Nineteenth-Century America," *Journal of American History* 66 (1980).

28. Stern, *Heads and Headlines*, 138; "The American Phrenological Journal for 1849," *American Phrenological Journal* 11 (1849): 11.

29. The firm then became known as Fowler and Wells; Stern, *Heads and Headlines*, 25–57.

30. "Phrenological Museum," *American Phrenological Journal* 30 (1859): 31.

31. Stern, *Heads and Headlines*, 59–61; John van Wyhe, *Phrenology and the Origins of Victorian Scientific Naturalism* (Aldershot: Ashgate, 2004), 183.

32. William Chambers employed a similar strategy in Britain; Fyfe, *Steam-Powered Knowledge*, 41–54.

33. See Aileen Fyfe, "Reading Natural History at the British Museum and the *Pictorial Museum*," in Aileen Fyfe and Bernard Lightman, eds., *Science in the Marketplace: Nineteenth-Century Sites and Experiences* (Chicago: University of Chicago Press, 2007), for a similar discussion of the relationship between physical and print spaces.

34. "A Shelf of Our Cabinet," *American Phrenological Journal* 35 (1862): 37.

35. Inside cover of May issue of the *American Phrenological Journal* 9 (1849).

36. "Miscellany," *American Phrenological Journal* 9 (1847): 101.

37. Eric Lupfer, "The Business of American Magazines," in Scott Casper, Jeffrey Groves, Stephen Nissenbaum, and Michael Winship, eds., *The History of the Book in America: The Industrial Book, 1840–1880* (Chapel Hill: University of North Carolina Press, 2007), 3:227; Jeffrey Groves, "Periodicals and Serial Publication," in Casper, Groves, Nissenbaum, and Winship, *The History of the Book in America*, 3:227.

38. "The Past and Future Course of the Journal," *American Phrenological Journal* 4 (1842): 315–316.

39. John Nerone, "Newspapers and the Public Sphere," in Casper, Groves, Nissenbaum, and Winship, *The History of the Book in America*, 3:233.

40. "The Past and Future Course of the Journal," 319.

41. "Notices of the Press," *American Phrenological Journal* 49 (1869): iii; "General Notices," *American Phrenological Journal* 13 (1851): 141.

42. Inside cover of May issue of the *American Phrenological Journal* 9 (1847).

43. Scott Casper, "The Census, the Post-Office, and Governmental Publishing," in Casper, Groves, Nissenbaum, and Winship, *The History of the Book in America*, 3:182–184.

44. Scott Casper, "Other Variations on the Trade," in Casper, Groves, Nissenbaum, and Winship, *The History of the Book in America*, 3:205.

45. "Business Notices," *American Phrenological Journal* 35 (1862): 141.

46. "To Readers and Friends," *American Phrenological Journal* 26 (1857): 110.

47. "Our Circular Prospectus," *American Phrenological Journal* 24 (1856): 127.

48. "A Few Editorial Notices of the Journal," *American Phrenological Journal* 15 (1852): 24.

49. "Salutatory," *American Phrenological Journal* 19 (1854): 15.

50. Loose printed prospectus for the *American Phrenological Journal*, BDSDS.1851, American Antiquarian Society, Worcester, MA.

51. Combe to Allen, 30 September 1839, MS7327, f. 18, Combe Papers.

52. "Lecturers," *American Phrenological Journal* 13 (1851): 22.

53. Nelson Sizer, *Forty Years in Phrenology* (New York: Fowler and Wells, 1882), 187; Stern, *Heads and Headlines*, 80.

54. "Prizes and Premiums," *American Phrenological Journal* 36 (1862): 133.

55. Stern, *Heads and Headlines*, 212.

56. "Introductory Statement," *American Phrenological Journal* 1 (1839): 7–8.

57. "Our Country and Its Prospective Greatness," *American Phrenological Journal* 13 (1854).

58. "Our Country: An Oration On Our Progress," *American Phrenological Journal* 49 (1869).

59. Donald Pease, "New Perspectives on U.S. Culture and Imperialism," in Amy Kaplan and Donald Pease, eds., *Cultures of United States Imperialism* (Durham, NC: Duke University Press, 1993).

60. Daniel Howe, *What Hath God Wrought? The Transformation of America, 1815–1848* (Oxford: Oxford University Press, 2007), 701–744.

61. Stern, *Heads and Headlines*, 172.

62. "A Rocky Mountain Book Store," *American Phrenological Journal* 47 (1868); Ottinger to Wells, November 1868, Fowler and Wells Family Papers, Cornell University Library (henceforth, Fowler Papers); Stern, *Heads and Headlines*, 216–217.

63. "Phrenology in California," *American Phrenological Journal* 23 (1856).

64. "Advertisements," *American Phrenological Journal* 22 (1855): 92; Stern, *Heads and Headlines*, 171–172.

65. "Phrenology in California," *American Phrenological Journal* 23 (1856).

66. Howe, *What Hath God Wrought?* 1–7, also puts "the communication revolution" at the center of nineteenth-century American history.

67. "Salutatory: Our New Year," *American Phrenological Journal* 19 (1854).

68. "The American Phrenological Journal for 1857," *The Liberator* 27 (1857).

69. "The Past and Future Course of this Journal," *American Phrenological Journal* 4 (1842).

70. "Dr. Buchanan in Florida," *American Phrenological Journal* 2 (1839).

71. Michael O'Brien, *Conjectures of Order: Intellectual Life and the American South, 1810–1860* (Chapel Hill: University of North Carolina Press, 2004), 1:426–430; "Mr. Fowler in the South," *American Phrenological Journal* 27 (1858).

72. Stern, *Heads and Headlines*, 205.

73. "Southward Ho!," *American Phrenological Journal* 41 (1865).

74. Iain Watts, "Philosophical Intelligence: Letters, Print, and Experiment during Napoleon's Continental Blockade," *Isis* 106 (2015), discusses the relationship between war and scientific publishing more generally.

75. "Money! Money! Money!," *American Phrenological Journal* 37 (1863).

76. "Not Received," *American Phrenological Journal* 38 (1863); "A Texas Editor on Phrenology," *American Phrenological Journal* 48 (1868).

77. "Our Journal in the Army," *American Phrenological Journal* 38 (1863).

78. "Opinions of the People," *American Phrenological Journal* 35 (1862).

79. Andrew to Wells, 22 April 1863, Fowler Papers.

80. On the relationship between the American Civil War and scientific publishing, see Bernard Lightman, "Spencer's American Disciples: Fiske, Youmans, and the Appropriation of the System," in Bernard Lightman, ed., *Global Spencerism: The Communication and Appropriation of a British Evolutionist* (Leiden: Brill, 2016).

81. "Southward Ho!," *American Phrenological Journal* 41 (1865).

82. "A Contrast: Massachusetts and South Carolina," *American Phrenological Journal* 44 (1866).

83. Howe, *What Hath God Wrought?* 704–707; Amy Kaplan, "'Left Alone with America': The Absence of Empire in the Study of American Culture," in Amy Kaplan and Donald Pease, eds., *Cultures of United States Imperialism* (Durham, NC: Duke University Press, 1993); Thomas Bender, "Historians, the Nation, and the Plenitude of Narratives," in Thomas Bender, ed., *Rethinking American History in a Global Age* (Berkeley: University of California Press. 2002).

84. John Perry, *Facing West: Americans and the Opening of the Pacific* (Westport, CT: Praeger, 1994), 168–172.

85. "From China," *American Phrenological Journal* 49 (1869).

86. "The Chinese Empire," *American Phrenological Journal* 46 (1867): 65–66; "Street Scenes in China," *Phrenological Journal and Science of Health* 50 (1870); "Sketches of Life in China," *Phrenological Journal and Science of Health* 50 (1870).

87. H. K. Barpujari, *The American Missionaries and North-East India, 1836–1900* (Guwahati: Spectrum, 1986), xi-lvii.

88. "Two Human Monsters," *American Phrenological Journal* 39 (1864).

89. "Fifty Years of Phrenology," *Phrenological Journal and Science of Health* 80 (1888): 19.

90. "The American Phrenological Journal for 1849," *American Phrenological Journal* 11 (1849): 10.

91. Advertisement opposite "Local Intelligence," *South African Quarterly Journal* 1 (1830): 225 and 463.

92. For Macartney's biography and details of phrenology in the Cape Colony, see Andrew Bank, "Of 'Native Skulls' and 'Noble Caucasians': Phrenology in Colonial South Africa," *Journal of Southern African Studies* 22 (1996); Bank, "Liberals and Their Enemies: Racial Ideology at the Cape of Good Hope, 1820–1850" (PhD diss., University of Cambridge, 1995), 313–346; Leigh Bregman, "'Snug Little Coteries': A History of Scientific Societies in Early Nineteenth-Century Cape Town, 1824–1835" (PhD diss., University College London, 2003), 103–104, 113.

93. Dubow, *A Commonwealth of Knowledge*, 1 and 40; Kock, *A Manner of Speaking*, 2, 51, and 65; Anna Smith, *The Spread of Printing: South Africa* (Amsterdam: Vangendt, 1971), 12.

94. McKenzie, "'Franklins of the Cape,'" 93; Kock, *A Manner of Speaking*, 22 and 93–94.

95. Bank, "Liberals and Their Enemies," 318–319.

96. Dubow, *A Commonwealth of Knowledge*, 36–46.

97. Jackie Loos, *Echoes of Slavery: Voices from South Africa's Past* (Cape Town: David Philip, 2004), 17–18; Timothy Keegan, *Colonial South Africa and the Origins of the Racial Order* (London: Leicester University Press, 1996), 50.

98. Dubow, *A Commonwealth of Knowledge*, 36.

99. *A Catalogue of the South African Public Library* (Cape Town: Gazette Office, 1829), 40 and 43.

100. Dubow, *A Commonwealth of Knowledge*, 36.

101. Henry Macartney, *Lectures Delivered in the Commercial Hall on the Subject of Life and Death* (Cape Town: George Greig, 1831), iii-viii and 9–24.

102. Bank, "Of 'Native Skulls,'" 390–392.

103. *The South African Commercial Advertiser*, 16 November 1825.

104. Bank, "Of 'Native Skulls,'" 392.

105. "A Catechism of Phrenology," *Cape of Good Hope Literary Gazette* 1 (1831): 175–176.

106. Martin Legassick and Robert Ross, "From Slave Economy to Settler Capitalism: The Cape Colony and Its Extensions, 1800–1854," in Carolyn Hamilton, Bernard Mbenga, and Robert Ross, eds., *The Cambridge History of South Africa* (Cambridge: Cambridge University Press, 2009), 1:255; Clifton Crais, *White Supremacy and Black Resistance: Pre-Industrial South Africa* (Cambridge: Cambridge University Press, 1992), 61–121.

107. Dubow, *A Commonwealth of Knowledge*, 19 and 60.

108. Kock, *A Manner of Speaking*, 79.

109. For Macartney in the Eastern Cape, see Bank, "Of 'Native Skulls,'" 396–399. For Godlonton and the *Graham's Town Journal*, see Kock, *A Manner of Speaking*, 8–9 and 81–85.

110. *Graham's Town Journal*, 26 November 1835.

111. Kock, *A Manner of Speaking*, 8 and 81–82.

112. Alfred Gordon-Brown, *The Settlers' Press: Seventy Years of Printing in Grahamstown* (Cape Town: Balkema, 1979), 11.

113. Kock, *A Manner of Speaking*, 23.

114. Crais, *White Supremacy*, 113–115 and 143; Norman Etherington, Patrick Harries, and Bernard Mbenga, "From Colonial Hegemonies to Imperial Conquest, 1840–1880," in Hamilton, Mbenga and Ross, *The Cambridge History of South Africa*, 1:321; and Legassick and Ross, "From Slave Economy to Settler Capitalism," 268.

115. Crais, *White Supremacy*, 127–133.

116. *Graham's Town Journal*, 23 January 1835 and 6 February 1835; Crais, *White Supremacy*, 132–133.

117. Bank, "Of 'Native Skulls,'" 390–391.

118. *Graham's Town Journal*, 10 December 1835.

119. *Graham's Town Journal*, 26 November 1835.

120. Bank, "Liberals and Their Enemies," 328–338.

121. *Graham's Town Journal*, 16 November 1843.

122. Bank, "Of 'Native Skulls,'" 398.

123. The District of Victoria was quickly abandoned, although the land was later permanently incorporated into Victoria East in 1847; Etherington, Harries, and Mbenga, "From Colonial Hegemonies to Imperial Conquest," 323.

124. *Graham's Town Journal*, 26 November 1835.

125. Etherington, Harries, and Mbenga, "From Colonial Hegemonies to Imperial Conquest," 358.

126. *The Natal Witness*, 9 August 1850 and 23 August 1850.

127. *The Natal Witness*, 29 March 1850.

128. Kock, *A Manner of Speaking*, 93 and 96.

129. *The Natal Witness*, 18 October 1850. The original Paris edition included lithographic illustrations of tribal leaders. However, these were not reproduced in the Cape edition; Thomas Arbousset, *Narrative of an Exploratory Tour to the North-East of*

the Colony of the Cape of Good Hope, trans. John Brown (Cape Town: A. S. Robertson, 1846).

130. *The Natal Witness*, 18 October 1850; "Moletsane, Abraham Makgothi," in Evert Verwey, ed., *New Dictionary of South African Biography* (Pretoria: HSRC Publishers, 1995).

131. Bank, "Of 'Native Skulls,'" 396–402.

132. *Sam Sly's African Journal*, 4 September 1845; Kock, *A Manner of Speaking*, 79.

133. *Sam Sly's African Journal*, 13 November 1845.

134. *Sam Sly's African Journal*, 14 August 1845.

135. *Sam Sly's African Journal*, 12 March 1846 and 18 September 1845.

136. Bank, "Of 'Native Skulls,'" 398.

137. Bank, "Of 'Native Skulls,'" 387–403.

138. Combe to Bose, 4 February 1848, MS7391, f. 297; Bose to Combe, 7 December 1848, MS7289, f. 80, Combe Papers.

139. For the broader history of Bengali phrenology, see Shruti Kapila, "Race Matters: Orientalism and Religion, India and Beyond, c. 1770–1880," *Modern Asian Studies* 41 (2007): 495–502; Gyan Prakash, *Another Reason: Science and the Imagination of Modern India* (Princeton, NJ: Princeton University Press, 1999), 56–57 and 74–80; "History of Phrenology in Calcutta," *Phrenological Magazine* 4 (1883): 453–458.

140. Combe, 7 December 1848, MS7289, f. 80, Combe Papers; Poonam Bala, *Imperialism and Medicine in Bengal: A Socio-Historical Perspective* (London: Sage, 1991), 72; Mel Gorman, "Introduction of Western Science into Colonial India: Role of the Calcutta Medical College," *Proceedings of the American Philosophical Society* 132 (1988).

141. Subir Raychoudhuri, "The Lost World of the Babus," in Sukanta Chaudhuri, ed., *Calcutta: The Living City* (Oxford: Oxford University Press, 1990), 1:70; Tithi Bhattacharya, *The Sentinels of Culture: Class, Education, and the Colonial Intellectual in Bengal, 1848–85* (New Delhi: Oxford University Press, 2005), 2–4. For the broader history and historiography of the "Bengal Renaissance," see David Kopf, *British Orientalism and the Bengal Renaissance* (Berkeley: University of California Press, 1969); Sumit Sarkar, "Calcutta and the 'Bengal Renaissance,'" in Sukanta Chaudhuri, ed., *Calcutta: The Living City* (Oxford: Oxford University Press, 1990), vol. 1.

142. Doss to Combe, 7 December 1848, MS7283, f. 63, Combe Papers.

143. For the history of the Bengali press in this period, see Anindita Ghosh, *Power in Print: Popular Publishing and the Politics of Language and Culture in a Colonial Society, 1778–1905* (New Delhi: Oxford University Press, 2006); Mrinal Chanda, *History of the English Press in Bengal, 1780–1857* (Calcutta: K. P. Bagchi, 1987); Smarajit Chakraborti, *The Bengali Press, 1818–1868* (Calcutta: Firma KLM, 1976).

144. Rajat Sanyal, *Voluntary Associations and the Urban Public Life in Bengal, 1815–1876* (Calcutta: Riddhi-India, 1980), 6 and 18–20; Christopher Bayly, *Recovering Liberties: Indian Thought in the Age of Liberalism and Empire* (Cambridge: Cambridge University Press, 2011), 109.

145. Prakash, *Another Reason*, 75–76.

146. Andrew Sartori, *Bengal in Global Concept History: Culturalism in the Age of Capital* (Chicago: University of Chicago Press, 2008), 114.

147. C. Mackenzie Brown, *Hindu Perspectives on Evolution: Darwin, Dharma, and Design* (London: Routledge, 2012), 97.

148. David Kopf, *The Brahmo Samaj and the Shaping of the Modern Indian Mind* (Princeton, NJ: Princeton University Press, 1979), 49–59; Chakraborti, *The Bengali Press*, 100–104; Prakash, *Another Reason*, 53–54.

149. J. Lourdusamy, *Science and National Consciousness in Bengal, 1870–1930* (New Delhi: Orient Longman, 2004), 47; Sanyal, *Voluntary Associations*, 111; Partha Chatterjee, *Texts of Power: Emerging Disciplines in Colonial Bengal* (Minneapolis: University of Minnesota Press, 1995), 12; Prakash, *Another Reason*, 54.

150. Ghosh, *Power in Print*, 95; Chatterjee, *Texts of Power*, 15.

151. For the connection between mental philosophy and Indian liberalism, see Bayly, *Recovering Liberties*, 63; Kapila, "Race Matters," 471–513.

152. "Our Plans and Objects," *The Pamphleteer* 1 (1850): 1–3.

153. On the relationship between nationalism and liberalism in Bengal and India, see Bayly, *Recovering Liberties*, 1; Sartori, *Bengal in Global Concept History*, 68–108. The Calcutta Medical College did provide an opportunity to forge connections beyond Bengal, with students arriving from Agra, Benares, Burma, and Ceylon; Home, Public, 14 September 1849, C, No. 7, National Archives of India, New Delhi; Home, Public, India Financials, 4 April 1848, India Office Records, E/4/796, ff. 68–69, British Library.

154. "Our Plans and Objects," *The Pamphleteer* 1 (1850): 1; "Phrenological Development of the Bengalees," *The Pamphleteer* 1 (1850): 66.

155. "Phrenological Development of the Bengalees," 66–69.

156. On the global outlook of Indian political thought, see Tapan Raychaudhuri, *Europe Reconsidered: Perceptions of the West in Nineteenth-Century Bengal* (New Delhi: Oxford University Press, 1988); Sartori, *Bengal in Global Concept History*; Bayly, *Recovering Liberties*.

157. Radhaballav Das, *General Reflections on Christianity* (Calcutta: Chunderseaker Bistoo, 1845), 202.

158. "The Prospect of the Age," *The Pamphleteer* 1 (1850): 84–89; "Phrenological Development of the Bengalees," 66–69.

159. "Education of the People," *The Pamphleteer* 1 (1850): 139–160.

160. Ghosh, *Power in Print*, 16; Sarkar, "Calcutta and the 'Bengal Renaissance,'" 105.

161. "History of Phrenology in Calcutta," *Phrenological Magazine* 4 (1883): 453–458, Sarkar, "Calcutta and the 'Bengal Renaissance,'" 102; Dipesh Chakrabarty, "*Adda*, Calcutta: Dwelling in Modernity," *Public Culture* 11 (1999).

162. "Education of the People," 139–160.

163. As Das recognized, Fox had read *The Constitution of Man* and was sympathetic to phrenological approaches to education; "Education of the People," 139–160.

164. "Education of the People," 139–160.

165. Daniel White, *From Little London to Little Bengal: Religion, Print, and Modernity in Early British India, 1793–1835* (Baltimore: John Hopkins University Press, 2013), 1–5.

166. Kopf, *The Brahmo Samaj*, xiii-xvii.

167. Prakash, *Another Reason*, 53, Lourdusamy, *Science and National Consciousness in Bengal*, 50; Kopf, *The Brahmo Samaj*, 50.

168. Brian Hatcher, *Idioms of Improvement: Vidyasagar and Cultural Encounter in Bengal* (Calcutta: Oxford University Press, 1996), 223–227; Sartori, *Bengal in Global Concept History*, 116.

169. Translation from Hatcher, *Idioms of Improvement*, 226–227.

170. Sartori, *Bengal in Global Concept History*, 114–116.

171. Das, *General Reflections*, 211.

172. Gorman, "Introduction of Western Science into Colonial India," 286–288; Samita Sen and Anirban Das, "A History of the Calcutta Medical College and Hospital, 1835–1936," in Uma Das Gupta, ed., *Science and Modern India: An Institutional History, 1784–1947* (New Delhi: Pearson Longman, 2011), 478; Home, Public, 3 October 1846, C, No. 3, National Archives of India, New Delhi.

173. "Plurality of the Cerebral Organs," *The Pamphleteer* 1 (1850): 112.

174. Das, *General Reflections*, iv-vii and 178–179.

175. "The Earth a Living Body," *The Pamphleteer* 1 (1850): 21–26.

176. On the Battala book market, see Ghosh, *Power in Print*, 107–151.

177. Hatcher, *Idioms of Improvement*, 215–222; Kopf, *The Brahmo Samaj*, 257. For evangelical publishing in Britain, see Aileen Fyfe, *Science and Salvation: Evangelical Popular Science Publishing in Victorian Britain* (Chicago: University of Chicago Press, 1994).

178. Chakraborti, *The Bengali Press*, 104.

179. "Education of the People," 157.

180. Bhattacharya, *The Sentinels of Culture*, 2, 4, and 70.

181. In 1857 the Purnochandradoy Press printed 8,450 books compared with the 55,000 printed by the Baptist Mission Press; Ghosh, *Power in Print*, 121.

182. *The New Calcutta Directory* (Calcutta: Military Orphan Press, 1856), pt. 9, 348; Nilmoni Basak, *Arabya Upanyas* (Calcutta: Purnochandradoy Press, 1850); Anindita Ghosh, "An Uncertain 'Coming of the Book': Early Print Cultures in Colonial India," *Book History* 6 (2003): 23–56.

183. Ghosh, *Power in Print*, 105.

184. "Education of the People," 158.

185. James Long, *The Proceedings and Transactions of the Bethune Society from November 10th 1859 to April 20th 1869* (Calcutta: Bishop's College Press, 1870), xx-xxii.

186. Radhaballav Das, *Manatatva Sarsangraha* (Calcutta: Purnochandradoy Press, 1849), i-iv. (I am extremely grateful to Sohini Chattopadhyay for her translation of this work and also to Shinjini Das, Ishan Mukherjee, and Rohan Deb Roy for discussing the Bengali literature further with me.)

187. Later Bengali phrenological charts do not feature the mustache, and more closely resemble those produced in Europe and the United States; for example, *Anubikshana* 1 (1875): 129; *Adrista* 1 (1896): 232.

188. Ghosh, *Power in Print*, 64 and 98.

189. "Plurality of the Cerebral Organs," 135.

190. "A Hindu of Small Cranium," *Bombay Times*, 13 February 1859; "Collet and Co's Advertisement," *Bombay Times*, 6 April 1839.

191. Bernard Lightman, *Victorian Popularizers of Science* (Chicago: University of Chicago Press, 2007), 11–18.

192. Lightman, *Victorian Popularizers*, 11–12; Jonathan Topham, "Introduction: Historicizing 'Popular Science,'" *Isis* 100 (2009).

CHAPTER SIX

1. "Written Descriptions from Daguerreotypes," *American Phrenological Journal* 24 (1856): 1.

2. "Character Sketches," *Phrenological Journal and Science of Health* 104 (1897): 96.

3. "Seven Modern Wonders," *Phrenological Journal and Science of Health* 52 (1871): 343.

4. "Character from Photographs," *Phrenological Journal and Science of Health* 89 (1890): 223.

5. "Answers to Correspondents," *Phrenological Magazine* 4 (1883): 175; "Answers to Correspondents," *Phrenological Magazine* 5 (1884): 131.

6. Lorraine Daston and Peter Galison, *Objectivity* (New York: Zone Books, 2007), 120–164; Phillip Prodger, *Darwin's Camera: Art and Photography in the Theory of Evolution* (Oxford: Oxford University Press, 2009), xxi-xxiv; Elizabeth Edwards, *Raw Histories: Photographs, Anthropology, and Museums* (Oxford: Berg, 2001), 133.

7. "Written Descriptions of Character from Likenesses," *American Phrenological Journal* 33 (1861): 100.

8. "History of Photography in America," *Phrenological Journal and Science of Health* 54 (1872): 298.

9. Gael Newton, "South-East Asia: Malay, Singapore, and Philippines," in John Hannavy, ed., *Encyclopedia of Nineteenth-Century Photography* (London: Routledge, 2008), 2:1315.

10. N. W. Beckwith, "Aborigines of the Philippine Archipelago," *Phrenological Journal and Life Illustrated* 50 (1870): 160; N. W. Beckwith, "Natives of the Philippine Islands—No. III," *Phrenological Journal and Life Illustrated* 53 (1871): 313–318.

11. Alfred Russel Wallace, "On the Varieties of Man in the Malay Archipelago," *Transactions of the Ethnological Society of London* 3 (1865): 196–215; Adrian Desmond and James Moore, *Darwin's Sacred Cause: Race, Slavery, and the Quest for Human Origins* (London: Penguin, 2009), 297–298 and 342–369; Martin Fichman, *An Elusive Victorian: The Evolution of Alfred Russel Wallace* (Chicago: University of Chicago Press,, 2004), 151–154.

12. Beckwith, "Aborigines of the Philippine Archipelago," 160.

13. Beckwith, "Natives of the Philippine Islands—No. III," 313–318.

14. "American Possessions in the Philippines," *Phrenological Journal and Science of Health* 107 (1899): 177.

15. "The Mahdi, Or Fighting Prophet of the Soudan," *Phrenological Journal and Science of Health* 80 (1885): 216–218.

16. "The Zulus and Their Phrenology," *Phrenological Magazine* 2 (1881): 138–145.

17. "The Todas," *Daily Telegraph*, 12 June 1895; *Sydney Morning Herald*, 21 January

1874; *Catalogue of the Library of the Mercantile Library Association of San Francisco* (San Francisco: Francis and Valentine, 1874), 651.

18. For the history of photography in colonial India, see Christopher Pinney, *Camera Indica: The Social Life of Indian Photographs* (London: Reakiton, 1992); John Falconer, "Photography in Nineteenth-Century India," in Christopher Bayly, ed., *The Raj and the British, 1600–1947* (London: National Portrait Gallery Publications, 1990). Pinney, *Camera*, 44, simply claims that it is "famously difficult to find documentation of the reception" of photographs. Anne Maxwell, *Colonial Photography and Exhibitions: Representations of the Native and the Making of European Identities* (London: Bloomsbury, 1999), x, examines photography in the context of exhibitions but admits her analysis of reception is based on crowd size and consumer demand rather than actual readings. Gary Sampson, "The Success of Samuel Bourne in India," *History of Photography* 16 (1992), provides the most detailed account of the reception of colonial photography.

19. This follows Edwards, *Raw Histories*, 13, arguing we need "social biographies of photographs as objects."

20. Edwards, *Raw Histories*, 143, also argues that treating photographs as material objects helps to trace the limits of colonial power.

21. For example, Pinney, *Camera Indica*, 50–51; Clare Anderson, *Legible Bodies: Race, Criminality, and Colonialism in South Asia* (Oxford: Berg, 2004), 86–88.

22. Deborah Sutton, *Other Landscapes: Colonialism and the Predicament of Authority in Nineteenth-Century South India* (Copenhagen: Nordic Institute of Asian Studies, 2009), 160–161; Dane Kennedy, *Magic Mountains: Hill Stations and the British Raj* (Berkeley: University of California Press, 1996), 70–57.

23. George Stocking, *Victorian Anthropology* (New York: Free Press, 1987), 154–155; Efram Sera-Shriar, *The Making of British Anthropology, 1813–1871* (London: Pickering and Chatto, 2013), 158.

24. Sera-Shriar, *The Making of British Anthropology*, 147–169; Adam Kuper, *The Invention of Primitive Society* (London: Routledge, 1988), 64–65.

25. Joy Harvey, "Darwin in a French Dress," in Eve-Marie Engels and Thomas Glick, eds., *The Reception of Charles Darwin in Europe* (London: Continuum, 2008), 2:368–370.

26. Sera-Shriar, *The Making of British Anthropology*, 157–159; Stocking, *Victorian Anthropology*, 247–249.

27. George Stocking, "What's in a Name? The Origins of the Royal Anthropological Institute (1837–71)," *Man* 6 (1971).

28. William Marshall, *A Phrenologist Amongst the Todas* (London: Longmans, Green and Co., 1873), vi-viii and 52.

29. The connection between evolution and phrenology can be traced back to Robert Chambers's *Vestiges of the Natural History of Creation*, published in 1844; James Secord, *Victorian Sensation: The Extraordinary Publication, Reception, and Secret Authorship of "Vestiges of the Natural History of Creation"* (Chicago: University of Chicago Press, 2000), 77–110.

30. Sutton, *Other Landscapes*, 19–21.

31. Kennedy, *Magic Mountains*, 92 and 109; Sutton, *Other Landscapes*, 103; Kavita Phillip, *Civilising Natures: Race, Resources, and Modernity in Colonial South India* (New Delhi: Orient Longman, 2003), 61–62.

32. Sutton, *Other Landscapes*, 26.

33. Sutton, *Other Landscapes*, 21–26 and 162–166.

34. Marshall, *A Phrenologist*, v and 94.

35. Sutton, *Other Landscapes*, 161, is one of few scholars to recognize Marshall's book as evidence of "the longevity of phrenology as an anthropometric science."

36. Henrika Kuklick, *The Savage Within: The Social History of British Anthropology, 1885–1945* (Cambridge: Cambridge University Press, 1991), 27 and 42.

37. Cadet Papers, India Office Records, L/MIL/9/203, ff. 142–147, British Library.

38. Erica Wald, *Vice in the Barracks: Medicine, Military, and the Making of Colonial India, 1780–1868* (Basingstoke: Palgrave, 2014), 86–89.

39. Janet Dewan, "Delineating Antiquities and Remarkable Tribes," *History of Photography* 16 (1992): 304; Falconer, "Photography in Nineteenth-Century India," 271.

40. Cadet Papers, India Office Records, L/MIL/9/203, ff. 142–147, Bengal Officers' Services, India Office Records, L/MIL/10/88, f. 104, British Library; "Marriages," *Gentleman's Magazine* 211 (1861): 557.

41. Marshall, *A Phrenologist*, 28.

42. Shruti Kapila, "Race Matters: Orientalism and Religion, India and Beyond, c. 1770–1880," *Modern Asian Studies* 41 (2007).

43. Bengal Service Army List, India Office Records, L/MIL/10/63, f.503, L/MIL/10/61, f.503, MIL/10/58, f.503, MIL/10/41, f. 503, British Library; Arun Kumar, "Thomason College of Engineering, Roorkee, 1847–1947," in Uma Das Gupta, ed., *Science and Modern India: An Institutional History, c. 1784–1947* (Delhi: Pearson Longman, 2011).

44. G. Thomas, "The 'Peccavi' Photographs," *History of Photography* 4 (1980): 49; Christopher Pinney, "Underneath the Banyan Tree: William Crooke and Photographic Depictions of Caste," in Elizabeth Edwards, ed., *Anthropology and Photography, 1860–1920* (New Haven, CT: Yale University Press, 1992), 166.

45. *Thomason Civil Engineering College Calendar* (Roorkee: Thomason Civil Engineering College Press, 1871), 52.

46. On the relationship between the physical sciences and anthropology, see Simon Schaffer, *From Physics to Anthropology and Back Again* (Cambridge: Prickly Pear Press, 1994).

47. Falconer, "Photography in Nineteenth-Century India," 267.

48. G. Thomas, "The Madras Photographic Society, 1856–61," *History of Photography* 16 (1992): 299–301.

49. Dewan, "Delineating Antiquities," 302 and 315.

50. Thomas, "The Madras Photographic Society," 299–301.

51. John Watson and John Kaye, *The People of India* (London: India Museum, 1875), vol. 8, documents the Madras Presidency including the Todas.

52. James Ryan, *Picturing Empire: Photography and the Vizualisation of the British Empire* (London: Reaktion, 1997), 155–158; Pinney, *Camera*, 34–45.

53. Marshall, *A Phrenologist*, v.

54. Bengal Officers' Service, India Office Records, L/MIL/10/88, f. 104, British Library.

55. Marshall, *A Phrenologist*, ix and 96–97.

56. Marshall, *A Phrenologist*, 32.

57. Marshall, *A Phrenologist*, 34.

58. Marshall, *A Phrenologist*, 88–89

59. Daston and Galison, *Objectivity*, 133; Prodger, *Darwin's Camera*, xxi.

60. Marshall, *A Phrenologist*, 32.

61. Marshall, *A Phrenologist*, 32–33.

62. Marshall, *A Phrenologist*, 12–13.

63. Marshall, *A Phrenologist*, 29–33. Anderson, *Legible Bodies*, 186–188, also notes the connection between uniformity and primitive culture in Marshall's argument.

64. The problem of photographic authority is discussed in Prodger, *Darwin's Camera*, xxi-xxiv and 220–224; Daston and Galison, *Objectivity*, 120–164; and Efram Sera-Shriar, "Anthropometric Portraiture and Victorian Anthropology: Situating Francis Galton's Photographic Work in the Late 1870s," *History of Science* 53 (2015).

65. Edwards, *Raw Histories*, 131–138.

66. Huxley to Granville, 12 August 1869, CO 854/10 f. 629, TNA.

67. John Lamprey, "On a Method of Measuring the Human Form, for the Use of Students in Ethnology," *Journal of the Ethnological Society of London* 1 (1869): 84–85; Frank Spencer, "Some Notes on the Attempt to Apply Photography to Anthropometry during the Second Half of the Nineteenth Century," in Edwards, *Anthropology and Photography*, 102.

68. Daston and Galison, *Objectivity*, 133–138.

69. F. Fisk Williams, *A Guide to the Indian Photographer* (Calcutta: R. C. Lepage, 1860), 7–8 and 18; Michael Pritchard, "Ottewill, Thomas," in John Hannavy, ed., *Encyclopedia of Nineteenth-Century Photography* (London: Routledge, 2008), 1:1033.

70. *A Catalogue of Photographic Apparatus and Chemicals* (Calcutta: R. C. Lepage, 1860), 8.

71. Williams, *A Guide to the Indian Photographer*, 18.

72. *A Catalogue of Photographic Apparatus*, 7.

73. *Hints to Travellers* (London: Royal Geographical Society, 1883), 245.

74. Edward Burnett Tylor, "Marshall's Todas of South India," *Nature* 9 (1873): 99–101.

75. Michael Gray, "Shades of Sepia: Photographic Processes," in Vidya Dehejia, ed., *India through the Lens: Photography, 1840–1911* (Washington, DC: Freer Gallery of Art, 2000), 295; Falconer, "Photography in Nineteenth-Century India," 269; Ryan, *Picturing Empire*, 29.

76. Ryan, *Picturing Empire*, 51.

77. Advertisement in Thomas Sutton, *A Dictionary of Photography* (London: S. Low and Son, 1858).

78. *A Catalogue of Photographic Apparatus*, 19.

79. *Hints to Travellers* (1883), 244; Brian Coe, *Cameras: From Daguerreotypes to Instant Pictures* (London: Marshall Cavendish, 1978), 26.

80. Kennedy, *Magic Mountains*, 91.

81. *Hints to Travellers*, 244.

82. Ryan, *Picturing Empire*, 51.

83. On the connection between the environment and race, particularly in India, see Mark Harrison, *Climates and Constitutions: Health, Race, Environment, and British Imperialism in India, 1600–1850* (Oxford: Oxford University Press, 1999); David Arnold, *Tropics and the Traveling Gaze: India, Landscape, and Science, 1800–1856* (New Delhi: Permanent Black, 2005), 110–146.

84. Marshall, *A Phrenologist*, 35, 87 and 93–94.

85. Ryan, *Picturing Empire*, 47–72; Sampson, "The Success of Samuel Bourne in India"; Garry Sampson, "Photographer of the Picturesque: Samuel Bourne," in Dehejia, *India through the Lens*.

86. For example, "Photographs of India," Y3022E, Royal Commonwealth Society Collections, University of Cambridge.

87. Ryan, *Picturing Empire*, 47; Sampson, "Photographer of the Picturesque," 164 and 171.

88. *Times of India*, 25 November 1874.

89. Prodger, *Darwin's Camera*, 13; Edwards, *Raw Histories*, 40; Maxwell, *Colonial Photography and Exhibitions*, 9; Ryan, *Picturing Empire*, 30.

90. Williams, *A Guide to the Indian Photographer*, 22; *A Catalogue of Photographic Apparatus*, 7.

91. Williams, *A Guide to the Indian Photographer*, 7–8.

92. *Hints to Travellers* (London: Royal Geographical Society, 1878), 51.

93. Williams, *A Guide to the Indian Photographer*, 18.

94. *Hints to Travellers* (1883), 245.

95. Williams, *A Guide to the Indian Photographer*, 21.

96. *Hints to Travellers* (1883), 249.

97. Prodger, *Darwin's Camera*, xxi-xxiv and 220–224; Daston and Galison, *Objectivity*, 120–164; Edwards, *Raw Histories*, 133.

98. Prodger, *Darwin's Camera*, xxii; Edwards, *Raw Histories*, 38; Ryan, *Picturing Empire*, 47.

99. The frontispiece for *A Phrenologist Amongst the Todas* appears in "Photographs of India," Y3022E, Royal Commonwealth Society Collections, University of Cambridge; "Samuel Bourne Photographs," RL.00130, David M. Rubenstein Rare Book and Manuscript Library, Duke University, Durham, NC.

100. Sampson, "The Success of Samuel Bourne in India," 337–347.

101. Marshall, *A Phrenologist*, 52–54.

102. Prodger, *Darwin's Camera*, 30–33, makes a similar argument in the case of Darwin, describing "illustration as strategy."

103. Ryan, *Picturing Empire*, 53.

104. Sutton, *Other Landscapes*, 21–26, 52, and 162–166.

105. Marshall, *A Phrenologist*, 57, 70–51, 86, and 109.

106. John Moore, *Celebration of Innovation: A History of Autotype, 1868–2005* (Wantage: Autotype International, 2005), 6–8; John Ward, "Autotype Fine Art Com-

pany," in John Hannavy, ed., *Encyclopedia of Nineteenth-Century Photography* (London: Routledge, 2008), 1:103.

107. *The Autotype Process* (London: Autotype Fine Art Company, 1871), 3 and 5.

108. Gordon Baldwin, *Looking at Photographs: A Guide to Technical Terms* (Malibu: J. Paul Getty Museum, 1991), 19–20.

109. *A Phrenologist Amongst the Todas* appears in at least three different contemporary cloth bindings, possibly reflecting different audiences imagined by the publisher. One features a quotation from the German philologist Max Müller in printed gold-leaf letters on the back cover.

110. *Athenaeum*, 29 November 1873, 680.

111. Sera-Shriar, *The Making of British Anthropology*, 147–148.

112. Stocking, *Victorian Anthropology*, 161–163; Kuklick, *The Savage Within*, 80.

113. Sera-Shriar, *The Making of British Anthropology*, 158.

114. Tylor, "Marshall's Todas of South India," 99–101. On Tylor's notion of "survivals," see Sera-Shriar, *The Making of British Anthropology*, 169; Kuper, *The Invention of Primitive Society*, 64–65.

115. Tylor, "Marshall's Todas of South India," 99.

116. "The Todas," *The Examiner*, 27 December 1873; "Who Are the Todas?," *Popular Science Review* 13 (1874).

117. Tylor, "Marshall's Todas of South India," 100.

118. Tylor, "Marshall's Todas of South India," 99–101.

119. Tylor, "Marshall's Todas of South India," 101.

120. Marshall, *A Phrenologist*, 32.

121. Tylor, "Marshall's Todas of South India," vii and 207.

122. "List of Presents," *Journal of the Anthropological Institute* 3 (1874): 307.

123. Fichman, *An Elusive Victorian*, 72–73.

124. [Alfred Russel Wallace], "A Primeval Race," *Athenaeum*, 15 November 1873, 624–625.

125. Fichman, *An Elusive Victorian*, 151–152.

126. [Wallace], "A Primeval Race," 624–625.

127. Peter Raby, *Alfred Russel Wallace: A Life* (Princeton, NJ: Princeton University Press, 2001), 176.

128. Alfred Russel Wallace, *Studies Scientific and Social* (London: Macmillan and Co., 1900), 1:473–486.

129. Charles Darwin also read *A Phrenologist Amongst the Todas* in this way, carefully annotating his own copy when revising *The Descent of Man* for a second edition in 1874 (held at the Darwin Collections, Cambridge University Library). Darwin cited Marshall's work in a new section on infanticide, comparing New Zealanders, the Todas, and Scottish islanders; Charles Darwin, *The Descent of Man*, 2nd ed. (London: John Murray, 1874), 255–258. I am grateful to Paul White for this reference.

130. Robert Stebbins, "France," in Thomas Glick, ed., *The Comparative Reception of Darwinism* (Chicago: University of Chicago Press, 1974), 117–163; Joy Harvey, "Darwin in a French Dress: Translating, Publishing, and Supporting Darwin in Nineteenth-Century France," in Eve-Marie Engels and Thomas Glick, eds., *The Reception of*

Charles Darwin in Europe (London: Continuum, 2008), 2:254–374; translation from *Revue Européenne* in Stebbins, "France," 122.

131. Alice Conklin, *In the Museum of Man: Race, Anthropology, and Empire in France, 1850–1950* (Ithaca, NY: Cornell University Press, 2013), 20–44.

132. Conklin, *In the Museum of Man*, 29; Martin Staum, *Labeling People: French Scholars on Society, Race, and Empire, 1815–1848* (Montreal: McGill-Queen's University Press, 2003), 126.

133. Conklin, *In the Museum of Man*, 29–30; Harvey, "Darwin in a French Dress," 355.

134. Jean Louis Armand de Quatrefages, "A Phrenologist Among the Todas," *Journal des Savants* (1873).

135. Elizabeth Williams, *The Physical and the Moral: Anthropology, Physiology, and Philosophical Medicine in France, 1750–1850* (Cambridge: Cambridge University Press, 2002), 257–258.

136. Williams, *The Physical and the Moral*, 269; Stebbins, "France," 151–153.

137. Jean Louis Armand de Quatrefages, *Hommes Fossiles et Hommes Sauvages* (Paris: Baillière, 1884), 505–506, 532, and 570 (translation my own, italics original).

138. "Chapon, Léon-Louis (1836–1900), Engraver," *Benezit Dictionary of Artists*, online edition; Quatrefages, *Hommes Fossiles*, ix (translation my own).

139. Quatrefages, *Hommes Fossiles*, 511 (translation my own).

140. Schaffer, *From Physics to Anthropology*, 5–54, for this argument in the Anglo-American context. Marie-Claude Mahias, "Le soleil noir des Nilgiri: L'astronomie, la photographie et l'anthropologie physique en Inde du Sud," *Gradhiva* 24 (1999): 44–56, makes a similar argument for the French case. I am grateful to Simon Schaffer for this reference.

141. Françoise Launay, *The Astronomer Jules Janssen: A Globetrotter of Celestial Physics*, trans. Storm Dunlop (New York: Springer, 2012), 63–69; Mahias, "Le soleil noir des Nilgiri," 45–54; Quatrefages, *Hommes Fossiles*, 503.

142. Jules Janssen's photographs, held today at the Musée de l'Homme, are reproduced in Mahias, "Le soleil noir des Nilgiri," 45–54.

143. Quatrefages, *Hommes Fossiles*, 512 (translation my own).

144. "Face—Slightly elongated, oval and of an agreeable shape without anything strange or striking (see fig. 180 and fig. 181)"; Quatrefages, *Hommes Fossiles*, 510 (translation my own).

145. Quatrefages, *Hommes Fossiles*, 512 (translation my own).

146. Quatrefages, *Hommes Fossiles*, 505 and 564–567 (translation my own).

147. Quatrefages, *Hommes Fossiles*, 565–567 (translation my own).

148. Quatrefages, *Hommes Fossiles*, vii (translation my own).

149. Daniel Headrick, *The Tentacles of Progress: Technology Transfer in the Age of Imperialism, 1850–1940* (Oxford: Oxford University Press, 1988), 20.

150. *The Pioneer*, 15 December 1873.

151. *Madras Mail*, 9 February 1874; *Madras Mail*, 14 March 1874; *The Pioneer*, 15 December 1873.

152. *Times of India*, 14 February 1874.

153. Anderson, *Legible Bodies*, 188; "Proceedings of the Asiatic Society of Bengal for December 1873," *Proceedings of the Asiatic Society of Bengal* (1873): 206, 155.

154. Henry Grigg, *Manual of the Nílagiri District in the Madras Presidency* (Madras: Government Press, 1880), 180–185.

155. Sutton, *Other Landscapes*, 26.

156. "Progress of Oriental Research in 1872–1873," *Indian Antiquary* 3 (1874): 325.

157. Gary, "Shades of Sepia," 297.

158. "Notice," *Indian Antiquary* 3 (1874).

159. Bijay Acharyya, *Codification in British India* (Calcutta: S. N. Banerji and Sons, 1914), v-vii and 324.

160. Richard Davenport-Hines, "Blavatsky, Helena Petrovna (1831–1891)," *Oxford Dictionary of National Biography* (2004), online edition.

161. Helena Blavatsky, *Isis Unveiled* (New York: J. W. Bouton, 1877), 2:612–613.

162. Helena Blavatsky, *The Modern Panarion* (London: Theosophical Publishing Company, 1895), 1:174–177.

163. Alfred Haddon's copy of *A Phrenologist Amongst the Todas* is held at the Haddon Library, University of Cambridge, classmark 231 M.

164. William Halse Rivers Rivers, *The Todas* (London: Macmillan, 1906), 14–15.

165. Rivers, *Todas*, 695.

166. Rivers, *Todas*, 195 and 695.

167. Pinney, *Camera*, 44; John Falconer, "'A Pure Labor of Love': A Publishing History of *The People of India*," in Eleanor Hight and Gary Sampson, eds., *Colonialist Photography: Imag(in)ing Race and Place* (London: Routledge, 2002), 77–79. Falconer's chapter is very much a publishing history without attention to reception. He simply repeats the oft-told story of Syed Ahmad Khan being shown a copy of *The People of India* in 1869, without reflecting on the diverse uses and readings of these photographs in the nineteenth century.

168. Joanna Scherer, "The Photographic Document: Photographs as Primary Data in Anthropological Enquiry," in Elizabeth Edwards, ed., *Anthropology and Photography, 1860–1920* (New Haven, CT: Yale University Press, 1992), 33.

169. Ryan, *Picturing Empire*, 19, argues the history of photography should be read as part of the history of evidence.

170. Edwards, *Raw Histories*, 27, seeks to understand how photographs are given "anthropological authority."

EPILOGUE

1. Christopher Reed, *Gutenberg in Shanghai: Chinese Print Capitalism, 1876–1937* (Vancouver: University of British Columbia Press, 2011), 77.

2. *Guxiangxue* was first published in 1919 by the Commercial Press in Shanghai. It went through at least four editions. All quotations in this chapter are from the fourth edition, published in 1923. Feng Pingsheng, *Guxiangxue* (Shanghai: Commercial Press, 1923). (I am extremely grateful to Kirie Stromberg for her translation of this source.)

3. Henry Cobbold, *Pictures of the Chinese, Drawn by Themselves* (London: John Murray, 1860), 137–143.

4. Robert Burts, *Around the World: A Narrative of a Voyage in the East India Squadron* (New York: Charles Francis, 1840), 2:263.

5. John Fairbank, *China: A New History* (Cambridge, MA: Harvard University Press, 1992), 235–281; Jonathan Spence, *The Search for Modern China* (London: W. W. Norton, 1990), 275–319; Sigrid Schmalzer, *The People's Peking Man: Popular Science and Human Identity in Twentieth-Century China* (Chicago: University of Chicago Press, 2008), 24–33; Ke Zunke and Li Bin, "Spencer and Science Education in China," in Bernard Lightman, ed., *Global Spencerism: The Communication and Appropriation of a British Evolutionist* (Leiden: Brill, 2016).

6. Yeh Hsueh and Benyu Guo, "China," in David Baker, ed., *The Oxford Handbook of the History of Psychology: Global Perspectives* (Oxford: Oxford University Press, 2011), 97.

7. E-Tu Zen Sun, "The Growth of an Academic Community, 1912–1949," in John Fairbank and Albert Feuerwerker, eds., *The Cambridge History of China* (Cambridge: Cambridge University Press, 1986), vol. 13, pt. 2, 368.

8. Geoffrey Blowers, Boris Tat Cheung, and Han Ru, "Emulation vs. Indigenization in the Reception of Western Psychology in Republican China: An Analysis of the Content of Chinese Psychology Journals (1922–1937)," *Journal of the History of the Behavioral Sciences* 45 (2009): 25–28.

9. Hsueh and Guo, "China," 94.

10. Hsueh and Guo, "China," 95–98.

11. Yang Haiyan, "Knowledge across Borders: The Early Communication of Evolution in China," in Bernard Lightman, Gordon McOuat, and Larry Stewart, eds., *The Circulation of Knowledge between Britain, India, and China: The Early-Modern World to the Twentieth Century* (Leiden: Brill, 2013).

12. Frank Dikötter, *Sex, Culture, and Modernity in China: Medical Science and the Construction of Sexual Identities in the Early Republican Period* (London: Hurst, 1995), 1–2, 38, and 90; advertisement in back of Feng Pingsheng, *Guxiangxue.*

13. Feng Pingsheng, *Guxiangxue,* 1–5 (translation by Kirie Stromberg).

14. Geoffrey Blowers, "Learning from Others: Japan's Role in Bringing Psychology to China," *American Psychologist* 55 (2000): 1433–36.

15. Hideki Nagamine, *Seisogaku Genron* (Tokyo: Senshindo, 1918).

16. William Kirby, *Germany and Republican China* (Stanford, CA: Stanford University Press, 1984), 8–17; Albert Feuerwerker, "The Foreign Presence in China," in John Fairbank, ed., *The Cambridge History of China: Republican China, 1912–1949* (Cambridge: Cambridge University Press, 1983), vol. 12, pt. 1.

17. Hsueh and Guo, "China," 9 and 35–36.

18. Dikötter, *Modernity in China,* 98; advertisement in back of Feng Pingsheng, *Guxiangxue.*

19. Yan Fu, *Qunxueyiyan* (Shanghai: Commercial Press, 1931). For the broader history of evolutionary theory in China, see Yang, "Knowledge across Borders"; Schmalzer, *The People's Peking Man,* 21–24.

20. Advertisement in back of Feng Pingsheng, *Guxiangxue.*

21. Fa-ti Fan, "Science and the Search for National Origins in the May Fourth Era," in Kai-Wing Chow, Tze-Ki Hon, Hung-Yok Ip, and Don Price, eds., *Beyond the May Fourth Paradigm: In Search of Chinese Modernity* (Lanham, MD: Lexington Books, 2008), 188–192.

22. Schmalzer, *The People's Peking Man*, 1–54.

23. Lung-Kee Sun, *The Chinese National Character: From Nationhood to Individuality* (London: M. E. Sharpe, 2002), 47.

24. Feng Pingsheng, *Guxiangxue*, 31 and 127 (translation by Kirie Stromberg).

25. Jouni-Matti Kuukkanen, "I Am Knowledge. Get Me Out of Here! On Localism and the Universality of Science," *Studies in the History and Philosophy of Science* 42 (2011).

26. For the development of this historiography, see Jan Golinski, *Making Natural Knowledge: Constructivism and the History of Science* (Cambridge: Cambridge University Press, 1998), 1–46.

27. Janet Browne, *Charles Darwin: The Power of Place* (London: Pimlico, 2002).

28. Steven Shapin and Simon Schaffer, *Leviathan and the Air-Pump: Hobbes, Boyle, and the Experimental Life* (Princeton, NJ: Princeton University Press, 2011), 225–282.

29. "If facts depend so much on these local features, how do they work elsewhere?"; Simon Schaffer, "Late Victorian Metrology and Its Instrumentation: A Manufactory of Ohms," in Robert Bud and Susan Cozzens, eds., *Invisible Connections: Instruments, Institutions, and Science* (Bellingham, WA: SPIE Optical Engineering Press, 1991), 23, asked this question directly at a relatively early stage. For an overview of this problem, see David Livingstone, *Putting Science in Its Place: Geographies of Scientific Knowledge* (Chicago: University of Chicago Press, 2003), 1–16 and 135–178.

30. Adi Ophir and Steven Shapin, "The Place of Knowledge: A Methodological Survey," *Science in Context* 4 (1991): 15–16.

31. Steven Shapin, "Pump and Circumstance: Robert Boyle's Literary Technology," *Social Studies of Science* 14 (1984); Simon Schaffer, "Astronomers Mark Time: Discipline and the Personal Equation," *Science in Context* 2 (1988).

32. Schaffer, "Late Victorian Metrology and Its Instrumentation"; Joseph O'Connell, "Metrology: The Creation of Universality by the Circulation of Particulars," *Social Studies of Science* 23 (1993). Many later attempts to study how science travels were inspired by Bruno Latour, *Science in Action: How to Follow Scientists and Engineers through Society* (Cambridge, MA: Harvard University Press, 1987).

33. Simon Schaffer's recent work on go-betweens in Asia and Africa provides an antidote to the earlier history of metrology grounded in European contexts alone; Simon Schaffer, "Golden Means: Assay Instruments and the Geography of Precision in the Guinea Trade," in Marie-Noëlle Bourguet, Christian Licoppe, and H. Otto Sibum, eds., *Instruments, Travel, and Science: Itineraries of Precision from the Seventeenth to the Twentieth Century* (London: Routledge, 2002); Schaffer, "The Asiatic Enlightenments of British Astronomy," in Simon Schaffer, Lissa Roberts, Kapil Raj, and James Delbourgo, eds., *The Brokered World: Go-Betweens and Global Intelligence, 1770–1820* (Sagamore Beach, MA: Science History Publications. 2009).

34. Most famously in Latour, *Science in Action*; Bruno Latour and Steve Woolgar, *Laboratory Life: The Construction of Scientific Facts* (Princeton, NJ: Princeton University Press, 1986).

35. Latour, *Science in Action*, 215–237.

36. "First consider the embeddedness of phenomena in localized technical and

cultural practices. Second, scrutinize how the content itself is mobilized to sustain reproduction of the phenomenon elsewhere"; Golinski, *Making Natural Knowledge*, 37.

37. In making this argument, I draw on insights from historical geography, particularly scholars influenced by Henri Lefebvre, *The Production of Space* (Oxford: Blackwell, 1991), trans. Donald Nicholson-Smith, such as Doreen Massey, *For Space* (London: Sage, 2009). For a helpful discussion of how to conceptualize "space" in the history of science, see Livingstone, *Putting Science in Its Place*, 5–12. Bruno Latour, *Reassembling the Social: An Introduction to Actor-Network-Theory* (Oxford: Oxford University Press, 2005), 159–246, can be read as making a similar claim. However, I prefer to establish this point empirically and without the theoretical baggage.

38. This distinction is built into most expressions of the problem of universality in science. For example, Ophir and Shapin, "The Place of Knowledge," 15: "How is it, if knowledge is indeed local, that certain forms of it appear global in domain of application?"

39. Shapin and Schaffer, *Leviathan and the Air-Pump*, xiii-xxi and 341–342.

40. For an overview of different approaches to global history, albeit largely focused on those developed by economic historians, see Maxine Berg, "Global History: Approaches and New Directions," in Maxine Berg, ed., *Writing the History of the Global: Challenges for the 21st Century* (Oxford: Oxford University Press, 2013).

41. Sanjay Subrahmanyam, "Connected Histories: Notes towards a Reconfiguration of Early Modern Eurasia," *Modern Asian Studies* 31 (1997); Claude Markovits, Jacques Pouchepadass, and Sanjay Subrahmanyam, eds., *Society and Circulation: Mobile People and Itinerant Cultures in South Asia, 1750–1950* (Delhi: Permanent Black, 2003).

42. Lynn Hunt, "The French Revolution in Global Context," in David Armitage and Sanjay Subrahmanyam, eds., *The Age of Revolutions in Global Context, c. 1760–1840* (Basingstoke: Palgrave Macmillan, 2010).

43. Maxine Berg, "In Pursuit of Luxury: Global History and British Consumer Goods in the Eighteenth Century," *Past & Present* 182 (2004); Prasannan Parthasarathi, *Why Europe Grew Rich and Asia Did Not: Global Economic Divergence, 1600–1850* (Cambridge: Cambridge University Press, 2011).

44. Anne Gerritsen and Giorgio Riello, eds., *The Global Lives of Things: The Material Culture of Connections in the Early Modern World* (London: Routledge, 2016).

45. The best introduction is Samuel Moyn and Andrew Sartori, "Approaches to Global Intellectual History," in Samuel Moyn and Andrew Sartori, eds., *Global Intellectual History* (New York: Columbia University Press, 2013).

46. Andrew Sartori, *Bengal in Global Concept History: Culturalism in the Age of Capital* (Chicago: University of Chicago Press, 2008); David Armitage, *The Declaration of Independence: A Global History* (Cambridge, MA: Harvard University Press, 2007).

47. Moyn and Sartori, "Approaches to Global Intellectual History," 5–17; Or Rosenboim, *The Emergence of Globalism: Visions of World Order in Britain and the United States, 1939–1950* (Princeton, NJ: Princeton University Press, 2017).

48. Vanessa Ogle, *The Global Transformation of Time, 1870–1950* (Cambridge, MA: Harvard University Press, 2015).

49. Technological challenges do come into Ogle's analysis at points—for instance,

in the introduction of daylight saving time; Ogle, *The Global Transformation of Time*, 64–74.

50. Lissa Roberts, "Situating Science in Global History: Local Exchanges and Networks of Circulation," *Itinerario* 33 (2009): 10.

51. Christopher Bayly, *The Birth of the Modern World, 1780–1914* (Oxford: Blackwell, 2004), 312–320; Jürgen Osterhammel, *The Transformation of the World: A Global History of the Nineteenth Century* (Princeton, NJ: Princeton University Press, 2014), 45–76, 779–781, 811–822.

52. David Stack, *Queen Victoria's Skull: George Combe and the Mid-Victorian Mind* (London: Hambledon Continuum, 2008), 175–176.

53. Ogle, *The Global Transformation of Time*, is exemplary in this respect, as is Alison Bashford and Joyce Chaplin, *The New Worlds of Thomas Robert Malthus: Rereading the "Principle of Population"* (Princeton, NJ: Princeton University Press, 2016).

54. Benedict Anderson, *Imagined Communities: Reflections on the Origin and Spread of Nationalism* (London: Verso, 1983); Bayly, *The Birth of the Modern World*, 199–219. For a critique of this historiography, see Partha Chatterjee, *The Nation and Its Fragments: Colonial and Postcolonial Histories* (Princeton, NJ: Princeton University Press, 1993), 3–13.

55. On the language of nations in the sciences, see Janet Browne, "A Science of Empire: British Biogeography before Darwin," *Revue d'Histoire des Sciences* 45 (1992).

56. "On the Coincidence Between the Natural Talents and Dispositions of Nations, and the Development of their Brains," *Phrenological Journal* 2 (1824–1825).

57. Christopher Bayly has done much to emphasize the longer history of nationalism, although not with reference to the history of science; Christopher Bayly, *The Origins of Nationality in South Asia: Patriotism and Ethical Government in the Making of Modern India* (Oxford: Oxford University Press, 1998).

58. Moyn and Sartori, "Approaches to Global Intellectual History," 5–17; Sanjay Subrahmanyam, "Global Intellectual History beyond Hegel and Marx," *History & Theory* 54 (2015): 126–37.

59. For example, Ogle, *The Global Transformation of Time*, 4; Rosenboim, *Globalism*, 1–23.

60. On the sciences as world history, see Sujit Sivasundaram, "Introduction: Global Histories of Science," *Isis* 101 (2010): 95–97.

61. On the relationship between global history and chronology, see Sebastian Conrad, *What Is Global History?* (Princeton, NJ: Princeton University Press, 2016), 141–161; Osterhammel, *The Transformation of the World*, 45–76.

62. James Secord, *Victorian Sensation: The Extraordinary Publication, Reception, and Secret Authorship of "Vestiges of the Natural History of Creation"* (Chicago: University of Chicago Press, 2000), 24–76; Aileen Fyfe and Bernard Lightman, "Science in the Marketplace: An Introduction," in Aileen Fyfe and Bernard Lightman, eds., *Science in the Marketplace: Nineteenth-Century Sites and Experiences* (Chicago: University of Chicago Press, 2007); Aileen Fyfe, "The Information Revolution," in David McKitterick, ed., *The Cambridge History of the Book in Britain* (Cambridge: Cambridge University Press, 2010), vol. 6; Bayly, *The Birth of the Modern World*, 19–20.

63. Wessel de Kock, *A Manner of Speaking: The Origins of the Press in South Africa* (Cape Town: Saayman and Weber, 1982), 93–94; Ulrike Stark, *An Empire of Books: The Naval Kishore Press and the Diffusion of the Printed Word in Colonial India* (Delhi: Permanent Black, 2008), 65–66.

64. Graham Shaw, "The British Book in India," in Michael Suarez and Michael Turner, eds., *The Cambridge History of the Book in Britain* (Cambridge: Cambridge University Press, 2009).

65. Jean Louis Armand de Quatrefages, *Histoire Générale des Races Humaines* (Paris: A. Hennuyer, 1889), 362–365 (translation my own).

66. Nancy Stepan, *The Idea of Race in Science: Great Britain, 1800–1960* (London: Macmillan, 1982), 1–110; Hannah Augstein, ed., *Race: The Origins of an Idea, 1760–1850* (Bristol: Thoemmes, 1996); George Stocking, "What's in a Name? The Origins of the Royal Anthropological Institute (1837–71)," *Man* 6 (1971).

67. I follow a number of recent critiques of the existing narrative, particularly Sadiah Qureshi, "Robert Gordon Latham, Displayed Peoples, and the Natural History of Race, 1854–1866," *Historical Journal* 54 (2011); Sujit Sivasundaram, "Race, Empire, and Biology before Darwin," in Denis Alexander and Ronald Numbers, eds., *Biology and Ideology from Descartes to Dawkins* (Chicago: University of Chicago Press, 2010); Shruti Kapila, "Race Matters: Orientalism and Religion, India and Beyond, c. 1770–1880," *Modern Asian Studies* 41 (2007).

68. "Miscellaneous Notices," *Phrenological Journal* 9 (1834–1836): 474.

69. Nagendranath Gupta, *Reflections and Reminiscences* (Bombay: Hind Kitabs, 1947), 55–60; *Times of India*, 3 September 1900.

70. Alfred Russel Wallace, *The Wonderful Century* (London: Swan Sonnenschein and Co., 1898), vii-ix, 1–11, 159–193, 381.

71. Eunice Johnson, *Timothy Richard's Vision: Education and Reform in China, 1880–1910* (Eugene, OR: Pickwick Publications, 2014), 169.

72. [Thomas Carlyle], "Signs of the Times," *Edinburgh Review* 49 (1829).

BIBLIOGRAPHY

Periodical and newspaper articles published before 1920 are cited in full in the notes and are not included below.

MANUSCRIPT SOURCES

Colonial Office Records. The National Archives, London.
Correspondence. Asiatic Society of Bengal, Kolkata.
Fowler and Wells Family Papers. Cornell University Library.
George Combe Papers. National Library of Scotland, Edinburgh.
George Murray Paterson Correspondence. College of Physicians of Philadelphia.
India Office Records. British Library.
James Burton Papers. British Library.
Johann Spurzheim Papers. Countway Medical Library, Harvard University.
Nahum Capen Papers. Countway Medical Library, Harvard University.
Narrative of the Escape of Eight Convicts from Macquarie Harbour. State Library of New South Wales, Australia.
National Archives of India. New Delhi.
Records of the Phrenological Society of Edinburgh. Edinburgh University Library.
Royal Commonwealth Society Archives. Cambridge University Library.
Samuel Bourne Photographs. David M. Rubenstein Rare Book and Manuscript Library, Duke University.
Samuel George Morton Papers. American Philosophical Society, Philadelphia.
Samuel George Morton Papers. Library Company of Philadelphia.
Samuel George Morton Papers. Princeton University.
Samuel Gridley Howe Papers. Houghton Library, Harvard University.

MUSEUM COLLECTIONS

Anatomy Department, University of Edinburgh.
Massachusetts Historical Society.

National Galleries of Scotland.
Science Museum, London.
Victoria and Albert Museum.
Warren Anatomical Museum, Harvard University.

PRINTED SOURCES

Acharya, Poromesh. "Education in Old Calcutta." In Sukanta Chaudhuri, ed., *Calcutta: The Living City.* Oxford: Oxford University Press, 1990.

Acharyya, Bijay. *Codification in British India.* Calcutta: S. N. Banerji and Sons, 1914.

Ackerknecht, Erwin. "P. M. A. Dumoutier et la Collection Phrénologique du Musée de l'Homme." *Bulletins et Mémoires de la Société d'Anthropologie de Paris* 7 (1956).

Alberti, Samuel. "Objects and the Museum." *Isis* 96 (2005).

———. "Owning and Collecting Natural History Objects in Nineteenth-Century Britain." In Marco Berretta, ed., *From Private to Public: Natural Collections and Museums.* Sagamore Beach, MA: Science History Publications, 2005.

Ambrose, Jennifer, and Erika Piola. "The First Fifty Years of Commercial Lithography in Philadelphia." In Erika Piola, ed., *Philadelphia on Stone: Commercial Lithography in Philadelphia, 1828–1878.* University Park: Penn State University Press, 2012.

Anderson, Benedict. *Imagined Communities: Reflections on the Origin and Spread of Nationalism.* London: Verso, 2006.

Anderson, Clare. *Legible Bodies: Race, Criminality, and Colonialism in South Asia.* Oxford: Berg, 2004.

Anderson, Robert. *Education and the Scottish People, 1750–1918.* Oxford: Clarendon Press, 1995.

Appadurai, Arjan. "How Histories Make Geographies." *Transcultural Studies* 1 (2010).

———. *Modernity at Large: Cultural Dimensions of Globalization.* Minneapolis: University of Minnesota Press, 1996.

———, ed. *The Social Life of Things: Commodities in Cultural Perspective.* Cambridge: Cambridge University Press, 1988.

Arbousset, Thomas. *Narrative of an Exploratory Tour to the North-East of the Colony of the Cape of Good Hope.* Trans. John Brown. Cape Town: A. S. Robertson, 1846.

Armistead, Wilson. *A Tribute for the Negro.* Manchester: W. Irwin, 1848.

Armitage, David. *The Declaration of Independence: A Global History.* Cambridge, MA: Harvard University Press, 2007.

Armitage, David, and Sanjay Subrahmanyam, eds. *The Age of Revolutions in Global Context, 1760–1840.* Basingstoke: Palgrave Macmillan, 2010.

Arnold, David. *Tropics and the Traveling Gaze: India, Landscape, and Science, 1800–1856.* Delhi: Permanent Black, 2005.

Arteaga, Juanma Sánchez, and Charbel Niño El-Hani. "Physical Anthropology and the Description of the 'Savage' in the Brazilian Anthropological Exhibition of 1882." *História, Ciências, Saúde-Manguinhos* 17 (2010).

Augstein, Hannah. *James Cowles Prichard's Anthropology: Remaking the Science of Man in Early Nineteenth-Century Britain.* Amsterdam: Editions Rodopi, 1999.

———, ed. *Race: The Origins of an Idea, 1760–1850*. Bristol: Thoemmes, 1996.

The Autotype Process. London: Autotype Fine Art Company, 1871.

Awful Calamities; or the Shipwrecks of December 1839. Boston: J. Howe, 1839.

Bala, Poonam. *Imperialism and Medicine in Bengal: A Socio-Historical Perspective*. London: Sage, 1991.

Baldwin, Gordon. *Looking at Photographs: A Guide to Technical Terms*. Malibu: J. Paul Getty Museum, 1991.

Bank, Andrew. "Liberals and Their Enemies: Racial Ideology at the Cape of Good Hope, 1820–1850" PhD diss., University of Cambridge, 1995.

———. "Of 'Native Skulls' and 'Noble Caucasians': Phrenology in Colonial South Africa." *Journal of Southern African Studies* 22 (1996).

Barnes, John, Bill Bell, Rimi Chatterjee, Wallace Kirsop, and Michael Winship. "A Place in the World." In Davud McKitterick, ed., *The Cambridge History of the Book in Britain, 1830–1914*. Cambridge: Cambridge University Press, 2009.

Barnhill, Georgia. "Transformations in Pictorial Printing." In Robert Gross and Mary Kelly, eds., *A History of the Book in America: An Extensive Republic, Print Culture, and Society in the New Nation, 1790–1840*. Chapel Hill: University of North Carolina Press, 2010.

Barpujari, H. K. *The American Missionaries and North-East India, 1836–1900*. Guwahati: Spectrum, 1986.

Barton, John, and Jennifer Phegley. "Introduction: 'An Age of Sensation . . . across the Atlantic.'" In John Barton, Jennifer Phegley, and Kristin Huston, eds., *Transatlantic Sensations*. Aldershot: Ashgate, 2010.

Basak, Nilmoni. *Arabya Upanyas*. Calcutta: Purnochandradoy Press, 1850.

Bashford, Alison, and Joyce Chaplin. *The New Worlds of Thomas Robert Malthus: Rereading the "Principle of Population."* Princeton, NJ: Princeton University Press, 2016.

Bayly, Christopher. *The Birth of the Modern World, 1780–1914: Global Connections and Comparisons*. Oxford: Blackwell, 2004.

———. *The Origins of Nationality in South Asia: Patriotism and Ethical Government in the Making of Modern India*. Oxford: Oxford University Press, 1998.

———. *Recovering Liberties: Indian Thought in the Age of Liberalism and Empire*. Cambridge: Cambridge University Press, 2011.

Beales, Derek. "The Idea of Reform in British Politics, 1829-1850." In Tim Blanning and Paul Wende, eds., *Reform in Great Britain and Germany, 1750–1850*. Oxford: Oxford University Press, 1999.

Bédollière, Émile de la. *La Morale en Action Illustrée*. Paris: Alphonse Henriot, 1837.

Bender, Thomas. "Historians, the Nation, and the Plenitude of Narratives." In Thomas Bender, ed., *Rethinking American History in a Global Age*. Berkeley: University of California Press, 2002.

Bennett, John, and Susan Rowley. *Uqalurait: An Oral History of Nunavut*. Montreal: McGill-Queen's University Press, 2008.

Berg, Maxine. "Global History: Approaches and New Directions." In *Writing the History of the Global: Challenges for the 21st Century*. Oxford: Oxford University Press, 2013.

———. "In Pursuit of Luxury: Global History and British Consumer Goods in the Eighteenth Century." *Past & Present* 182 (2004).

Bernstein, Robin. "Dances with Things: Material Culture and the Performance of Race." *Social Text* 101 (2009).

Bhattacharya, Tithi. *The Sentinels of Culture: Class, Education, and the Colonial Intellectual in Bengal 1848–85*. New Delhi: Oxford University Press, 2005.

Blackburn, Anne. *Locations of Buddhism: Colonialism and Modernity in Sri Lanka.* Chicago: University of Chicago Press, 2010.

Blair, Sandy. "'The convict press': Edward Smith Hall and *The Sydney Monitor.*" In Denis Cryle, ed., *Disreputable Profession: Journalists and Journalism in Colonial Australia.* Rockhampton: Central Queensland University Press, 1997.

Blanckaert, Claude. "A Jamaican Émigré as the 'Father of Ethnology in France.'" In George Stocking, ed., *Bones, Bodies, and Behavior: Essays in Behavioral Anthropology.* Madison: University of Wisconsin Press, 1988.

Blavatsky, Helena. *Isis Unveiled.* New York: J. W. Bouton, 1877.

———. *The Modern Panarion.* London: Theosophical Publishing Company, 1895.

Bljj, Harm de. *The Power of Place: Geography, Destiny, and Globalization's Rough Landscape.* Oxford: Oxford University Press, 2009.

Blowers, Geoffrey. "Learning from Others: Japan's Role in Bringing Psychology to China." *American Psychologist* 55 (2000).

Blowers, Geoffrey, Boris Tat Cheung, and Han Ru. "Emulation vs. Indigenization in the Reception of Western Psychology in Republican China: An Analysis of the Content of Chinese Psychology Journals (1922–1937)." *Journal of the History of the Behavioral Sciences* 45 (2009).

Blum, Ann. *Picturing Nature: American Nineteenth-Century Zoological Illustration.* Princeton, NJ: Princeton University Press, 1993.

Branson, Susan, and Leslie Patrick. "Étrangers dans un pays étrange: Saint-Domingan Refugees of Color in Philadelphia." In David Geggus, ed., *The Impact of the Haitian Revolution in the Atlantic World.* Columbia: University of South Carolina Press, 2001.

Bravo, Michael. *The Accuracy of Ethnoscience: A Study of Inuit Cartography and Cross-Cultural Commensurability.* Manchester: Department of Social Anthropology, University of Manchester, 1996.

———. "Ethnographic Navigation and the Geographical Gift." In David Livingstone and Charles Withers, eds., *Geography and Enlightenment.* Chicago: University of Chicago Press, 1996.

———. "Ethnological Encounters." In Nick Jardine, James Secord, and Emma Spary, eds., *Cultures of Natural History.* Cambridge: Cambridge University Press, 1994.

———. "Geographies of Exploration and Improvement: William Scoresby and Arctic Whaling, 1782–1822." *Journal of Historical Geography* (2006).

Bregman, Leigh. "'Snug Little Coteries': A History of Scientific Societies in Early Nineteenth-Century Cape Town, 1824–1835." PhD diss., University College London, 2003.

Bridges, Frederick. *Phrenology Made Practical and Popularly Explained.* London: Sampson, Low and Co., 1857.

———. *The Popular Manual of Phrenology.* London: George Philip and Son, 1860.

Broussais, François-Joseph-Victor. *Cours de Phrénologie.* Paris: Baillière, 1836.

Brown, C. Mackenzie. *Hindu Perspectives on Evolution: Darwin, Dharma, and Design.* London: Routledge, 2012.

Brown, Chandos. "A Natural History of the Gloucester Sea Serpent: Knowledge, Power, and the Culture of Science in Antebellum America." *American Quarterly* 42 (1990).

Browne, Janet. *Charles Darwin: The Power of Place.* London: Verso, 2002.

———. "A Science of Empire: British Biogeography before Darwin." *Revue d'Histoire des Sciences* 45 (1992).

Burns, Arthur, and Joanna Innes. "Introduction." In Arthur Burns and Joanna Innes, eds., *Rethinking the Age of Reform: Britain 1780–1850.* Cambridge: Cambridge University Press, 2003.

Burts, Robert. *Around the World: A Narrative of a Voyage in the East India Squadron.* New York: Charles Francis, 1840.

Byron, George. *Voyage of HMS Blonde to the Sandwich Islands.* London: John Murray, 1826.

Calcutta Annual Register and Directory. Calcutta: Scott and Company, 1831.

Caldwell, Charles. *The Autobiography of Charles Caldwell.* Philadelphia: Lippincott, Grambo and Co., 1855.

Cantor, Geoffrey. "A Critique of Shapin's Social Interpretation of the Edinburgh Phrenology Debate." *Annals of Science* 32 (1975).

Cantor, Geoffrey, and Sally Shuttleworth. "Introduction." In Geoffrey Cantor and Sally Shuttleworth, eds., *Science Serialized: Representation of the Sciences in Nineteenth-Century Periodicals.* Cambridge, MA: MIT Press, 2004.

Carmichael, Andrew. *A Memoir of the Life and Philosophy of Spurzheim.* Boston: Marsh, Capen and Lyon, 1833.

Casid, Jill. *Sowing Empire: Landscape and Colonization.* Minneapolis: University of Minnesota Press, 2005.

Casper, Scott. "The Census, the Post-Office, and Governmental Publishing." In Scott Casper, Jeffrey Groves, Stephen Nissenbaum, and Michael Winship, eds., *The History of the Book in America: The Industrial Book, 1840–1880.* Chapel Hill: University of North Carolina Press, 2007.

———. "Other Variations on the Trade." In Scott Casper, Jeffrey Groves, Stephen Nissenbaum, and Michael Winship, eds., *The History of the Book in America: The Industrial Book, 1840–1880.* Chapel Hill: University of North Carolina Press, 2007.

Catalogo da Exposição Medica Brasileira. Rio de Janeiro: Typographia Nacional, 1884.

Catalogue of the Calcutta Public Library. Calcutta: Saunders and Cones, 1846.

Catalogue of the Glasgow Mechanics' Institution Library. Glasgow: George Troup, 1848.

A Catalogue of the Library of the College of St. Margaret and St. Bernard. London: S. and R. Bentley, 1827.

Catalogue of the Library of the Mechanics' Institute, of the City of New-York. New York: A. Baptist Jr, 1844.

Catalogue of the Library of the Mercantile Library Association of San Francisco. San Francisco: Mercantile Library Association, 1874.

Catalogue of the Library of the Royal College of Surgeons in London. London: Richard Taylor, 1831.

Catalogue of Madame Sohier's Waxworks Exhibition. Melbourne: Wilson and MacKinnon, 1866.

The Catalogue of the Melbourne Public Library for 1861. Melbourne: Clarson, Shallard and Co, 1861.

Catalogue of the Mercantile Library in New York. New York: Baker, Godwin and Company, 1850.

A Catalogue of Photographic Apparatus and Chemicals. Calcutta: R. C. Lepage, 1860.

A Catalogue of Phrenological Specimens Belonging to the Boston Phrenological Society. Boston: John Ford, 1835.

A Catalogue of the South African Public Library. Cape Town: Gazette Office, 1829.

Catalogue of the Works in the Library of the Sydney Mechanics' School of Arts. Sydney: Caxton Steam Machine Printing Office, 1869.

Chakrabarty, Dipesh. "*Adda*, Calcutta: Dwelling in Modernity." *Public Culture* 11 (1999).

Chakraborti, Smarajit. *The Bengali Press, 1818–1868*. Calcutta: Firma KLM, 1976.

Chanda, Mrinal Kanti. *History of the English Press in Bengal, 1780–1857*. Calcutta: K. P. Bagchi, 1987.

"Chapon, Léon-Louis (1836–1900), Engraver." *Benezit Dictionary of Artists*, online edition.

Chase, Malcolm. *Chartism: A New History*. Manchester: Manchester University Press, 2007.

Chatterjee, Partha. *The Nation and Its Fragments: Colonial and Postcolonial Histories*. Princeton, NJ: Princeton University Press, 1993.

———. *Texts of Power: Emerging Disciplines in Colonial Bengal*. Minneapolis: University of Minnesota Press, 1995.

Chen, Songchuan. "An Information War Waged by Merchants and Missionaries at Canton: The Society for the Diffusion of Useful Knowledge in China, 1834–1839." *Modern Asian Studies* 46 (2012).

Clark, Jerrey. *The Shawnee*. Lexington: University Press of Kentucky, 2007.

A Classified Catalogue of the Mercantile Library of San Francisco. San Francisco: Mercantile Library Association, 1861.

Cliff, Alice. "Coming Home—Bally's Miniature Phrenological Specimens." *Science Museum Group Journal* 1 (2015).

Clifford, Timothy. "The Plaster Shops of the Rococo and Neo-Classical Era in Britain." *Journal of the History of Collections* 4 (1992).

Cobbold, Robert. *Pictures of the Chinese, Drawn by Themselves*. London: John Murray, 1860.

Coe, Brian. *Cameras: From Daguerreotypes to Instant Pictures*. London: Marshall Cavendish, 1978.

Cohen, William B. *The French Encounter with Africans: White Response to Blacks, 1530–1880*. Bloomington: Indiana University Press, 1980.

Colbert, Charles. *A Measure of Perfection: Phrenology and the Fine Arts in America*. Chapel Hill: University of North Carolina Press, 1997.

Colla, Elliot. *Conflicted Antiquities: Egyptology, Egyptomania, Egyptian Modernity.* Durham, NC: Duke University Press, 2007.

Colligan, Mimi. "Kreitmayer, Maximilian Ludwig (Max) (1830–1906)." In Christopher Cunneen, Jill Roe, Beverley Kingston, and Stephen Garton, eds., *Australian Dictionary of Biography (Supplement), 1580–1980.* Melbourne: Melbourne University Publishing, 2005.

Collini, Stefan. "The Idea of 'Character' in Victorian Political Thought." *Transactions of the Royal Historical Society* 35 (1985).

Combe, Andrew. *Principles of Physiology.* Edinburgh: Maclachlan and Stewart, 1841.

Combe, George. *The Constitution of Man.* Edinburgh: John Anderson, 1828.

———. *The Constitution of Man.* Edinburgh: Maclachlan and Stewart, 1835.

———. *Elements of Phrenology.* Edinburgh: Maclachlan and Stewart, 1850.

———. *Essays on Phrenology.* Edinburgh: Bell and Bradfute, 1819.

———. *Lectures on Phrenology.* London: Simpkin, Marshall and Co., 1839.

———. *Notes on the United States of North America.* Edinburgh: Maclachlan and Stewart, 1841.

———. *Outlines of Phrenology.* London: Simpkin, Marshall and Co., 1835.

———. *A System of Phrenology.* Edinburgh: Maclachlan and Stewart, 1843.

Conklin, Alice. *In the Museum of Man: Race, Anthropology, and Empire in France, 1850–1950.* Ithaca, NY: Cornell University Press, 2013.

Conrad, Sebastian. *What Is Global History?* Princeton, NJ: Princeton University Press, 2016.

Cooper, Frederick. *Colonialism in Question: Theory, Knowledge, History.* Berkeley: University of California Press, 2005.

Cooter, Roger. *The Cultural Meaning of Popular Science: Phrenology and the Organization of Consent in Nineteenth-Century Britain.* Cambridge: Cambridge University Press, 1984.

Costa, Carlos. *Catalogo Systematico da Bibliotheca da Faculdade de Medicina do Rio de Janeiro.* Rio de Janeiro: Imprensa Nacional, 1892.

Crais, Clifton. *White Supremacy and Black Resistance: Pre-Industrial South Africa.* Cambridge: Cambridge University Press, 1992.

Crass, Barbara. "Gender in Inuit Burial Practices." In Alison Rautman, ed., *Reading the Body: Representations and Remains in the Archaeolgical Record.* Philadelphia: University of Pennsylvania Press, 2000.

Credland, Arthur. *The Hull Whaling Trade.* Beverley: Hutton, 1995.

Curl, James. *Egyptomania: The Egyptian Revival, a Recurring Theme in the History of Taste.* Manchester: Manchester University Press, 1994.

———. *The Victorian Celebration of Death.* Newton Abbot: David and Charles, 1972.

Dain, Bruce. *A Hideous Monster of the Mind: American Race Theory in the Early Republic.* Cambridge, MA: Harvard University Press, 2002.

Darwin, Charles. *The Descent of Man.* London: John Murray, 1874.

Das, Cally Coomar. *General Reflections on Christianity.* Calcutta: Chunderseaker Bistoo, 1845.

Das, Radhaballav. *Manatatva Sarsangraha.* Calcutta: Purnochandradoy Press, 1849.

Daston, Lorraine, and Peter Galison. *Objectivity.* New York: Zone Books, 2007.

Davenport-Hines, Richard. "Blavatsky, Helena Petrovna (1831–1891)." *Oxford Dictionary of National Biography* (2004), online edition.

David, Robert. *The Arctic in the British Imagination, 1818–1914.* Manchester: Manchester University Press, 2001.

Davies, John. *Phrenology: Fad and Science.* New Haven, CT: Yale University Press, 1955.

Davis, Joseph. *Thesaurus Craniorum.* London, 1867.

Davy, John. *An Account of the Interior of Ceylon, and of its Inhabitants.* London: Longman, Hurst, Rees, Orme, and Brown, 1821.

Delessert, Benjamin. *Les Bons Exemples.* Paris: G. Kugelmann, 1858.

Desmond, Adrian, and James Moore. *Darwin's Sacred Cause: Race, Slavery, and the Quest for Human Origins.* London: Penguin, 2009.

Dewan, Janet. "Delineating Antiquities and Remarkable Tribes." *History of Photography* 16 (1992).

Dierks, Konstantin. "The Familiar Letter and Social Refinement in America, 1750–1800." In David Barton and Nigel Hall, eds., *Letter Writing as a Social Practice.* Amsterdam: John Benjamins Publishing Company, 2000.

Dikötter, Frank. *Sex, Culture, and Modernity in China: Medical Science and the Construction of Sexual Identities in the Early Republican Period.* London: Hurst, 1995.

Douglass, Frederick. *The Claims of the Negro Ethnologically Considered.* Rochester, NY: Lee, Mann and Co., 1854.

———. "Letter from the Editor." In *The Works of James McCune Smith: Black Intellectual and Abolitionist.* Oxford: Oxford University Press, 2007.

———. *The Life and Times of Frederick Douglass.* Boston: De Wolfe and Fiske Co 1882.

Drescher, Seymour. "British Way, French Way: Opinion Building and Revolution in the Second French Slave Emancipation." *American Historical Review* 96 (1991).

Drummond, David. *Objections to Phrenology: Being the Substance of a Series of Papers Communicated to the Calcutta Phrenological Society.* Calcutta: Printed for the Author, 1829.

Dubow, Saul. *A Commonwealth of Knowledge: Science, Sensibility, and White South Africa, 1820–2000.* Oxford: Oxford University Press, 2006.

Duncan, James. *In the Shadows of the Tropics: Climate, Race, and Biopower in Nineteenth-Century Ceylon.* Aldershot: Ashgate, 2007.

Eber, Dorothy. *Encounters on the Passage: Inuit Meet the Explorers.* Toronto: University of Toronto Press, 2008.

The Edinburgh Almanack. Edinburgh: Oliver and Boyd, 1828.

Edwards, Elizabeth. *Raw Histories: Photographs, Anthropology, and Museums.* Oxford: Berg, 2001.

Eley, Geoff. "Historicizing the Global, Politicizing Capital: Giving the Present a Name." *History Workshop Journal* 63 (2007).

État Présent de la Noblesse Française. Paris: Librairie Bachelin-Deflorenne, 1866.

Etherington, Norman, Patrick Harries, and Bernard Mbenga. "From Colonial Hegemonies to Imperial Conquest, 1840–1880." In Carolyn Hamilton, Bernard Mbenga, and

Robert Ross, eds., *The Cambridge History of South Africa*. Cambridge: Cambridge University Press, 2009.

Fabian, Ann. *The Skull Collectors: Race, Science, and America's Unburied Dead*. Chicago: University of Chicago Press, 2010.

Fairbank, John. *China: A New History*. Cambridge, MA: Harvard University Press, 1992.

Falconer, John. "Photography in Nineteenth-Century India." In Christopher Bayly, ed., *The Raj and the British, 1600–1947*. London: National Portrait Gallery Publications, 1990.

———. "'A Pure Labor of Love': A Publishing History of *The People of India*." In Eleanor Hight and Gary Sampson, eds., *Colonialist Photography: Imag(in)ing Race and Place*. London: Routledge, 2002.

Fan, Fa-ti. "The Global Turn in the History of Science." *East Asian Science, Technology and Society* 6 (2012).

———. "Science and the Search for National Origins in the May Fourth Era." In Kai-Wing Chow, Tze-Ki Hon, Hung-Yok Ip, and Don Price, eds., *Beyond the May Fourth Paradigm: In Search of Chinese Modernity*. Lanham, MD: Lexington Books, 2008.

Feng Pingsheng. *Guxiangxue*. Shanghai: Commercial Press, 1923.

Fetherling, George. *The Gold Crusades: A Social History of Gold Rushes, 1849–1929*. Toronto: University of Toronto Press, 1988.

Feuerwerker, Albert. "The Foreign Presence in China." In John Fairbank, ed., *The Cambridge History of China: Republican China, 1912–1949*, vol. 12, pt. 1. Cambridge: Cambridge University Press, 1983.

Fichman, Martin. *An Elusive Victorian: The Evolution of Alfred Russel Wallace*. Chicago: University of Chicago Press, 2004.

First Report of the Proceedings, Recommendations, and Transactions of the British Association for the Advancement of Science. York: Thomas Wilson and Sons, 1832.

Fladeland, Betty. *Men and Brothers: Anglo-American Antislavery Cooperation*. Urbana-Champaign: University of Illinois Press, 1972.

Fossati, Giovanni. *Manuel Pratique de Phrénologie*. Paris: Baillière, 1845.

Fowler, Orson. *Phrenology and Physiology Explained and Applied to Education and Self-Improvement*. New York: O. S. and L. N. Fowler, 1843.

Frederiksen, Rune, and Eckart Marchan. "Introduction." In Rune Frederiksen and Eckart Marchand, eds., *Plaster Casts: Making, Collecting, and Displaying from Classical Antiquity to the Present*. Berlin: Walter de Gruyter, 2010.

Fyfe, Aileen. "The Information Revolution." In David McKitterick, ed., *The Cambridge History of the Book in Britain*. Cambridge: Cambridge University Press, 2010.

———. "Reading Natural History at the British Museum and the *Pictorial Museum*." In Aileen Fyfe and Bernard Lightman, eds., *Science in the Marketplace: Nineteenth-Century Sites and Experiences*. Chicago: University of Chicago Press, 2007.

———. *Science and Salvation: Evangelical Popular Science Publishing in Victorian Britain*. Chicago: University of Chicago Press, 1994.

———. *Steam-Powered Knowledge: William Chambers and the Business of Publishing, 1820–1860*. Chicago: University of Chicago Press, 2012.

Fyfe, Aileen, and Bernard Lightman. "Science in the Marketplace: An Introduction." In Aileen Fyfe and Bernard Lightman, eds., *Science in the Marketplace: Nineteenth-Century Sites and Experiences.* Chicago: University of Chicago Press, 2007.

Gange, David. *Dialogues with the Dead: Egyptology in British Culture and Religion, 1822–1922.* Oxford: Oxford University Press, 2013.

Garrett, Aaron. "Mind and Matter." In James Harris, ed., *The Oxford Handbook of British Philosophy in the Eighteenth Century.* Oxford: Oxford University Press, 2013.

Gascoigne, John. *The Enlightenment and the Origins of European Australia.* Cambridge: Cambridge University Press, 2002.

Gatrell, Vic. *The Hanging Tree: Execution and the English People, 1770–1868.* Oxford: Oxford University Press, 1994.

Gérards, Emile. *Paris souterrain.* Paris: Garnier Frères, 1908.

Gerritsen, Anne, and Giorgio Riello, eds. *The Global Lives of Things: The Material Culture of Connections in the Early Modern World.* London: Routledge, 2016.

Ghosh, Anindita. *Power in Print: Popular Publishing and the Politics of Language and Culture in a Colonial Society, 1778–1905.* New Delhi: Oxford University Press, 2006.

———. "An Uncertain 'Coming of the Book': Early Print Cultures in Colonial India." *Book History* 6 (2003).

Giustino, David. *Conquest of Mind: Phrenology and Victorian Social Thought.* London: Croom Helm, 1975.

Goff, Victoria. "Convicts and Clerics: Their Roles in the Infancy of the Press in Sydney, 1803–1840." *Media History* 4 (1998).

Golinski, Jan. *Making Natural Knowledge: Constructivism and the History of Science.* Cambridge: Cambridge University Press, 1998.

Gordon-Brown, Alfred. *The Settlers' Press: Seventy Years of Printing in Grahamstown.* Cape Town: Balkema, 1979.

Gorman, Mel. "Introduction of Western Science into Colonial India: Role of the Calcutta Medical College." *Proceedings of the American Philosophical Society* 132 (1988).

Gould, Stephen. *The Mismeasure of Man.* New York: Norton, 1981.

Gray, Michael. "Shades of Sepia: Photographic Processes." In Vidya Dehejia, ed., *India through the Lens: Photography, 1840–1911.* Washington, DC: Freer Gallery of Art, 2000.

Green, Andy. *Education and State Formation: The Rise of Education Systems in England, France, and the USA.* London: Palgrave, 1992.

Green, James. "The Rise of Book Publishing." In Robert Gross and Mary Kelly, eds., *A History of the Book in America: An Extensive Republic, Print Culture, and Society in the New Nation, 1790–1840.* Chapel Hill: University of North Carolina Press, 2010.

Grigg, Henry. *Manual of the Nílagiri District in the Madras Presidency.* Madras: Government Press, 1880.

Groseclose, Barbara. *British Sculpture and the Company Raj.* Newark, DE: University of Delaware Press, 1995.

Groves, Jeffrey. "Periodicals and Serial Publication." In Scott Casper, Jeffrey Groves, Stephen Nissenbaum, and Michael Winship, eds., *The History of the Book in America: The Industrial Book, 1840–1880.* Chapel Hill: University of North Carolina Press, 2007),

Gupta, Nagendranath. *Reflections and Reminiscences.* Bombay: Hind Kitabs, 1947.

Hall, Nigel. "The Materiality of Letter Writing: A Nineteenth-Century Perspective." In David Barton and Nigel Hall, eds., *Letter Writing as a Social Practice.* Amsterdam: John Benjamins Publishing Company, 2000.

Hamilton, Cynthia. "'Am I Not a Man and a Brother?' Phrenology and Anti-slavery." *Slavery & Abolition* 29 (2008).

———. "Hercules Subdued: The Visual Rhetoric of the Kneeling Slave." *Slavery and Abolition* 34 (2013).

Harbsmeier, Michael. "Bodies and Voices from Ultima Thule: Inuit Explorations of the Kablunat from Christian IV to Knud Rasmussen." In Michael Bravo and Sverker Sörlin, eds., *Narrating the Arctic: A Cultural History of Nordic Scientific Practices.* Canton, MA: Science History Publications, 2002.

Harlow, Luke. "Neither Slavery nor Abolitionism: James M. Pendleton and the Problem of Christian Conservative Antislavery in 1840s Kentucky." *Slavery and Abolition* 27 (2006).

Harrison, Mark. *Climates and Constitutions: Health, Race, Environment, and British Imperialism in India, 1600–1850.* Oxford: Oxford University Press, 1999.

Hart-Davis, Duff. *Audubon's Elephant.* London: Weidenfeld, 2003.

Harvey, Joy. "Darwin in a French Dress: Translating, Publishing, and Supporting Darwin in Nineteenth-Century France." In Eve-Marie Engels and Thomas Glick, eds., *The Reception of Charles Darwin in Europe.* London: Continuum, 2008.

Haskell, Thomas. "Capitalism and the Origins of the Humanitarian Sensibility, Part 1." *American Historical Review* 90 (1985).

Hatcher, Brian. *Idioms of Improvement: Vidyasagar and Cultural Encounter in Bengal.* Calcutta: Oxford University Press, 1996.

Haynes, Christine. *Lost Illusions: The Politics of Publishing in Nineteenth-Century France.* Cambridge, MA: Harvard University Press, 2010.

Headrick, Daniel. *The Tentacles of Progress: Technology Transfer in the Age of Imperialism, 1850–1940.* Oxford: Oxford University Press, 1988.

Heaton, Herbert. "The Early Tasmanian Press, and Its Struggle for Freedom." *Papers and Proceedings of the Royal Society of Tasmania* 42 (1916).

Hideki, Nagamine. *Seisogaku Genron.* Tokyo: Senshindo, 1918.

Hindus, Michael. *Prison and Plantation: Crime, Justice, and Authority in Massachusetts and South Carolina, 1767–1878.* Chapel Hill: University of North Carolina Press, 1980.

Hints to Travellers. London: Royal Geographical Society, 1878.

Hints to Travellers. London: Royal Geographical Society, 1883.

Hoare, Merval. *Norfolk Island: An Outline of Its History, 1774–1977.* Brisbane: University of Queensland Press, 1969.

Hodges, Sarah. "Global Menace." *Social History of Medicine* 25 (2012).

Hollis, Patricia. "Anti-slavery and British Working Class Radicalism in the Years of Freedom." In Christine Bolt and Seymour Drescher, eds., *Anti-slavery, Religion and Reform: Essays in Memory of Roger Anstey.* Folkstone: Dawson, 1980.

Hopwood, Nick. "Visual Standards and Disciplinary Change: Normal Plates, Tables, and Stages in Embryology." *History of Science* 43 (2005).

Howe, Daniel. *What Hath God Wrought? The Transformation of America, 1815–1848.* Oxford: Oxford University Press, 2007.

Hsueh, Yeh, and Benyu Guo. "China." In David Baker, ed. *The Oxford Handbook of the History of Psychology: Global Perspectives.* Oxford: Oxford University Press, 2011.

Hughes, Robert. *The Fatal Shore: A History of the Transportation of Convicts to Australia, 1787–1868.* London: Collins Harvill, 1986.

Hunt, Alfred. *Haiti's Influence on Antebellum America: Slumbering Volcano in the Caribbean.* Lafayette: Louisiana State University Press, 1988.

Hunt and Co's Directory and Topography for the Cities of Gloucester & Bristol. London: B. W. Gardner, 1849.

Hutz, Claudio Simon, Gustavo Gauer, and William Barbosa Gomes. "Brazil." In David Baker, ed., *The Oxford Handbook of the History of Psychology: Global Perspectives.* Oxford: Oxford University Press, 2012.

Huzzey, Richard. "The Moral Geography of British Anti-slavery Responsibilities." *Transactions of the Royal Historical Society* 22 (2012).

Infant Education. Edinburgh: William and Robert Chambers, 1837.

Innes, Joanna. "'Reform' in English Public Life: The Fortunes of a Word." In Arthur Burns and Joanna Innes, eds., *Rethinking the Age of Reform: Britain, 1780–1850.* Cambridge: Cambridge University Press, 2003.

Jacyna, L. "The Physiology of Mind, the Unity of Nature, and the Moral Order in Victorian Thought." *British Journal for the History of Science* 14 (1981).

Jaffrelot, Christophe. "The Idea of the Hindu Race in the Writings of Hindu Nationalist Ideologues in the 1920s and 1930s." In Peter Robb, ed., *The Concept of Race in South Asia.* Oxford: Oxford University Press, 1995.

Jasanoff, Maya. *Edge of Empire: Conquest and Collecting in the East, 1750–1850.* London: Harper Perennial, 2006.

Jennings, Lawrence. *French Anti-slavery: The Movement for the Abolition of Slavery in France, 1802–1848.* Cambridge: Cambridge University Press, 2000.

Jerkic, Sonja. "Burials and Bones: A Summary of Burial Patterns and Human Skeletal Research in Newfoundland and Labrador." *Newfoundland Studies* 9 (1993).

John, Richard. *Spreading the News: The American Postal System from Franklin to Morse.* Cambridge, MA: Harvard University Press, 1995.

Johns, Adrian. "The Physiology of Reading." In Marina Frasca-Spada and Nick Jardine, eds., *Books and the Sciences in History.* Cambridge: Cambridge University Press, 2000.

Johnson, Eunice. *Timothy Richard's Vision: Education and Reform in China, 1880–1910.* Eugene, OR: Pickwick Publications, 2014.

Johnston, Anna. *The Paper War: Morality, Print Culture, and Power in Colonial New South Wales.* Perth: University of Western Australia Press, 2011.

Jones, Gareth Stedman. *Languages of Class: Studies in English Working Class History, 1832–1982*. Cambridge: Cambridge University Press, 1983.

Jones, Silas. *Practical Phrenology.* Boston: Russell, Shattuck and Williams, 1836.

Joyce, Barry. *The Shaping of American Ethnography: The Wilkes Exploring Expedition, 1838–1842*. Lincoln: University of Nebraska Press, 2001.

Kaestle, Carl, and Maris Vinovskis. *Education and Social Change in Nineteenth-Century Massachusetts*. Cambridge: Cambridge University Press, 1980.

Kapila, Shruti. "Race Matters: Orientalism and Religion, India and Beyond, c. 1770–1880." *Modern Asian Studies* 41 (2007).

Kaplan, Amy. "'Left Alone with America': The Absence of Empire in the Study of American Culture." In Amy Kaplan and Donald Pease, eds., *Cultures of United States Imperialism*. Durham, NC: Duke University Press, 1993.

Karsh, Efraim, and Inari Karsh. *Empires of the Sand: The Struggle for Mastery in the Middle East, 1789–1923*. Cambridge, MA: Harvard University Press, 1999.

Keegan, Timothy. *Colonial South Africa and the Origins of the Racial Order.* London: Leicester University Press, 1996.

Kennedy, Dane. *Magic Mountains: Hill Stations and the British Raj.* Berkeley: University of California Press, 1996.

Kidd, James. *The Forging of Races: Race and Scripture in the Protestant Atlantic World, 1600–2000*. Cambridge: Cambridge University Press, 2006.

Kirby, William. *Germany and Republican China*. Stanford, CA: Stanford University Press, 1984.

Klancher, Jon. *The Making of English Reading Audiences, 1790–1832*. Madison: University of Wisconsin Press, 1987.

Kock, Wessel. *A Manner of Speaking: The Origins of the Press in South Africa*. Cape Town: Saayman and Weber, 1982.

Kopf, David. *The Brahmo Samaj and the Shaping of the Modern Indian Mind.* Princeton, NJ: Princeton University Press, 1979.

———. *British Orientalism and the Bengal Renaissance.* Berkeley: University of California Press, 1969.

Kopytoff, Igor. "The Cultural Biography of Things: Commodification as Process." In Arjan Appadurai, ed., *The Social Life of Things: Commodities in Cultural Perspective.* Cambridge: Cambridge University Press, 1988.

Kuklick, Henrika. *The Savage Within: The Social History of British Anthropology, 1885–1945*. Cambridge: Cambridge University Press, 1991.

Kumar, Arun. "Thomason College of Engineering, Roorkee, 1847-1947." In Uma Das Gupta, ed., *Science and Modern India: An Institutional History, c. 1784–1947*. Delhi: Pearson Longman, 2011.

Kuper, Adam. *The Invention of Primitive Society.* London: Routledge, 1988.

Kuukkanen, Jouni-Matti. "I Am Knowledge. Get Me Out of Here! On Localism and the Universality of Science." *Studies in the History and Philosophy of Science* 42 (2011).

Lachance, Paul. "Repercussions of the Haitian Revolution in Louisiana." In David Geggus, ed., *The Impact of the Haitian Revolution in the Atlantic World.* Columbia: University of South Carolina Press, 2001.

Lambert, David. *Mastering the Niger: James MacQueen's African Geography and the Struggle over Atlantic Slavery.* Chicago: University of Chicago Press, 2013.

Langer, Rita. *Buddhist Rituals of Death and Rebirth: Contemporary Sri Lankan Practice and Its Origins.* London: Routledge, 2007.

Latour, Bruno. *Science in Action: How to Follow Scientists and Engineers through Society.* Cambridge, MA: Harvard University Press, 1987.

Latour, Bruno, and Steve Woolgar. *Laboratory Life: The Construction of Scientific Facts.* Princeton, NJ: Princeton University Press, 1986.

Laugrand, Frédérick, and Jarich Oosten. "Canicide and Healing: The Position of the Dog in the Inuit Cultures of the Canadian Arctic." *Anthropos* 97 (2002).

Launay, Françoise. *The Astronomer Jules Janssen: A Globetrotter of Celestial Physics.* Trans. Storm Dunlop. New York: Springer, 2012.

Leaney, Enda. "Phrenology in Nineteenth-Century Ireland." *New Hibernia Review* 10 (2006).

Lefebvre, Henri. *The Production of Space.* Trans. Donald Nicholson-Smith. Oxford: Oxford University Press, 1991.

Legassick, Martin, and Robert Ross. "From Slave Economy to Settler Capitalism: The Cape Colony and Its Extensions, 1800–1854." In Carolyn Hamilton, Bernard Mbenga, and Robert Ross, eds., *The Cambridge History of South Africa.* Cambridge: Cambridge University Press, 2009.

Levere, Trevor. *Science and the Canadian Arctic: A Century of Exploration, 1818–1918.* Cambridge: Cambridge University Press, 2004.

Lightman, Bernard. "Spencer's American Disciples: Fiske, Youmans, and the Appropriation of the System." In Bernard Lightman, ed., *Global Spencerism: The Communication and Appropriation of a British Evolutionist.* Leiden: Brill, 2016.

———. *Victorian Popularizers of Science.* Chicago: University of Chicago Press, 2007.

Livingstone, David. *Putting Science in Its Place: Geographies of Scientific Knowledge.* Chicago: University of Chicago Press, 2003.

Long, James. *The Proceedings and Transactions of the Bethune Society from November 10th 1859 to April 20th 1869.* Calcutta: Bishop's College Press, 1870.

Loos, Jackie. *Echoes of Slavery: Voices from South Africa's Past.* Cape Town: David Philip, 2004.

Lourdusamy, J. *Science and National Consciousness in Bengal, 1870–1930.* New Delhi: Orient Longman, 2004.

Ludlow, Helen. "The Government Teacher as Mediator of a 'Superior' Education in Colesberg, 1849–1858." *Historia* 57 (2012).

Lupfer, Eric. "The Business of American Magazines." In Scott Casper, Jeffrey Groves, Stephen Nissenbaum, and Michael Winship, eds., *The History of the Book in America: The Industrial Book, 1840–1880.* Chapel Hill: University of North Carolina Press, 2007.

Lyon, George. *A Brief Narrative of an Unsuccessful Attempt to Reach Repulse Bay.* London: John Murray, 1825.

———. *The Private Journal of Captain G.F. Lyon During the Recent Voyage of Under Captain Parry.* London: John Murray, 1824.

Macartney, Henry. *Lectures Delivered in the Commercial Hall on the Subject of Life and Death*. Cape Town: George Greig, 1831.

Macleod, Roy. "On Visiting the 'Moving Metropolis': Reflections on the Architecture of Imperial Science." In Nathan Reingold and Marc Rothenberg, eds., *Scientific Colonialism*. Washington, DC: Smithsonian Institution Press, 1987.

Magendie, François. *Précis Élémentaire de Physiologie*. Paris: Méquignon-Marvis, 1825.

Mahar, William. *Behind the Burnt Cork Mask: Early Blackface Minstrelsy and Antebellum American Popular Culture*. Urbana-Champaign: University of Illinois Press, 1999.

Mahias, Marie-Claude. "Le soleil noir des Nilgiri: L'astronomie, la photographie et l'anthropologie physique en Inde du Sud." *Gradhiva* 24 (1999).

Malone, Peter. "How the Smiths Made a Living." In Rune Frederiksen and Eckart Marchand, eds., *Plaster Casts: Making, Collecting, and Displaying from Classical Antiquity to the Present*. Berlin: Walter de Gruyter, 2010.

Markovits, Claude, Jacques Pouchepadass, and Sanjay Subrahmanyam, eds. *Society and Circulation: Mobile People and Itinerant Cultures in South Asia, 1750–1950*. Delhi: Permanent Black, 2003.

Marshall, Henry. *Ceylon: A General Description of the Island and its Inhabitants, with an Historical Sketch of the Conquest of the Colony by the English*. London: W. H. Allen, 1846.

Marshall, William Elliot. *A Phrenologist Amongst the Todas*. London: Longmans and Green, 1873.

Massey, Doreen. *For Space*. London: Sage, 2009.

Maxwell, Anne. *Colonial Photography and Exhibitions: Representations of the Native and the Making of European Identities*. London: Bloomsbury, 1999.

May, Samuel. *Some Recollections of the Antislavery Conflict*. Boston: Fields, Osgood and Co., 1869.

McAleese, Kevin. "The Reinterment of Thule Inuit Burials and Associated Artifacts." *Études Inuit Studies* 22 (1998).

McCandless, Peter. "Mesmerism and Phrenology in Antebellum Charleston: 'Enough of the Marvellous.'" *Journal of Southern History* 58 (1992).

McKenzie, Kirsten. "'Franklins of the Cape': *The South African Commercial Advertiser* and the Creation of a Colonial Public Sphere, 1824–1854." *Kronos* 25 (1998).

McKeown, Adam. "Periodizing Globalization." *History Workshop Journal* 63 (2007).

McLaren, Angus. "A Prehistory of the Social Sciences: Phrenology in France." *Comparative Studies in Society and History* 23 (1981).

Meigs, James. *Catalogue of Human Crania, in the Collection of the Academy of Natural Sciences of Philadelphia*. Philadelphia: J.B. Lippincott and Co., 1857.

Melossi, Dario, and Massimo Parvani. *The Prison and the Factory: Origins of the Penitentiary System*. Trans. Glynis Cousin. London: Macmillan, 1981.

Meranze, Michael. *Laboratories of Virtue: Punishment, Revolution, and Authority in Philadelphia, 1760–1835*. Chapel Hill: University of North Carolina Press, 1996.

Metcalf, Thomas. *Ideologies of the Raj*. Cambridge: Cambridge University Press, 1994.

Midgley, Clare. *Women against Slavery: The British Campaigns, 1780–1870.* London: Routledge, 1995.

"Moletsane, Abraham Makgothi." In Evert Verwey, ed., *New Dictionary of South African Biography.* Pretoria: HSRC Publishers, 1995.

Moore, John. *Celebration of Innovation: A History of Autotype, 1868–2005.* Wantage: Autotype International, 2005.

Morrell, Jack, and Arnold Thackray. *Gentlemen of Science: Early Years of the British Association for the Advancement of Science.* Oxford: Clarendon Press, 1981.

Morton, Samuel George. *Crania Americana.* Philadelphia: J. Dobson, 1839.

Moser, Stephanie. *Wondrous Curiosities: Ancient Egypt at the British Museum.* Chicago: University of Chicago Press, 2006.

Moyn, Samuel, and Andrew Sartori. "Approaches to Global Intellectual History." In Samuel Moyn and Andrew Sartori, eds., *Global Intellectual History.* New York: Columbia University Press, 2013.

Nair, Savithri. "'Bungallee House Set on Fire by Galvanism': Natural and Experimental Philosophy as Public Science in a Colonial Metropolis (1794–1806)." In Bernard Lightman, Gordon McOuat, and Larry Stewart, eds., *The Circulation of Knowledge between Britain, India, and China: The Early-Modern World to the Twentieth Century.* Leiden: Brill, 2013.

Nerone, John. "Newspapers and the Public Sphere." In Scott Casper, Jeffrey Groves, Stephen Nissenbaum, and Michael Winship, eds., *The History of the Book in America: The Industrial Book, 1840–1880.* Chapel Hill: University of North Carolina Press, 2007.

The New Calcutta Directory. Calcutta: Military Orphan Press, 1856.

Newman, Richard. *The Transformation of American Abolitionism: Fighting Slavery in the Early Republic.* Chapel Hill: University of North Carolina Press, 2002.

Newton, Gael. "South-East Asia: Malay, Singapore, and Philippines." In John Hannavy, ed., *Encyclopedia of Nineteenth-Century Photography.* London: Routledge 2008.

Nicholson, Peter. *The New and Improved Practical Builder.* London: Thomas Kelly, 1837.

Nott, Josiah. *Types of Mankind.* Philadelphia: J. B. Lippincott and Co., 1854.

O'Brien, Michael. *Conjectures of Order: Intellectual Life and the American South, 1810–1860.* Chapel Hill: University of North Carolina Press, 2004.

O'Connell, Joseph. "Metrology: The Creation of Universality by the Circulation of Particulars." *Social Studies of Science* 23 (1993).

Ogle, Vanessa. *The Global Transformation of Time, 1870–1950.* Cambridge, MA: Harvard University Press, 2015.

O'Hanlon, Michael. "Introduction." In Michael O'Hanlon and Robert Welsch, eds., *Hunting the Gatherers: Ethnographic Collectors and Agency in Melanesia, 1870s–1930s.* New York: Berghahn Books, 2000.

O'Neil, Anthony. *Catalogue of Casts of Skulls of Different Nations, selected from the Museum of the Phrenological Society.* Edinburgh, 1834.

Ophir, Adi, and Steven Shapin. "The Place of Knowledge: A Methodological Survey." *Science in Context* 4 (1991).

Orange, A. D. "The Beginnings of the British Association, 1831–1851." In Roy MacLeod and Peter Collins, eds., *The Parliament of Science: The British Association for the Advancement of Science, 1831–1981*. Northwood: Science Reviews, 1981.

Osterhammel, Jürgen. *The Transformation of the World: A Global History of the Nineteenth Century*. Princeton, NJ: Princeton University Press, 2014.

Parry, William. *Journal of a Second Voyage for the Discovery of a North-West Passage from the Atlantic to the Pacific*. London: John Murray, 1824.

Parthasarathi, Prasannan. *Why Europe Grew Rich and Asia Did Not: Global Economic Divergence, 1600–1850*. Cambridge: Cambridge University Press, 2011.

Pearce, Susan. *On Collecting: An Investigation into Collecting in the European Tradition*. London: Routledge, 1994.

Pease, Donald. "New Perspectives on U.S. Culture and Imperialism." In Amy Kaplan and Donald Pease, eds., *Cultures of United States Imperialism*. Durham, NC: Duke University Press, 1993.

Perry, John. *Facing West: Americans and the Opening of the Pacific*. Westport, CT: Praeger, 1994.

Philadelphia Directory. Philadelphia: A. McElroy, 1839.

Phillip, Kavita. *Civilising Natures: Race, Resources, and Modernity in Colonial South India*. New Delhi: Orient Longman, 2003.

Pingsheng, Feng. *Guxiangxue*. Shanghai: Commercial Press, 1923.

Pinney, Christopher. *Camera Indica: The Social Life of Indian Photographs*. London: Reaktion, 1992.

———. "Underneath the Banyan Tree: William Crooke and Photographic Depictions of Caste." In Elizabeth Edwards, ed., *Anthropology and Photography, 1860–1920*. New Haven, CT: Yale University Press, 1992.

Popkin, Jeremy. *You Are All Free: The Haitian Revolution and the Abolition of Slavery*. Cambridge: Cambridge University Press, 2010.

Prakash, Gyan. *Another Reason: Science and the Imagination of Modern India*. Princeton, NJ: Princeton University Press, 1999.

Pratt, Mary Louise. *Imperial Eyes: Travel Writing and Transculturation*. London: Routledge, 1992.

Prichard, James Cowles. *Researches into the Physical History of Mankind*. London: John and Arthur Arch, 1826.

———. *Researches into the Physical History of Mankind*. London: John and Arthur Arch, 1836.

———. *A Treatise on Insanity and Other Disorders Affecting the Mind*. London: Sherwood, Gilbert and Piper, 1835.

Pritchard, Michael. "Ottewill, Thomas." In John Hannavy, ed., *Encyclopedia of Nineteenth-Century Photography*. London: Routledge, 2008.

Prodger, Phillip. *Darwin's Camera: Art and Photography in the Theory of Evolution*. Cambridge: Cambridge University Press, 2009.

Quatrefages, Jean Louis Armand de. *Histoire Générale des Races Humaines*. Paris: A. Hennuyer, 1889.

———. *Hommes Fossiles et Hommes Sauvages*. Paris: Baillière, 1884.

Quigley, Christine. *Skulls and Skeletons: Human Bone Collections and Accumulations*. Jefferson, NC: McFarland, 2001.

Qureshi, Sadiah. "Displaying Sara Baartman, the 'Hottentot Venus.'" *History of Science* 42 (2004).

———. *Peoples on Parade: Exhibitions, Empire, and Anthropology in Nineteenth-Century Britain.* Chicago: University of Chicago Press, 2011.

———. "Robert Gordon Latham, Displayed Peoples, and the Natural History of Race, 1854–1866." *Historical Journal* 54 (2011).

Raby, Peter. *Alfred Russel Wallace: A Life.* Princeton, NJ: Princeton University Press, 2001.

Raj, Kapil. *Relocating Modern Science: Circulation and the Construction of Knowledge in South Asia and Europe, 1650–1900.* New Delhi, 2006.

Raychaudhuri, Tapan. *Europe Reconsidered: Perceptions of the West in Nineteenth-Century Bengal.* New Delhi: Oxford University Press, 1988.

Raychoudhuri, Sukanta. "The Lost World of the Babus." In Sukanta Chaudhuri, ed., *Calcutta: The Living City.* Oxford: Oxford University Press, 1991.

Redman, Samuel. *Bone Rooms: From Scientific Racism to Human Prehistory in Museums.* Cambridge, MA: Harvard University Press, 2016.

Reed, Christopher. *Gutenberg in Shanghai: Chinese Print Capitalism, 1876–1937.* Vancouver: University of British Columbia Press, 2011.

Reid, Donald. *Whose Pharaohs? Archaeology, Museums, and Egyptian National Identity from Napoleon to World War I.* Berkeley: University of California Press, 2002.

Renneville, Marc. *Le langage des crânes: Un histoire de la phrénologie.* Paris: Institut d'édition Sanofi-Synthélabo, 2000.

———. "Un musée d'anthropologie oublié: Le cabinet phrénologique de Dumoutier." *Bulletins et Mémoires de la Société d'Anthropologie de Paris* 10 (1998).

The Report of the Eleventh Meeting of the British Association for the Advancement of Science held at Plymouth. London: John Murray, 1842.

Report of the Ninth Meeting of the British Association for the Advancement of Science held in Birmingham. London: John Murray, 1840.

Revista da Exposição Anthropologica Brazileira. Rio de Janeiro: Typographia de Pinheiro, 1882.

Rhinevault, Carney, and Tatiana Rhinevault. *Hidden History of the Mid-Hudson Valley.* Charleston, SC: History Press, 2011.

Richardson, Robert. *Travels along the Mediterranean and Parts Adjacent.* London: T. Cadell, 1822.

Rivers, William Halse Rivers. *The Todas.* London: Macmillan, 1906.

Robb, Peter. "Introduction: The Concept of Race in South Asia." In Peter Robb, ed., *The Concept of Race in South Asia.* Oxford: Oxford University Press, 1995.

Robben, Antonius. "Death and Anthropology: An Introduction." In Antonius Robben, ed., *Death, Mourning, and Burial: A Cross-Cultural Reader.* Malden, MA: Blackwell, 2004.

Roberts, Lissa. "Situating Science in Global History: Local Exchanges and Networks of Circulation." *Itinerario* 33 (2009).

Robertson, Roland. *Globalization: Social Theory and Global Culture.* London: Sage, 1992.

Rochette, Marc. "Dumont d'Urville's Phrenologist: Dumoutier and the Aesthetics of Races." *Journal of Pacific History* 38 (2003).

Rogers, Beverley. "Unwrapping the Past: Egyptian Mummies on Show." In Joe Kember, John Plunkett, and Jill Sullivan, eds., *Popular Exhibitions, Science, and Showmanship, 1840–1910*. London: Pickering and Chatto, 2012.

Rogers, John. "Caste as a Social Category and Identity in Colonial Lanka." *Indian Economic and Social History Review* 97 (2006).

———. "Early British Rule and Social Classification in Lanka." *Modern Asian Studies* 38 (2004).

Roque, Ricardo. *Headhunting and Colonialism: Anthropology and the Circulation of Human Skulls in the Portuguese Empire, 1870–1930*. Basingstoke: Palgrave Macmillan, 2010.

Rosenboim, Or. *The Emergence of Globalism Visions of World Order in Britain and the United States, 1939–1950*. Princeton, NJ: Princeton University Press, 2017.

Ross, William. *Arctic Whalers, Icy Seas: Narratives of the Davis Strait Whale Fishery*. Toronto: Irwin, 1985.

Rothman, David. *The Discovery of the Asylum: Social Order and Disorder in the New Republic*. Boston: Little Brown, 1971.

Rudiments of Knowledge. Edinburgh: William and Robert Chambers, 1838.

Rudwick, Martin. "The Emergence of a Visual Language for Geological Science, 1760–1840." *History of Science* 14 (1976).

Ruffle, John. "The Journeys of Lord Prudhoe and Major Orlando Felix in Egypt, Nubia, and the Levant, 1826–1829." In Janet Starkey and Paul Starkey, eds., *Travellers in Egypt*. London: I. B. Tauris, 1998.

———. "Lord Prudhoe and Major Felix, *Hiéroglyphiseurs Décidés*." In Jason Thompson, ed., *Egyptian Encounters*. Cairo: American University in Cairo Press, 2002.

Rusert, Britt. *Fugitive Science: Empiricism and Freedom in Early African American Culture*. New York City: New York University Press, 2017.

———. "The Science of Freedom: Counterarchives of Racial Science on the Antebellum Stage." *African American Review* 45 (2012).

Ryan, James. *Picturing Empire: Photography and the Vizualisation of the British Empire*. London: Reaktion, 1997.

Sahlins, Marshall. *Islands of History*. Chicago: University of Chicago Press, 1985.

Said, Edward. *Orientalism*. London: Routledge and Kegan Paul, 1978.

Sampson, Gary. "Photographer of the Picturesque: Samuel Bourne." In Vidya Dehejia, ed., *India through the Lens: Photography, 1840–1911*. Washington, DC: Freer Gallery of Art, 2000.

———. "The Success of Samuel Bourne in India." *History of Photography* 16 (1992).

Sanders, Scott. "Antioch College: Establishing the Faith." In James Hodges, James O'Donnell, and John Oliver, eds., *Cradles of Conscience: Ohio's Independent Colleges and Universities*. Kent, OH: Kent State University Press, 2003.

Sanyal, Rajat. *Voluntary Associations and the Urban Public Life in Bengal, 1815–1876*. Calcutta: Riddhi-India, 1980.

Sarkar, Sumit. "Calcutta and the 'Bengal Renaissance.'" In Sukanta Chaudhuri, ed., *Calcutta: The Living City*. Oxford: Oxford University Press, 1991.

Sartori, Andrew. *Bengal in Global Concept History: Culturalism in the Age of Capital.* Chicago: University of Chicago Press, 2008.

Sattin, Anthony. *Lifting the Veil: British Society in Egypt, 1768–1956.* London: J. M. Dent, 1988.

Sayyid Marsot, Afaf Lutfi al-. *Egypt in the Reign of Muhammad Ali.* Cambridge: Cambridge University Press, 1984.

Schaffer, Simon. "The Asiatic Enlightenments of British Astronomy." In Simon Schaffer, Lissa Roberts, Kapil Raj, and James Delbourgo, eds., *The Brokered World: Go-Betweens and Global Intelligence, 1770–1820.* Sagamore Beach, MA: Science History Publications, 2009.

———. "Astronomers Mark Time: Discipline and the Personal Equation." *Science in Context* 2 (1988).

———. *From Physics to Anthropology and Back Again.* Cambridge: Prickly Pear Press, 1994.

———. "Golden Means: Assay Instruments and the Geography of Precision in the Guinea Trade." In Marie-Noëlle Bourguet, Christian Licoppe, and H. Otto Sibum, eds., *Instruments, Travel, and Science: Itineraries of Precision from the Seventeenth to the Twentieth Century.* London: Routledge, 2002.

———. "Late Victorian Metrology and Its Instrumentation: A Manufactory of Ohms." In Robert Bud and Susan Cozzens, eds., *Invisible Connections: Instruments, Institutions, and Science.* Bellingham, WA: SPIE Optical Engineering Press, 1991.

Scherer, Joanna. "The Photographic Document: Photographs as Primary Data in Anthropological Enquiry." In Elizabeth Edwards, ed., *Anthropology and Photography, 1860–1920.* New Haven, CT: Yale University Press, 1992.

Schmalzer, Sigrid. *The People's Peking Man: Popular Science and Human Identity in Twentieth-Century China.* Chicago: University of Chicago Press, 2008.

Schultz, Lucille. "Letter-Writing Instruction in 19th-Century Schools in the United States." In David Barton and Nigel Hall, eds., *Letter Writing as a Social Practice.* Amsterdam: John Benjamins Publishing Company, 2000.

Schwarcz, Lilia Moritz. "Predictions Are Always Deceptive: João Baptista de Lacerda and His White Brazil." *Historia, Ciencias, Saude—Manguinhos* 18 (2011).

Scott, Donald. "The Popular Lecture and the Creation of a Public in Mid-Nineteenth-Century America." *Journal of American History* 66 (1980).

Secord, Anne. "Corresponding Interests: Artisans and Gentlemen in Nineteenth-Century Natural History." *British Journal for the History of Science* 27 (1994).

Secord, James. "Knowledge in Transit." *Isis* 95 (2004).

———. "Monsters at the Crystal Palace." In Soraya de Chadarevian and Nick Hopwood, eds., *Models: The Third Dimension of Science.* Stanford, CA: Stanford University Press, 2004.

———. "Progress in Print." In Marina Frasca-Spada and Nick Jardine, eds., *Books and the Sciences in History.* Cambridge: Cambridge University Press, 2000.

———. *Victorian Sensation: The Extraordinary Publication, Reception, and Secret Authorship of "Vestiges of the Natural History of Creation"*. Chicago: University of Chicago Press, 2000.

———. *Visions of Science: Books and Readers at the Dawn of the Victorian Age*. Oxford: Oxford University Press, 2014.

Sen, Samita, and Anirban Das. "A History of the Calcutta Medical College and Hospital, 1835–1936." In Uma Das Gupta, ed., *Science and Modern India: An Institutional History, 1784–1947*. New Delhi: Pearson Longman, 2011.

Sera-Shriar, Efram. "Anthropometric Portraiture and Victorian Anthropology: Situating Francis Galton's Photographic Work in the Late 1870s." *History of Science* 53 (2015).

———. *The Making of British Anthropology, 1813–1871*. London: Pickering and Chatto, 2013.

Shapin, Steven. "Phrenological Knowledge and the Social Structure of Early Nineteenth-Century Edinburgh." *Annals of Science* 32 (1975).

———. "The Politics of Observation: Anatomy and Social Interests in the Edinburgh Phrenology Disputes." In Roy Wallis, ed., *On the Margins of Science: The Social Construction of Rejected Knowledge*. Keele: University of Keele Press, 1979.

———. "Pump and Circumstance: Robert Boyle's Literary Technology." *Social Studies of Science* 14 (1984).

Shapin, Steven, and Simon Schaffer. *Leviathan and the Air-Pump: Hobbes, Boyle, and the Experimental Life*. Princeton, NJ: Princeton University Press, 2011.

Shaw, Graham. "The British Book in India." In Michael Suarez and Michael Turner, eds., *The Cambridge History of the Book in Britain*. Cambridge: Cambridge University Press, 2009.

———. "Calcutta: The Birthplace of the Indian Lithographed Book." *Journal of the Printing Historical Society* 27 (1998).

Silva, Colvin de. *Ceylon under the British Occupation, 1795–1833*. Colombo: Colombo Apothecaries'' Co., 1953.

Sivasundaram, Sujit. "Buddhist Kingship, British Archaeology, and Historical Narratives in Sri Lanka c. 1750–1850." *Past & Present* 197 (2010).

———. "'A Christian Benares': Orientalism, Science, and the Serampore Mission of Bengal." *Indian Economic and Social History Review* 44 (2007).

———. "Ethnicity, Indigeneity, and Migration in the Advent of British Rule to Sri Lanka." *American Historical Review* 115 (2010).

———. "Introduction: Global Histories of Science." *Isis* 101 (2010).

———. *Islanded: Britain, Sri Lanka, and the Bounds of an Indian Ocean Colony*. Chicago: University of Chicago Press, 2013.

———. "Race, Empire, and Biology before Darwin." In Denis Alexander and Ronald Numbers, eds., *Biology and Ideology from Descartes to Dawkins*. Chicago: University of Chicago Press, 2010.

———. "Sciences and the Global: On Methods, Questions, and Theory." *Isis* 101 (2010).

Sizer, Nelson. *Forty Years in Phrenology*. New York: Fowler and Wells, 1882.

Skidmore, Thomas. *Black into White: Race and Nationality in Brazilian Thought*. Durham, NC: Duke University Press, 1993.

Smith, Anna. *The Spread of Printing: South Africa*. Amsterdam: Vangendt, 1971.

Smith, James McCune. "The Heads of Colored People." In *The Works of James McCune Smith: Black Intellectual and Abolitionist*. Oxford: Oxford University Press, 2007.

———. "Outside Barbarians." In *The Works of James McCune Smith: Black Intellectual and Abolitionist*. Oxford: Oxford University Press, 2007.

Smith, John. *Select Discourses on the Functions of the Nervous System in Opposition to Phrenology*. New York: D. Appleton and Co., 1840.

Soler, Mariano Cubí y. *La Frenologia i Sus Glorias*. Barcelona: Imprenta Hispana de V. Castaños, 1853.

Spence, Jonathan. *The Search for Modern China*. London: W. W. Norton, 1990.

Spencer, Frank. "Some Notes on the Attempt to Apply Photography to Anthropometry during the Second Half of the Nineteenth Century." In Elizabeth Edwards, ed., *Anthropology and Photography, 1860–1920*. New Haven, CT: Yale University Press, 1992.

Sprod, Dan. *Alexander Pearce of Macquarie Harbour*. Hobart: Cat and Fiddle Press, 1977.

Spurzheim, Johann. *Phrenology, or, The Doctrine of the Mind*. London: Charles Knight, 1825.

———. *The Physiognomical System of Drs. Gall and Spurzheim*. London: Baldwin, Cradock and Joy, 1815.

Stack, David. *Queen Victoria's Skull: George Combe and the Mid-Victorian Mind*. London: Hambledon Continuum, 2008.

Stanton, William. *The Leopard's Spots: Scientific Attitudes towards Race in America, 1815–59*. Chicago: University of Chicago Press, 1960.

Stark, Ulrike. *An Empire of Books: The Naval Kishore Press and the Diffusion of the Printed Word in Colonial India*. Delhi: Permanent Black, 2008.

Stauffer, John. "Introduction." In *The Works of James McCune Smith: Black Intellectual and Abolitionist*. Oxford: Oxford University Press, 2007.

Staum, Martin. *Labeling People: French Scholars on Society, Race, and Empire, 1815–1848*. Montreal: McGill-Queen's University Press, 2003.

———. "Physiognomy and Phrenology at the Paris Athénée." *Journal of the History of Ideas* 56 (1995).

Stebbins, Robert. "France." In Thomas Glick, ed., *The Comparative Reception of Darwinism*. Chicago: University of Chicago Press, 1974.

Steel, Wilfred. *The History of the London and North Western Railway*. London: Railway and Travel Monthly, 1914.

Stepan, Nancy. *"The Hour of Eugenics": Race, Gender, and Nation in Latin America*. Ithaca, NY: Cornell University Press, 1991.

———. *The Idea of Race in Science: Great Britain, 1800–1960*. London: Macmillan, 1982.

Stern, Madeleine. *Heads and Headlines: The Phrenological Fowlers*. Norman: University of Oklahoma Press, 1971.

Stocking, George, ed. *Objects and Others: Essays on Museums and Material Culture*. Madison: University of Wisconsin Press, 1985.

———. *Victorian Anthropology*. London: Free Press, 1987.

———. "What's in a Name? The Origins of the Royal Anthropological Institute." *Man* 6 (1971).

Straton, James. *Contributions to the Mathematics of Phrenology*. Aberdeen: William Russel, 1845.

Streets, Heather. *Martial Races: The Military, Race, and Masculinity in British Imperial Culture, 1857–1914*. Manchester: Manchester University Press, 2004.

Stronach, George. "Marshall, Henry (1775–1851)." *Oxford Dictionary of National Biography* (2005), online edition.

Subrahmanyam, Sanjay. "Connected Histories: Notes towards a Reconfiguration of Early Modern Eurasia." *Modern Asian Studies* 31 (1997).

———. "Global Intellectual History beyond Hegel and Marx." *History & Theory* 54 (2015).

Sun, Lung-Kee. *The Chinese National Character: From Nationhood to Individuality.* London: M. E. Sharpe, 2002.

A Supplement to the Appendix of Captain Parry's Voyage for the Discovery of a North-West Passage in the Years 1819–1820. London: John Murray, 1824.

Sutton, Deborah. *Other Landscapes: Colonialism and the Predicament of Authority in Nineteenth-Century South India.* Copenhagen: Nordic Institute of Asian Studies, 2009.

Sutton, Thomas. *A Dictionary of Photography.* London: S. Low and Son, 1858.

Tagore, Jyotirindranath. *Twenty-Five Collotypes.* Hammersmith: Emery Walker Limited, 1914.

Tansill, Charles. *The United States and Santo Domingo, 1798–1873.* Baltimore: John Hopkins University Press, 1938.

Thackray, John. "R. I. Murchison's *Silurian System* (1839)." *Archives of Natural History* 9 (1978).

Thomas, G. "The Madras Photographic Society, 1856–61." *History of Photography* 16 (1992).

———. "The 'Peccavi' Photographs." *History of Photography* 4 (1980).

Thomason Civil Engineering College Calendar. Roorkee: Thomason Civil Engineering College Press, 1871.

Thompson, Jason. *Sir Gardner Wilkinson and His Circle.* Austin: University of Texas Press, 1992.

Tilley, Helen. "Racial Science, Geopolitics, and Empires: Paradoxes of Power." *Isis* 105 (2014).

Tomek, Beverly. *Colonization and Its Discontents: Emancipation, Emigration, and Antislavery in Antebellum Pennsylvania.* New York: New York University Press, 2011.

Tomlinson, Stephen. *Head Masters: Phrenology, Secular Education, and Nineteenth-Century Social Thought.* Tuscaloosa: University of Alabama Press, 2005.

Topham, Jonathan. "Anthologizing the Book of Nature: The Circulation of Knowledge and the Origins of the Scientific Journal in Late Georgian Britain." In Bernard Lightman, Gordon McOuat, and Larry Stewart, eds., *The Circulation of Knowledge between Britain, India, and China: The Early-Modern World to the Twentieth Century.* Leiden: Brill, 2013.

———. "Introduction: Historicizing 'Popular Science.'" *Isis* 100 (2009).

———. "Science, Print, and Crossing Borders: Importing French Science Books into Britain 1789–1815." In David Livingstone and Charles Withers, eds., *Geographies of Nineteenth-Century Science.* Chicago: University of Chicago Press, 2011.

———. "Scientific Publishing and the Reading of Science in Nineteenth-Century Britain: A Historiographical Survey and Guide to Sources." *Studies in History and Philosophy of Science Part A* 31 (2000).

Turley, David. *The Culture of English Antislavery, 1780–1860.* London: Routledge, 2003.

Turnball, Paul. "'Rare Work amongst the Professors': The Capture of Indigenous Skulls within Phrenological Knowledge in Early Colonial Australia." In Barbara Creed and Jeanette Horn, eds., *Body Trade: Captivity, Cannibalism, and Colonialism in the Pacific.* New York: Routledge, 2001.

Twyman, Michael. "Putting Philadelphia on Stone." In Erika Piola, ed., *Philadelphia on Stone: Commercial Lithography in Philadelphia, 1828–1878.* University Park: Penn State University Press, 2012.

Vilela, Ana Maria Jacó. "History of Psychology in Brazil: A Narrative from Its Teaching." *Psicologia: Ciência e Profissão* 32 (2012).

Vimont, Joseph. *Traité de Phrénologie Humaine et Comparée.* Paris: Baillière, 1832.

Waddell, David. "International Politics and Latin American Independence." In Leslie Bethell, ed., *The Cambridge History of Latin America: From Independence to 1870.* Cambridge: Cambridge University Press, 1985.

Wagner, Kim. "Confessions of a Skull: Phrenology and Colonial Knowledge in Early Nineteenth-Century India." *History Workshop Journal* 69 (2010).

Wald, Erica. *Vice in the Barracks: Medicine, Military, and the Making of Colonial India, 1780–1868.* Basingstoke: Palgrave Macmillan, 2014.

Walker, Charles. *Smoldering Ashes: Cuzco and the Creation of Republican Peru, 1780–1840.* Durham, NC: Duke University Press, 1999.

Wallace, Alfred Russel. *Studies Scientific and Social.* London: Macmillan and Co., 1900.

———. *The Wonderful Century.* London: Swan Sonnenschein and Co., 1898.

Ward, John. "Autotype Fine Art Company." In John Hannavy, ed., *Encyclopedia of Nineteenth-Century Photography.* London: Routledge 2008.

Watson, Hewett. *Statistics of Phrenology.* London: Longman, Rees, Orme, Brown, Green, and Longman, 1836.

Watson, John Forbes, and John William Kaye, eds., *The People of India.* London: India Museum, 1875.

Watts, Iain. "Philosophical Intelligence: Letters, Print, and Experiment during Napoleon's Continental Blockade." *Isis* 106 (2015).

———. "'We want no authors': William Nicholson and the Contested Role of the Scientific Journal in Britain, 1797–1813." *British Journal for the History of Science* 47 (2014).

Wende, Peter. "1848: Reform or Revolution in Germany and Great Britain." In Tim Blanning and Peter Wende, eds., *Reform in Great Britain and Germany, 1750–1850.* Oxford: Oxford Unviersity Press, 1999.

White, Daniel. *From Little London to Little Bengal: Religion, Print, and Modernity in Early British India, 1793–1835.* Baltimore: John Hopkins University Press, 2013.

White, Paul. "Lives and Letters: Correspondence and Public Character in the Nineteenth Century." In Rosalind Crone, David Gange, and Katy Jones, eds., *New Perspectives in British Cultural History.* Cambridge: Cambridge Scholars Publishing, 2007.

Whitridge, Peter. "Landscapes, Houses, Bodies, Things: 'Place' and the Archaeology of Inuit Imaginaries." *Journal of Archaeological Method and Theory* (2004).

Whittier, John. *The Writings of John Greenleaf Whittier.* London: Macmillan, 1888.

Wickramasinghe, Nira. "The Return of Keppetipola's Cranium: Authenticity in a New Nation." *Economic and Political Weekly* 32 (1997).

Wilde, William. *Narrative of a Voyage to Madeira, Teneriffe and Along the Shores of the Mediterranean.* Dublin: William Curry and Co., 1840.

Williams, Elizabeth. *The Physical and the Moral: Anthropology, Physiology, and Philosophical Medicine in France, 1750–1850.* Cambridge: Cambridge University Press, 2002.

Williams, F. Fisk. *A Guide to the Indian Photographer.* Calcutta: R. C. Lepage, 1860.

Windler, Erica. "Madame Durocher's Performance: Cross-Dressing, Midwifery, and Authority in Nineteenth-Century Rio de Janeiro, Brazil." In William French and Katherine Bliss, eds., *Gender, Sexuality, and Power in Latin America since Independence.* Lanham, MA: Rowman and Littlefield, 2007.

Withers, Charles. *Geography and Science in Britain, 1831–1939: A Study of the British Association for the Advancement of Science.* Manchester: Manchester University Press, 2010.

Withers, Charles, and David Livingstone. "Thinking Geographically about Nineteenth-Century Science." In Charles Withers and David Livingstone, eds., *Geographies of Nineteenth-Century Science.* Chicago: University of Chicago Press, 2011.

Wyhe, John van. "The Authority of Human Nature: The Schädellehre of Franz Joseph Gall." *British Journal for the History of Science* 35 (2002).

———. "The Diffusion of Phrenology through Public Lecturing." In Aileen Fyfe and Bernard Lightman, eds., *Science in the Marketplace: Nineteenth-Century Sites and Experiences.* Chicago: University of Chicago Press, 2007.

———. *Phrenology and the Origins of Victorian Scientific Naturalism.* Aldershot: Ashgate, 2004.

———. "Was Phrenology a Reform Science? Towards a New Generalization for Phrenology." *History of Science* 42 (2004).

Yan, Fu. *Qunxueyiyan.* Shanghai: Commercial Press, 1931.

Yang, Haiyan. "Knowledge across Borders: The Early Communication of Evolution in China." In Bernard Lightman, Gordon McOuat, and Larry Stewart, eds., *The Circulation of Knowledge between Britain, India, and China: The Early-Modern World to the Twentieth Century.* Leiden: Brill, 2013.

Zen, Sun E-tu. "The Growth of an Academic Community, 1912–1949." In John Fairbank and Albert Feuerwerker, eds., *The Cambridge History of China.* Cambridge: Cambridge University Press, 1986.

Zunke, Ke, and Li Bin. "Spencer and Science Education in China." In Bernard Lightman, ed., *Global Spencerism: The Communication and Appropriation of a British Evolutionist.* Leiden: Brill, 2016.

INDEX

Note: References to figures are denoted by an "f" in italics following the page number.

Abel, Clarke, 138. *See also* Calcutta Phrenological Society
Aberdeen, 64. *See also* Britain
abolitionists, 123–27, 303n64, 308n210; British, 127; and evolutionary theory, 301n16; female, 124–26. *See also* anti-slavery movement; slavery
Aborigines, 13, 51, 285n176. *See also* Australia
Aborigines' Protection Society, 86, 106. *See also* Aborigines
Académie des Sciences, 109, 226, 229. *See also* Paris; science
Academy of Natural Sciences of Philadelphia, 94, 106. *See also* Philadelphia; science
Acharyya, Bijay Kisor, 236–37
Adam, John, 140. *See also* Medical and Physical Society of Calcutta
Adams, President John Quincy, 123
advertising, 108–9; for periodicals, 158; phrenological, 155–56, 169, 186–87, 234, 290n92. *See also* periodicals
Africa, 3, 13–14, 37, 130, 141; people of, 15; racial character of, 16; slaves from, 123; southern, 198. *See also* Africans; North Africa; South Africa; West Africa
African Americans, 15, 128–131; convicts of the, 134, 146; trials of the, 16, 73–76. *See also* Africa; race; United States
African Repository, 73. *See also* periodicals
Africans, 13, 15, 37, 56–66, 71, 73, 119–22, 126–31, 173–76. *See also* Africa; race
Ahmad, Muhammad, 197. *See also* Sudan
Allahabad, 234. *See also* India
Al-Tahtawi, Rifa'a, 40. *See also* Egypt
American Colonization Society (ACS), 72, 126–29. *See also* Africa; slavery
American Journal of Science, 96, 100, 103, 105–8, 113. *See also* periodicals; science
American Phrenological Journal, 11–12, 73–75, 105, 115, 151, 154–68, 156*f*, 191, 193–94; "Pictorial Prospectus" for the 159*f*; Portraits of "Nena Sahib" and "The King of Oude" (1864) in the, 167*f*. *See also* periodicals; phrenology
Americas, 3, 12; African slaves in the, 13. *See also* Canada; North America; South America; United States
Amherst College, 154
Amsterdam, 1. *See also* Europe
anatomy, 4; measurements of, 106. *See also* craniology; physiology
Anatomy Act (1832), 32. *See also* politics
Anderson, Benedict, 254
animals, 48, 119
Annals of Phrenology, 53, 65–66, 120, 151–53. *See also* Boston; periodicals
Anthropological Society, 200. *See also* anthropology; London
anthropology, 4, 20, 112, 231; in Brazil, 110–11; colonial, 203; evolutionary, 201; historical, 49, 278n8; in India, 200–22, 235, 238; photography in, 215, 325n170; physical, 111; physical sciences and, 320n46; professionalization of, 23, 279n24. *See also* ethnology; phrenology

Antioch College, 142. *See also* education; Mann, Horace
antiquities: collection of, 36, 39–40; and the modern state, 40; studies of, 39–40
antislavery movement, 58, 62, 65, 68–72, 124–31; 125*f*. *See also* abolitionists; reform; slavery
Arbousset, Thomas: *Relation d'un Voyage d'Exploration au Nord-Est de la Colonie du Cap de Bonne Espérance* (1842) of, 176. *See also* Cape Colony; travel
Arctic, 8, 14, 20, 41–49; folkloric studies of the, 114; Inuit settlements in the, 41–42; Inuit skulls collected in the, 18, 20–21, 36, 41–50; missionary work in the, 285n165; race in, 21. *See also* Inuit; Northwest Passage
Armistead, William: *A Tribute for the Negro* (1848) of, 64. *See also* race
Arthur, Frederick, 209. *See also* photography
Asia, 3–4, 12, 16, 20, 141; eastern, 222; race in, 21; southern, 222. *See also* East Asia; Pacific; South Asia; Southeast Asia
Asiatic Society of Bengal, 113. *See also* India
Astronomical Society, 4. *See also* London; astronomy
astronomy, 88, 229–30. *See also* science
atlas: ethnological, 204; expensive, 105; phrenological, 108, 110–11; scientific, 112–13, 294n42. *See also* books; publishing
Auburn Prison, 133–36, 146. *See also* prisons; United States
Audubon, John James, 93; *Birds of America* (1827–38) of, 93. *See also* illustration; natural history
Australia, 6, 240; Chinese immigrants in, 51; gold rush in, 53; phrenology in, 286n1. *See also* Aborigines; New Holland; Pacific
Australian, 151. *See also* Australia; newspapers
Autotype Printing and Publishing Company, 217–18, 228, 236. *See also* printing; publishing

Baartman, Sara, 62
Bahia, 110; Faculty of Medicine in, 110–11. *See also* Brazil
Baltimore, 70–72. *See also* United States
Baltimore Leader, 157. *See also* Baltimore; newspapers
Bancroft, George, 65
Barbados, 15–16. *See also* Americas
Barbados Globe, 150. *See also* newspapers
Beatson, James, 140. *See also* East India Company
Belfast Natural History Society, 40. *See also* Ireland
Belgium, 123. *See also* Europe
Belin, Eustache, 55–66, 60*f*, 68–76, 74*f*; bust of, 68, 76, 261; name of, 288n34. *See also* Haiti
Bell, John, 106, 137–39, 141. *See also* periodicals; Philadelphia Phrenological Society
Bengal, 136–45, 169, 178–90, 316n153. *See also* India
Bengal Hurkaru, 138. *See also* newspapers-
Berlin, 1. *See also* Germany
Bengali, language, 186–88. *See also* Bengal
Bethune Society, 188. *See also* India
Biassou, Georges, 57. *See also* Haiti
Bible, 34, 39
Bidyakalpadrum, 179. *See also* India; periodicals; Young Bengal
Birmingham, 85–86, 88–89, 93. *See also* Britain
Bissette, Cyrille, 68. *See also* antislavery movement
Blavatsky, Helena, 237–39; *Isis Unveiled* (1877) of, 237; *The Modern Panarion* (1895) of, 237
Blumenbach, Johann, 83, 88, 91, 106. *See also* phrenology; race
Bolding, John, 75–76
Bombay, 3, 77, 79, 203, 212, 234, 236. *See also* India
Bombay Times, 15, 203. *See also* periodicals
Bonaparte, Napoleon, 33–34, 57. *See also* France
books: British, 225; cheap, 80–83; European, 170; expensive, 108–9; exports of, 96–97; global circulation of, 81–82; illustrated, 17; imports of, 96–97; lost at sea, 84; photography in, 198; phrenological, 2, 10–11, 78–114, 159, 162, 178, 202; steam-printed, 18. *See also* atlas; booksellers; illustration; libraries; lithography; printing; publishing; stereotype
booksellers, 157. *See also* books; publishing
Boston, 5, 51–53, 65, 69, 71, 81, 83, 90, 122, 141–42, 151; periodicals in, 97, 106, 153; publishing in, 152. *See also* United States

Boston Phrenological Society, 65, 120, 134. *See also* Boston; phrenology
Boston Prison Discipline Society, 146. *See also* Boston; prisons
Bourne, Samuel, 199, 210, 212–17, 228, 240; "Panoramic View from a Spur of Dodabet" of, 216*f*. *See also* India; photography
Boyer, Jean-Pierre, 72–73. *See also* Haiti
Boyle, Robert, 249, 251. *See also* science
brain, 1–2, 12; American Indian, 101*f*, 102, 107; European, 101*f*, 102, 107; organization of the, 35, 141. *See also* craniology; mind; phrenology; skull
Brazil: abolitionist movement in, 111; physicians in, 114; science in, 112; sources in, 299n204; teeth in, 109–12
Brazilian Anthropological Exposition, 111. *See also* Brazil
Bridges, Frederick: *The Popular Manual of Phrenology* (1860) of, 64
Briggs, Captain Thomas, 203. *See also* photography
Brigham, Amariah, 19
Bristol, 53, 59, 61, 93, 113, 131. *See also* Britain
Britain, 1–9, 12, 15, 22, 36, 42–43, 52–53, 61–65, 74–79, 86–88, 93–100, 104–8, 116, 130–37, 147, 150, 153–54, 190–91, 229, 257; attitudes to burial in, 46–47; editions of scientific works in, 82; educational systems of, 307n184; industrial towns of, 86; penny post in, 115; postal service in, 143; reading of the *Crania Americana* in, 90; reforms in, 169; school reform in, 143; scientific audiences in, 309n4; scientific establishment of, 89; working class in, 80, 123. *See also* British Empire; Europe
British and Foreign Medical Review, 96, 107. *See also* periodicals
British Association for the Advancement of Science, 85–90, 93–94, 97, 109, 113; organizing committee of the, 87. *See also* Britain; science
British Empire, 12, 23–30, 56, 87–88, 193, 203; photography in the, 225; slavery in the, 118; war in the, 280n50. *See also* Britain; empire; science
British Guyana, 150. *See also* South America
British Journal of Photography, 212. *See also* periodicals; photography
British Museum, 34, 36. *See also* Britain; museum
Broca, Pierre Paul, 225–26. *See also* anatomy; anthropology; physicians
Broussais, François-Joseph-Victor, 11, 58–59, 61–69, 76; *Cours de Phrénologie* (1836) of, 79. *See also* phrenology; slavery
Buchanan, David Dale, 176. *See also Natal Witness*; newspapers
Buchanan, Joseph, 163. *See also* phrenology
Bulletin de la Société de Géographie, 58. *See also* periodicals
Bull Ring Riots, 85. *See also* Britain
burial, 20, 32, 42, 49; Christian views of, 44–48; Inuit views of, 42–47. *See also* culture; death; skull
Burts, Robert, 243
busts, 8, 52, 63, 66, 67*f*, 68, 76–77, 137, 155, 158–59, 291n111; Egyptian, 39; list history of, 66; plaster, 14, 16–18, 66, 76–77, 257; proslavery inscriptions and, 69. *See also* casts; craniology; material culture; phrenology; skull
Buxton, Thomas, 127. *See also* antislavery movement
Byron, George, 19

Cairo, 33–35, 40. *See also* Egypt
Calcutta, 9–11, 15, 77, 79–80, 136–45, 151, 168, 178–92, 202–3, 235–37, 240, 261; booksellers in, 234; photography in, 208–10, 212; publishing in, 179. *See also* India
Calcutta Medical College, 178, 183, 185. *See also* Calcutta
Calcutta Phrenological Society, 8, 11, 53, 137–38, 178–79, 181–88, 190–91. *See also* Calcutta; phrenology
Calcutta Public Library, 79–80. *See also* libraries
Caldwell, Charles, 9, 12–13, 52, 118–26, 163, 255; letter of, 121*f*, 122; "New Views of Penitentiary Discipline and Moral Education and Reformation of Criminals" (1831–32) of, 146. *See also* phrenology; race
Calthorpe Islands, 46. *See also* Arctic
Canada, 116. *See also* Americas; North America
Canadian Phrenological and Psychological Magazine, 151. *See also* Canada; periodicals

Canton, 4. *See also* China
Cape Colony, 1–2, 9, 23–24, 151, 154; Boors in the, 176; British in the, 169, 176; Hottentot skulls collected in the, 20; phrenological settlers in the, 168–78, 313n92; schools of the, 141. *See also* Cape Town; South Africa
Capen, Nahum, 81–82. *See also* publishing
Cape of Good Hope Literary Gazette, 170–72. *See also* periodicals
Cape Town, 1, 4, 10, 78, 80, 168–73, 176–78. *See also* South Africa
capitalism, 132; of print, 254
Carey, William, 139. *See also* missionaries
Carlyle, Thomas, 34
casts, 51–77, 60*f*; manufacture of, 81, 289n56; plaster, 162, 250, 260, 292n120. *See also* busts; material culture; phrenology; skull; stereotype
Catechism of Phrenology, A (1831), 171. *See also* Edinburgh; phrenology
Catholic Telegraph, 73. *See also* periodicals
Ceylon: British power in, 24–33, 41, 280n50; collection of skulls in, 20–33, 36, 40. *See also* Kandyans; South Asia
Chambers, Robert: *Vestiges of the Natural History of Creation* (1844) of, 3, 34, 79, 185
Chambers's Edinburgh Journal, 96, 108–9. *See also* periodicals
Channing, William Ellery, 122, 126. *See also* antislavery movement
Chapman, Maria Weston, 122–23. *See also* antislavery movement
Chapon, Léon-Louis, 228. *See also* photography; phrenology
character, 2, 116–18, 145, 147; African, 56, 58–63, 72, 76, 120, 122, 129–30, 177, 261; African American, 129; Chinese, 247; European, 25, 43, 63; Indian, 165, 181; Inuit, 42–44; Kandyan, 24–29; language of, 122; national, 91; Native American, 120; in political thought, 118; racial, 20; of slaveholders, 122; Toda, 231, 235; Xhosa, 174, 178. *See also* human nature; phrenology
Chartists, 116–17, 123. *See also* politics; reform
Child, David Lee, 68–69. *See also* antislavery movement
Child, Lydia Maria, 70. *See also* antislavery movement
China, 2, 9, 14, 18, 154, 261; New Culture Movement in, 243, 246; phrenology in, 241–49, 256; Republican, 245, 249, 256; Revolution of 1911 in, 243, 246, 257; sketch of a servant boy of, 166*f*; skulls in, 165. *See also* Asia
Christian Examiner, 97, 106. *See also* periodicals
Christianity, 184–85, 255. *See also* religion
civilization: classical, 34; Egyptian, 33–35, 40; Peruvian, 103
Clay, Senator Henry, 126. *See also* politics
Coates, Benjamin Hornor, 100. *See also* physicians
Cobbold, Richard, 241–42. *See also* missionaries
Cole, Sir Galbraith Lowry, 169. *See also* Cape Colony
Collins, John, 92–93, 97. *See also* lithography; publishing
colonialism, 9, 15, 20, 41; British, 116, 139; European, 61; French, 61, 68; photography and, 199; violence of, 35. *See also* empire; imperialism; slavery; violence
Combe, Andrew: *Principles of Physiology* (1838) of, 79. *See also* physiology
Combe, George, 1–9, 12–16, 20–23, 42, 50–69, 75, 82–109, 115, 119–36, 121*f*, 141–49, 153, 159, 249, 254, 276n42; *The Constitution of Man* (1828) of, 2, 9, 16, 79–83, 98, 108, 116, 124, 130–32, 135, 142–44, 155, 184, 241, 254, 256–57, 273n3, 316n163; *Elements of Phrenology* (1835) of, 79, 82, 98, 170, 202; *Essays on Phrenology* (1819) of, 78–80, 137; *Lectures on Phrenology* (1839) of, 67*f*; *Moral Philosophy* (1842) of, 143; *Notes on the United States of North America* (1841) of, 120, 122, 128, 134–35, 178; *Outlines of Phrenology* (1824) of, 98; *Remarks on National Education* (1845) of, 146; *System of Phrenology* (1834) of, 77, 79, 81–82, 108, 120, 188, 245. *See also* phrenology
Commercial Press, 241, 244–45, 261. *See also* China; publishing
communications, 2–5, 257, 274n14; global, 77, 181–82; material realities of, 9, 145; technologies of, 18, 77. *See also* culture; letters; newspapers; publishing; society; transportation
Cooper, Frederick, 5

Cooter, Roger, 2–3, 5–6
Copenhagen, 1. *See also* Europe
Cosimi, Cosimo, 53. *See also* casts
Cox, Abram, 65. *See also* casts; museum
Cox, Robert, 23, 42, 63. *See also* Edinburgh Phrenological Society
Craig, Edward, 33
Craigie, David, 107. *See also* periodicals
Cramer, Charles, 114. *See also* geology
craniology, 1–2, 210–17; collection of, 10; philosophical, 106; plates of, 87*f*, 88, 93. *See also* anatomy; brain; busts; phrenology; skull
criminology, 110. *See also* science
culture: American, 112; British, 32; civilized aspects of, 31–32; primitive, 321n63; print, 154, 159, 170, 179; reading by phrenologists of, 41; scientific, 170. *See also* communications; libraries; religion; society
Cuvier, George, 4, 58, 62, 289n49; *Histoire Naturelle des Mammifères* (1824–42) of, 58. *See also* anatomy; race
Cuyer, Édouard, 226, 228. *See also* photography; phrenology

Daguerre, Louis, 193, 202. *See also* photography; printing
Daily Chronicle, 150. *See also* newspapers
Dain, Bruce, 84. *See also* history
Darwin, Charles, 87, 170, 213, 222, 225, 231, 249–51, 301n16; *The Descent of Man* (1871) of, 200, 218–19, 323n129; *The Origin of Species* (1859) of, 2, 222–23, 225. *See also* natural history
Das, Cally Coomar, 178–79, 180*f*, 182–85, 255, 261; *General Reflections on Christianity* (1845) of, 185. *See also* Calcutta; *The Pamphleteer*
Das, Radhaballav, 316n163; *Manatatva Sarsangraha* (1849) of, 187*f*, 188, 189*f*, 190. *See also* phrenology
Daston, Lorraine, 112–13
Davy, John, 25, 27; *An Account of the Interior of Ceylon* (1821) of, 25, 26*f*
death, 20, 24, 42, 49; cultural significance of, 278n4; Inuit view of, 41–47; Nubian dance of, 38*f*; phrenological fascination with, 40; rituals of, 20, 29–31, 45–48, 285n172. *See also* burial; culture
Delhi, 212. *See also* India
democracy, 161. *See also* politics
Denmark, 52. *See also* Europe
de Quatrefages, Jean Louis Armand, 198, 225–26, 228–31, 233–34, 239; *Charles Darwin et ses Précurseurs Français: Étude sur le Transformisme* (1870) of, 225; *Histoire Générale des Races Humaines* (1889) of, 258, 259*f*; *Hommes Fossiles et Hommes Sauvages* (1884) of, 225–26, 227*f*, 228, 228*f*, 230*f*, 232*f*, 233–34; *L'Unité de l'Espèce Humaine* (1861), 225. *See also* anthropology
Dessalines, Jean-Jacques, 57. *See also* Haiti
Dickens, Charles, 134
Dinwiddie, James, 138. *See also* science
Doolittle, Justus, 164–65, 166*f*, 257. *See also* missionaries
Doornik, Elisa, 9
Douglass, Frederick, 129–30, 255; *Life and Times of Frederick Douglass* (1881) of, 130. *See also* antislavery movement
Drummond, David, 10, 140. *See also* Calcutta; schools
Dublin, 124. *See also* Ireland
Dumoutier, Pierre, 12, 59, 61–63. *See also* phrenology
D'Urban, Benjamin, 174. *See also* Cape Colony
Dutt, Akshay Kumar, 184–85, 255; *Bahya Bastur Sahit Manab Prakritir Sambandha Bicar* (1851, 1853) of, 184. *See also Tattvabodhini Patrika*

East Asia, 243, 245–46. *See also* Asia
Eastern State Penitentiary, 134–36. *See also* prisons; United States
East India Company, 9–10, 12, 15, 50, 77, 79, 137–38, 140, 169, 178, 181, 183, 201, 203, 245, 261; Seminary of the, 202. *See also* Britain
East Timor, 13. *See also* Asia
Eclectic Journal of Medicine, 106. *See also* periodicals
Edinburgh, 5, 9, 12, 16, 27–37, 41, 52–56, 64, 73–80, 91, 115, 122, 126, 130–31, 136, 152–53, 299n190. *See also* Britain; Scotland
Edinburgh Medical and Surgical Journal, 107–8. *See also* periodicals
Edinburgh New Philosophical Journal, 107. *See also* periodicals

Edinburgh Phrenological Society, 20–23, 33–36, 42, 45, 50–54, 63–68, 82, 90–91, 115, 173; catalogue of the, 23, 63, 278n3; collection of skulls of the, 285n176; museum of the, 89. *See also* Edinburgh; phrenology
Edinburgh Review, 10, 145. *See also* periodicals
education, 116; national, 182–84; the penny post and, 115, 142; phrenological approach to, 136–45, 316n163; physical, 140–41; rural, 136–45. *See also* reform; schools
Edwards, William Frédéric, 113. *See also* ethnology
Egypt, 13, 20, 23, 32–41, 282n101, 284n137; collection of heads in, 32–41; grave robbers in, 39, 45; Greek and Roman accounts of, 35–37; hieroglyphics of, 34, 36, 38; modern, 34, 38, 40–41, 49; mummies of, 32–40; Ottoman, 33–34, 39; temples of, 39. *See also* Africa; antiquities; Ottoman Empire
empire, 3, 10, 91, 283n128; American, 56, 168; British, 12, 23–30, 56, 87–88; colonial cities of, 8; European, 9; French, 56. *See also* British Empire; colonialism; imperialism
environment, 37, 41–49. *See also* geography
epistemology, 8. *See also* knowledge; philosophy; scienceEstlin, John, 54. *See also* phrenology
Ethnological Society, 114, 194, 200, 207. *See also* ethnology; London
ethnology: American, 85, 106; anatomical aspects of, 200; British, 85–93, 95–96; collecting in, 279n25; as a discipline, 88–90; in India, 198–222, 235; reception of the *Crania Americana* in, 108; in South America, 109; visual language of, 102. *See also* anthropology; phrenology; Prichard, James Cowles; race; skull
eugenics, 110. *See also* race; science
Europe, 1–4, 8–10, 12, 15, 39, 100, 111, 133–34, 141, 147, 151, 160, 171; death rituals of, 47–48; phrenological lectures in, 50; publishing in, 96–97, 153; race in, 21; radicalism in, 169; reception of the *Crania Americana* in, 90, 96, 101, 105; reformism in, 20; working class in, 123
Fabian, Ann, 84. *See also* history
Fairbairn, John, 170, 173–74, 177–78. *See also* liberalism; *South African Commercial Advertiser*
Falconer, John, 199, 239
Felix, Orlando, 32–39, 49. *See also* Egypt
Fergusson, James: *Rude Stone Monuments in All Countries* (1872) of, 234
First Anglo-Burmese War, 139. *See also* Britain; India
First Anglo-Sikh War, 203. *See also* Britain; India
Flanders, J. J., 109
Flourens, Marie-Jean-Pierre, 225; *Examen du Livre de M. Darwin* (1864) of, 225. *See also* physiology
folios. *See* books
Follitt, John, 16
Forbes, John, 107. *See also* periodicals
Forten, James, 129. *See also* antislavery movement
Fourier, Joseph, 33
Fowler, Lorenzo, 2, 10, 53, 55, 73–76, 80, 124, 154–65, 168–69, 193–96, 257; *Self-Instructor in Phrenology* (1875) of, 10. *See also American Phrenological Journal*; periodicals; publishing
Fowler, Orson, 2, 10, 53, 55, 73–76, 80, 115, 154–65, 168–69, 193–96, 257; *Fowler on Matrimony* (1841) of, 179; *Self-Instructor in Phrenology* (1875) of, 10; *Synopsis of Phrenology* (1848) of, 202. *See also American Phrenological Journal*; periodicals; publishing
Fox, William Johnson, 182; *Lectures, Addressed Chiefly to the Working Classes* (1846) of, 182–83. *See also* education; reform
France, 2, 6, 53, 74, 76, 110, 115, 134, 153, 225, 229; *Les Bons Exemples* (1858) of, 58; *Description de l'Égypte* (1809–29) of, 33–34; economy of, 61; global empire of, 33; government of, 153, 229; July Monarchy in, 116, 225; Orléanist regime in, 61; Third Republic in, 200. *See also* Europe
Franco-Prussian War, 222. *See also* France; Prussia
Frederick Douglass' Paper, 129. *See also* periodicals
Freedom's Journal, 70. *See also* periodicals
French Revolution, 57. *See also* France

Freud, Sigmund, 244. *See also* psychology
Friends' Intelligencer, 72. *See also* periodicals
Fugitive Slave Act (1850), 75–76. *See also* slavery
Fuller, John, 94. *See also* publishing

Galison, Peter, 112–13
Gall, Franz Joseph, 1–2, 80, 83, 226. *See also* phrenology
Gall, Franz Joseph, and Spurzheim, Johann: *Anatomie et Physiologie du Système Nerveux* (1810) of, 83, 108, 110. *See also* atlas
Gamarra, Agustín, 103. *See also* Peru
Garrison, William Lloyd, 71, 130, 146, 162. *See also* abolitionists; antislavery movement
Gentleman's Journal, 150. *See also* London; periodicals
geography, 5, 18, 77, 112, 116; of Egypt, 37; historical, 328n37; human, 20–21, 86; Inuit knowledge of, 45–48; moral, 118, 123; political, 8, 117; of science, 8, 249, 256; social, 8. *See also* environment
geology, 88, 114. *See also* science
Georgia Citizen, 157. *See also* newspapers
Germany, 6, 82, 147, 245–46. *See also* Europe
Gilchrist, John, 139
Glasgow, 32, 54, 65. *See also* Scotland
Glasgow Mechanics' Institute, 109. *See also* Glasgow
Gloucester sea serpent, 96
Godlonton, Robert, 172, 174, 191. *See also* *Graham's Town Journal*
Graham's Town Journal, 8, 172, 174, 178; "Government Notice" following the 1834–35 Xhosa War (1835) in the, 175*f*. *See also* newspapers
Great Moon Hoax, 96
Gregory, William, 63. *See also* Edinburgh Phrenological Society
Grigg, Henry: *Manual of the Nílagiri District in the Madras Presidency* (1880) of, 235

Haeckel, Ernst, 244–46; *Miracle of Life* (1904) of, 245–46
Haiti: independence of, 72; memories of, 56–62; progress of, 70; refugees from, 71, 73; violence in, 56–57. *See also* Americas; colonialism; slavery
Haitian Revolution (1791), 56, 62, 64–71. *See also* Haiti
Harcourt, William, 87–88. *See also* British Association for the Advancement of Science
Hawaii, 19–20, 164. *See also* United States
Hawaiian Spectator, 152. *See also* Hawaii; periodicalsHecker, Friedrich, 147. *See also* revolution
Heidelberg, 153. *See also* Germany
Herschel, John, 170. *See also* science
Hicks, Reverend W. W., 165, 191. *See also* missionaries
Hill, Rowland, 115
Hinduism, 184–85, 255. *See also* religion
historiography, 14, 109, 116, 151, 239, 327n26; of science, 116, 249, 275nn28–29. *See also* history; history of science
history: of African geography, 286n179; of anthropology, 114; of the book, 79, 84–85, 112; of colonial photography, 220, 239; of communication, 257; global, 109, 154, 191, 222, 241, 253–54, 256, 274n25, 275nn28–29, 328n40; imperial, 42; intellectual, 15–16, 253–54, 256; of mankind, 21; of metrology, 327n33; oral, 48, 285n175; periodization in the study of, 274n15; of photography, 199, 319n18, 325n169; political, 116, 254; of political thought, 116–17; of publishing, 84–85, 325n167; racial, 110, 116, 200; of scientific atlases, 113. *See also* historiography; history of science
history of science, 2–6, 10, 18, 49, 79, 84, 112, 171, 249–54, 329n57; external, 251; geography of the, 112–13; global, 3–6, 8, 13–18; internal, 251. *See also* historiography; history; science
Hitchcock, Romyn: *The Ainos of Yezo* (1892) of, 222
Hobart Town Courier, 152. *See also* periodicals
Hodges, Sarah, 5, 275n28
Hodgkin, Thomas, 86, 93
Holland, 123. *See also* Europe
Howe, Samuel Gridley, 65, 120, 134–35, 142–44, 301n16. *See also* Boston Phrenological Society; education
human nature: genealogy of, 48; investigation of, 6. *See also* character; race
Hume, David, 79. *See also* philosophy

Hunt, James, 200, 259. *See also* anthropology
Hunt, Lynn, 253. *See also* history
Hurlbut, Elisha, 135–36
Hussy, William Cobb, 10. *See also* phrenology
Huxley, Thomas, 207–8, 220, 240; *Evolution and Ethics* (1893) of, 244. *See also* anatomy

Illustrated Australian Phrenological, Physiognomical and Hygienic Magazine, 151. *See also* Australia; periodicals
illustration: in books, 17, 99, 111; natural history, 97; in a review, 101–2, 105. *See also* books; images; lithography; publishing; woodcuts
images: production of, 112–13. *See also* illustration; lithography; photography; woodcuts
imperialism, 15, 151, 154, 196; American, 164, 196; British, 15, 33, 196; cultural, 33; French, 33; nationalism and, 161. *See also* colonialism; empire; nationalism; slavery
Imperial Mineralogical Academy, 114. *See also* Russia
India, 2, 6, 9, 11–12, 15, 18, 52, 77, 115, 130–34, 154, 179, 229, 316n153, 322n83; British colonialism in, 116; educational policy in, 182–83; government of, 203, 235; miniatures on ivory from, 167*f*; natives in, 234–40; North, 32, 133; photography in, 199, 202–10, 319n18; phrenology in, 234–40; race in, 21; Rebellion of 1857 in, 168, 203; rural schools in, 136–45; ships in, 178–92; Siege of Delhi in, 203; society in colonial, 178–92; South, 28; Thugs in, 20, 32, 134; widow burning in, 20. *See also* South Asia
Indian Antiquary, 236. *See also* periodicals
Indian Journal of Medical and Physical Science, 77. *See also* periodicals
Indian Removal Act (1830), 54. *See also* United States
indigenous peoples, 111. *See also* anthropology
industry, 91. *See also* science
Inuit: burial of the, 44–45, 285n176; collection of the skulls of the, 18, 20–21, 36, 41–50; death rituals of the, 45–47; dogs in the life of the, 47; geographic knowledge of the, 46–48; human genealogy of the, 48; mourning rituals of the, 47; religious beliefs of the, 45–46; stone cairns (*inuksuit*) of the, 45–48. *See also* Arctic
Ireland, 2, 10, 34; Catholic emancipation in, 116; poverty in, 124. *See also* Britain

Jackson, President Andrew, 106
Jamaica, 146. *See also* Americas; slavery
James De Ville, 137. *See also* busts; casts; London
Jameson, Robert, 107. *See also* natural history
Janet, Paul: *La Famille. Leçons de Philosophie Morale* (1856) of, 58. *See also* France
Janssen, Jules, 229–31, 232*f*. *See also* astronomy
Japan, 2, 193, 222, 233, 240, 245–46; the Ainus of, 233. *See also* Asia
Java, 9. *See also* Asia
Jenkins, Edward, 155. *See also* New York City; publishing
Journal de la Société Phrénologique de Paris, 8, 62, 64–65, 152–53. *See also* Paris; periodicals
Journal des Savants, 225. *See also* periodicals

Kandyans, 23–32, 41; Buddhism of the, 29–31, 281n71, 281n74. *See also* Ceylon; Keppetipola
Kant, Immanuel, 65. *See also* philosophy
Kashmir, 212. *See also* India
Kennedy, James, 107
Keppetipola, 23–24, 26–32, 41; skull of, 25, 26*f*, 29, 32–33, 41, 49, 279n28. *See also* Ceylon; Kandyans
knowledge: colonial, 49; phrenological, 64; racial, 51; scientific, 86; transatlantic exchange of, 299n190. *See also* epistemology; science
Knox, Robert, 259. *See also* anthropology
Kreitmayer, Maximilian, 53–54. *See also* busts

Lacerda, João Baptista, 111–12, 114, 300n214. *See also* Brazil; physicians
Ladies' Repository, 106. *See also* periodicals
Lamarck, Jean-Baptiste, 58, 225. *See also* natural history; race
Lamprey, John, 207. *See also* ethnology

language: of nations in the sciences, 329n55; of phrenology, 116, 122, 129, 134, 140; and race, 88; of revolution, 147; structuralist, 10; visual, 88, 90–91, 102
Le Caméléon, 69. *See also* Paris; periodicals
Lee, Hannah, 71. *See also* antislavery movement
Leeds Anti-Slavery Association, 64. *See also* Britain; slavery
Le Journal du Havre, 69. *See also* Paris; periodicals
Le Tour du Monde, 229. *See also* periodicals
letters, 8, 14–15, 17, 115–49; manuals on the writing of, 303n58; as material objects, 117, 145; as a practice of reform, 117–18. *See also* communications; material culture
liberalism, 117, 171; Indian, 316n151; nationalism and, 316n153; popular science and, 171. *See also* nationalism; political thought
Liberator, 129, 162. *See also* antislavery movement; press
Liberia, 126–28. *See also* Africa
libraries, 79–80; colonial, 178; of mechanics, 109; medical, 110; of the schools, 143–44. *See also* books; material culture
Lightman, Bernard: *Victorian Popularizers of Science* (2007) of, 190–91. *See also* science
Linnean Society, 109. *See also* London; science
lithography: cost of, 99; of the *Crania Americana*, 87*f*, 90–93, 97, 99–101, 105, 111, 113, 261; limestone for, 97; in Philadelphia, 93, 97, 99. *See also* illustration; images; material culture; printing; publishing; woodcuts
Liverpool, 3–4, 99. *See also* Britain
Lockyer, Norman, 229. *See also* astronomy
London, 4–5, 15–18, 34, 51, 77, 98–99, 109, 112, 154, 169–70, 193, 207, 240; publishing in, 79, 99, 234, 238; working-class audiences in, 191. *See also* Britain; publishing
London and Foreign Quarterly Review, 96. See also periodicals
London and North Western Railway, 4. *See also* transportation
London Stereoscopic Company, 213. *See also* photography
Longmans and Green, 217–18, 220. *See also* publishing
Lord Prudhoe (Algernon Percy), 35–40. *See also* Egypt
Louverture, Toussaint, 57, 70. *See also* Haiti
Lovett, William, 116. *See also* politics
Lubbock, John: *Origin of Civilisation and the Primitive Condition of Man* (1870) of, 200. *See also* anthropology
Lucassi, Charles, 53. *See also* casts
Lucknow, 212. *See also* India
Lyell, Charles: *Principles of Geology* (1830) of, 34
Lyon, Captain Edward, 203. *See also* photography
Lyon, Captain George, 41–47; "An Esquimaux Grave," in *A Brief Narrative of an Unsuccessful Attempt to Reach Repulse Bay* (1825) of, 46*f*. *See also* Arctic

Macartney, Henry, 169–78; advertisement for the phrenological lecture of, 175*f*. *See also* Cape Colony; phrenology
Mackenzie, George, 49–50; *Illustrations of Phrenology* (1820) of, 170, 178. *See also* phrenology
Mackenzie, William, 50. *See also* East India Company
Maconochie, Alexander, 131–36, 145–46; *The Management of Prisons in the Australian Colonies* (1838) of, 135–36. *See also* prisons
Madras, 77, 199, 201–3, 209–10, 234, 236, 240; publishing in, 238. *See also* India
Madras Photographic Society, 203. *See also* photography
Magendie, François, 10. *See also* physiology
Mann, Horace, 136, 142–45, 153. *See also* education; reform
Mantell, Gideon, 96. *See also* geology
Maori, 14. *See also* Pacific; race
Marshall, Henry, 23–32
Marshall, William Elliot, 202–22, 226, 228–29, 233–35, 240; *A Phrenologist amongst the Todas* (1873) of, 198–240, 205*f*, 213*f*, 214*f*, 224*f*, 231, 260, 314n129, 320n35, 323n109, 323n129. *See also* phrenology
Marsh, Capen and Lyon, 81–82, 153. *See also* publishing
Martineau, Harriet: *Society in America* (1837) of, 125. *See also* abolitionists

Maryland Colonization Society, 72–73. *See also* slavery
Massachusetts Board of Education, 143–44. *See also* education
Massachusetts School Library, 143–44. *See also* education; libraries
Massachusetts State Prison, 146. *See also* prisons
material culture, 8–16, 200, 249, 260; global history of, 12; and performance, 15–16; and possession, 22; and racial politics, 13; of time, 254. *See also* books; busts; casts; materiality; museum; phrenology; skull
material exchange, 1, 8–9, 51–77; global, 18, 79, 199, 210; transatlantic, 78–112. *See also* material culture; materiality; phrenology
materialism: Lamarckian, 200; nineteenth-century, 12; phrenological, 141, 169, 181; in religion, 184. *See also* philosophy; phrenology
materiality, 37, 41, 117; of letters, 124, 145; of the mind, 91, 141; and practice, 275n29. *See also* material culture; materialism; material exchange
Mauritius, 28. *See also* Africa
Maxwell, W. D., 15–16, 150–51. *See also* phrenology
Medical and Physical Society of Calcutta, 140. *See also* Calcutta
Medico-Chirurgical Review, 107–8. *See also* periodicals
Melbourne, 51–54. *See also* Australia
Melbourne Public Library, 80. *See also* libraries
Mercantile Library of New York, 66, 79–80. *See also* libraries
Mercantile Library of San Francisco, 79. *See also* libraries
Mercer, Charles, 126. *See also* politics
metaphysics, 86. *See also* philosophy
mind: material cultures of the, 8–10; materiality of the, 91, 141; sciences of the, 116. *See also* brain; materiality; phrenology
Minerva, 152. *See also* periodicals
missionaries: in the Arctic, 285n165; in the Cape Colony, 176; in China, 154, 164–65, 191; in India, 139, 165, 168, 183, 191. *See also* religion
Mittermaier, Karl Joseph Anton von, 136
modernism, 243. *See also* nationalism; political thought
Moffat, Alexander, 21. *See also* surgery
Mohammad Ali, 34, 38–41, 282n101. *See also* Egypt
Moletsane, 176–77. *See also* Cape Colony
Morrison, John, 135. *See also* prisons
Morton, James, 95. *See also* Morton, Samuel George
Morton, Samuel George, 4, 9–10, 50, 83–104, 106–11, 113, 297n137; *Crania Americana* (1839) of, 9, 21, 50, 83–114, 84*f*, 87*f*, 92*f*, 94*f*, 95*f*, 220, 261; editions of the *Crania Americana* of, 296n104; printing of the *Crania Americana* of, 297n133. *See also* craniology; ethnology; lithography; phrenology; publishing; skull
Mott, Lucretia, 124–29, 125*f*, 146, 301n16. *See also* abolitionists
Murchison, Roderick, 86; *Silurian System* (1839) of, 113. *See also* British Association for the Advancement of Science; geology
Murray, John, 42. *See also* Arctic
museum, 20–24; phrenological, 51, 59, 89; plaster casts in the, 292n120; private status of, 22; public status of, 22. *See also* busts; casts; material culture; skull
Muséum d'Histoire Naturelle, 225, 229, 231, 233, 240. *See also* France; museum

Napoleonic Wars, 53. *See also* Europe
Natal Witness, 176–77. *See also* Cape Colony; newspapers
nation: identities of the, 151; India as a, 179, 181; theory of the, 255; understandings of, 52. *See also* nationalism; race
nationalism, 84, 110, 112, 151, 154, 160, 163, 181, 190, 329n57; black, 129; Chinese, 243; colonial, 255; education and, 184; European, 255; and imperialism, 161; Indian, 236, 241; and liberalism, 316n153. *See also* imperialism; liberalism; nation; political thought
Native Americans, 13–14, 120. *See also* Morton, Samuel George; race
Nat Turner rebellion (1831), 71
natural history: American, 87, 99, 104; collections of, 279n20; the *Crania Americana* as a work of, 107–8; and natural philosophy, 91; specimens of, 41; study

of mankind in, 89. *See also* illustration; naturalists; phrenology; race; science
naturalists, 58, 100–1; American, 86, 93; European, 96; French, 58, 61. *See also* natural history
natural theology, 106. *See also* philosophy
Nena Sahib, 165, 167–68. *See also* India
New Haven, 100, 105. *See also* United States
New Holland, 21. *See also* Australia
New Orleans, 70, 118. *See also* United States
New Orleans Lyceum, 118–19. *See also* New Orleans
New South Wales, 4, 51, 77, 79, 131–33, 151; prisons of, 116, 131–36, 146. *See also* Australia
newspapers, 4, 96; abolitionist, 291n114; South American, 299n202. *See also* communications; periodicals; press; publishingNew York Anti-Slavery Office, 73. *See also* antislavery movement
New York City, 1–2, 10–15, 50–54, 66, 70–76, 80–81, 96, 115, 129, 156–59, 161, 168, 193, 196, 248. *See also* publishing; United States
New Yorker, 68. *See also* periodicals
New York Mechanics' Institute, 109. *See also* New York City
New York Mercantile Association, 83. *See also* New York City
Nicholson's Journal, 4–5. *See also* periodicals
Nilgiri Hills, 199–204, 209–18, 220–21, 229, 231, 235, 237–40. *See also* India; Marshall, William Elliot
Norfolk Island, 131–35, 137, 151. *See also* prisons
North Africa: race in, 21; slave trade in, 35. *See also* Africa
North America, 134. *See also* Americas
North American Review, 9, 91, 99, 103, 106, 157. *See also* periodicals
Northwest Passage, 23, 41–42, 48. *See also* Arctic
Nottingham, 64. *See also* Britain
Nubia, 38–39; death dance in, 38*f*. *See also* Africa

O'Connell, Joseph, 250
Ogle, Vanessa: *The Global Transformation of Time* (2015) of, 253–54
Ohio, 106. *See also* United States
O'Neil, Anthony, 52–54, 64. *See also* casts
O'Neil, Luke, 52–54, 64. *See also* casts
Orientalism, 283n103. *See also* Egypt
Ottewill, Thomas, 208, 208*f*. *See also* photography
Ottoman Empire, 33–35; military bodyguards of the, 38. *See also* Egypt
Owen, Robert Dale, 64, 182; *Tracts on Republican Government and National Education* (1840) of, 183. *See also* politics; reform

Pacific, 3, 9–10, 12–13, 42, 52, 130, 134, 164. *See also* Asia; Southeast Asia
Pacific Islanders, 13. *See also* Pacific; race
Pamphleteer, 179, 181–83, 185–87, 190; advertisement for Radhaballav Das, *Manatatva Sarsangraha* (1850) of the, 187*f*; title page (1850) of the, 180*f*. *See also* periodicals; phrenology
Paris, 2, 4–5, 9–12, 18, 51–62, 76, 80, 83, 228; an American in, 68–73; colonial photography in, 231; Faculty of Medicine in, 110; National Assembly in, 57; phrenology in, 58–59, 66, 226; publishing in, 79, 83, 152. *See also* France
Paris Phrenological Society, 11, 59. *See also* Paris; phrenology
Parley's Magazine, 73. *See also* periodicals
Parry, Captain William, 41–50. *See also* Arctic
Paterson, George Murray, 12, 137–42; "On the Phrenology of Hindostan" (1826) of, 140. *See also* education; schools; surgery
Patrocínio, José do, 110–11. *See also* Brazil; race
Pearce, Alexander, 10; the skull of, 11*f*
Pearce, Susan, 49
Pedro II, 110. *See also* Brazil
Pennsylvania Freeman, 69, 71–73. *See also* antislavery movement; periodicals
Penny Magazine, 4. *See also* periodicals
People of India, The (1868–75), 204, 239. *See also* atlas; photography
Périer, Honorine, 123, 303n54. *See also* Spurzheim, Johann
periodicals, 4–5, 8, 11–12, 73, 99, 150–92, 198; cheap, 108; European, 170; foreign articles in, 309n9; phrenological, 178; scientific, 100, 107. *See also* books; communications; newspapers; publishing

Perkins School for the Blind, 142. *See also* Boston; education
Peru, 103, 109. *See also* South America
Petit, Jahangir Bomanji, 261. *See also* India; phrenology
Pettigrew, Thomas, 40. *See also* surgery
Philadelphia, 3–4, 10–11, 50–56, 69–70, 76, 79, 86, 89, 91, 94, 98–99, 112; abolitionists in, 70–72, 124; lithography in, 93, 97; periodicals in, 72, 97, 106; publishing in, 83, 93, 104–5, 107, 113, 155; race riots in, 146; unrest in, 126. *See also* United States
Philadelphia Museum, 66. *See also* museum; Philadelphia
Philadelphia Phrenological Society, 124. *See also* Philadelphia; phrenology
Philippines, 164, 194–96. *See also* Asia; Southeast Asia
Phillips, John, 94–95. *See also* Morton, Samuel George
Philosophical Society of Australasia, 4. *See also* philosophy; Sydney
Philosophical Transactions, 4. *See also* periodicals; Royal Society
philosophy: Baconian, 86; materialist, 4, 256; mental, 6, 316n151; moral, 12; natural, 91; Newtonian, 86; of phrenology, 143. *See also* metaphysics; natural theology; phrenology
Photographic Society of Bengal, 203. *See also* Bengal; photography
photography, 193–240; anthropological, 240; anthropometric, 207, 233; authority of, 321n64; carbon process of, 217–22; collodion, 209–10; colonial, 198, 203–4, 208–13, 216, 231, 235, 240; commercial, 213–15; the daguerreotype in, 203, 209; as image production, 113; in India, 199, 202–22, 216*f*, 319n18; as a material object, 319n20; mountain, 214; the Ottewill Folding Box Camera in, 208*f*; phrenological, 16–18, 198, 202, 205*f*, 222, 225, 237, 239; portable darkroom in, 209–10; primitive, 202–10, 205*f*; reproduction of, 18, 226; and Rouch's Registered Portable Dark Operating Chamber, 211*f*; visual technology of, 193. *See also* images; material culture
Phrenological Association, 88. *See also* phrenology
Phrenological Journal, 6, 9, 11, 21–29, 35–45, 50–54, 63–64, 73, 77, 88–89, 98, 107–8, 115, 122, 127, 134–36, 146, 151–53, 159, 178, 194, 196–97, 255. *See also* periodicals; phrenology
Phrenological Journal and Life Illustrated: "A Negrito" (1870) in the, 195*f*. *See also* periodicals; phrenology
Phrenological Journal and Science of Health: Photograph of Malay Chiefs, Mindanao (1899) in the, 197*f*; Sketch of a "Chinese Servant Boy" (1870) in the, 166*f*. *See also* periodicals; phrenology
Phrenological Magazine, 197–98. *See also* periodicals; phrenology
Phrenological Museum of Melbourne, 51–52. *See also* phrenology
phrenology: ancient Egyptian heads in, 40–41; Bengali, 178–92, 202, 251, 315n139; British, 202; business empire of, 154–68; in California, 161; chart of, 153, 189*f*, 242*f*, 317n187; in China, 241–49; and the colonial state, 204; conservative, 123; converts to, 83; critics of, 15, 103, 107, 129–30, 145, 206; demise of, 273n6; as a discipline, 89–90, 112; evolutionary theory and, 199, 210–13, 225–26, 319n29; fugitive, 73–77; and gender, 188; global history of, 1–6, 8, 13–16, 18, 51–77; goal of, 21; human origins in works of, 201–2; international popularity of, 12, 51; lecturing in, 290n78; measurements of, 206, 226; museums of, 51, 59, 89; new audiences for, 80–81, 170; paraphernalia, 52; physical condition of human remains in, 41; on the plantation, 118–24; politics of, 5–6, 86; as popular spectacle, 159, 183; reform and, 115–49; as a science, 2, 91, 118, 150; shows of Egyptian mummies in, 40; social history of, 5; theories of race in, 51–77; and the Third Republic, 222–34. *See also* anthropology; books; brain; busts; casts; Combe, George; craniology; ethnology; material culture; material exchange; mind; natural history; philosophy; photography; physiology; prisons; race; science; skull; slavery; violence
physicians, 1, 4, 9, 12, 15, 52, 54, 58, 85, 100, 103, 110; provincial, 86; in South America, 109. *See also* surgery

physiology, inquiry of, 102. *See also* anatomy; craniology; phrenology
Piddington, Henry, 10
Pingsheng, Feng: *Guxiangxue* (1923) of, 241, 242*f*, 244, 246–49, 247*f*, 248*f*, 258, 264, 325n2. *See also* China; phrenology
Pinney, Christopher, 199, 239
Pinto, Antonio Pereira d'Araújo, 110. *See also* Brazil; phrenology
political thought: global, 183; history of, 116–17; Indian, 183, 316n156; nineteenth-century, 118, 183, 255. *See also* politics
politics, 91; change in, 147; global, 2, 118; imperial, 49; of the mind, 116; of oppression, 16; of phrenological photography, 196; of phrenology, 5–6; postwar, 9; of publishing, 168; racial, 13, 15–16; radical, 222; reformist vision of, 4, 115–49; science and, 18, 116, 249. *See also* democracy; political thought; reform; revolution; society
Popular Science Review, 219. *See also* periodicals
press: American, 96; Bengali, 315n143; British, 99; foreign, 143; freedom of the, 153, 169, 171; periodical, 150–92. *See also* newspapers; periodicals; publishing
Prichard, James Cowles, 85–91, 93–98, 102, 106–7, 111, 113, 259–60; Dedication in the "foreign edition" of the *Crania Americana* (1839) to, 95–96, 95*f*; ethnological project of, 85–91, 95; "The Extinction of Some Varieties of the Human Race" (1839) of, 85; philological interests of, 91; *Researches into the Physical History of Mankind* (1813, 1826, 1836) of, 88, 95, 95*f*, 102; *Treatise on Insanity* (1835) of, 89. *See also* ethnology; race
Pringle, Thomas, 78, 80, 173. *See also* antislavery movement; phrenology
printing: and circulating, 87; copies in, 217–22; industrial world and, 83; material culture of, 169; political culture of, 169; practices of, 85, 93; technologies of, 81, 217; transatlantic, 105. *See also* books; illustration; industry; lithography; press; publishing
prisons: American, 136; colonial, 136; discipline in the, 116, 131, 134–36; global, 131–36; reform of the, 115, 118, 131–36. *See also* reform
Prussia, 76, 116, 134. *See also* Europe
pseudo-science, 10. *See also* science
psychology, 243–44; national, 246
publishing, 2–4, 8, 77–81, 93, 317n181; American, 81, 85, 96–97, 152; British, 81, 96–99; cheap, 80–81; European, 96–97, 152; evangelical, 80, 317n177; fractured environment of, 163; as an international business, 81; local networks of, 105; politics of, 168; scientific, 80, 258, 292n12, 312n74, 312n80; technologies of, 168; transatlantic, 84–85, 98, 104–5, 109, 292n12; unscrupulous, 9. *See also* books; illustration; lithography; newspapers; periodicals; press; printing; steam press
Purnochandradoy Press, 187. *See also* India; press; publishing

Quakers, 72. *See also* antislavery movement

race: Brazilian, 111; European, 35; history of, 55, 277n70, 277n74; phrenological theories of, 17–21, 49–77, 259; politics of, 57; science of, 13–15, 110–11; slavery and, 68; theories of, 23, 51–77, 119; on the world stage, 13–18. *See also* ethnology; human nature; language; phrenology; science; skin color
Rait, James, 141. *See also* schools
Ramses II, 36, 39; Ramesseum of, 39. *See also* Egypt
reform, 115–49; abolitionism as, 308n210; in Britain, 169; educational, 178, 182; global, 118, 145; of labor, 123; phrenological principles in, 132–36; as a political ideology, 308n211; religious, 178, 182. *See also* education; politics; prisons; slavery; society
Reform Act (1832), 5, 116. *See also* Britain; reform
religion: controversy in, 86, 144, 184; feelings of, 45; materialist critique of, 185; natural theology in, 45; perfectibility of man in place of, 179. *See also* culture; missionaries; schools
reviews, 99–106, 297n141; American, 105; British, 105; reception of, 105–9. *See also* periodicals; publishing
revolution, 116, 147; political language of, 147. *See also* politics
Revue Européenne, 223. *See also* periodicals

Richardson, Robert: *Travels along the Mediterranean* (1822) of, 35
Rio de Janeiro, 110–12; Faculty of Medicine in, 110; Museu Nacional in, 111–12. *See also* Brazil
Rivers, William Halse Rivers, 198–99, 238–39. *See also* anthropology
Roberts, Henry: *Indian Exchange Tables* (1855) of, 234
Rodrigues, Raimundo Nina, 111. *See also* Brazil; physicians
Rosetta Stone, 34. *See also* Britain; Egypt
Ross, David, 138. *See also* science
Ross, John, 42. *See also* Arctic
Rouch, William, 210; Portable Dark Operating Chamber of, 210, 211*f*. *See also* photography
Routledge, George, 79. *See also* publishing
Roy, Rammohun, 53–55, 55*f*, 59, 62–63, 77, 287n19. *See also* India
Royal College of Surgeons, 83. *See also* Britain
Royal Geographical Society, 106; *Hints to Travellers* (1878) of the, 209, 214. *See also* Britain
Royal Society, 4, 109, 250. *See also* London; science
Royal Society of Northern Antiquaries, 114. *See also* Copenhagen
Rudiments of Knowledge (1838), 142. *See also* books; phrenology; schools
Ruschenberger, William, 94–95. *See also* science; surgery
Russia, 114. *See also* Asia; Europe

Sächsische Vaterlands-Blätter, 136. *See also* periodicals
Sahlins, Marshall, 20
Salt, Henry, 39. *See also* Egypt
Salt Lake City, 161. *See also* United States
Sam Sly's African Journal, 177. *See also* periodicals
San Francisco, 79. *See also* United States
Santos, José Manuel de Castro, 110. *See also* Brazil; phrenology
Say, Thomas: *American Entomology* (1817) of, 113. *See also* natural history
Schaffer, Simon, 250–52; *Leviathan and the Air Pump* (1985) of, 251
Scherer, Joanna, 239
Schoelcher, Victor, 58, 61. *See also* naturalists; science
schools: in Bengal, 116, 136–45; phrenological, 137, 142; phrenological textbooks in the, 143–44. *See also* education
science: African American, 277n76; American, 96–97, 110; circulation of, 9; commonwealth of, 88; disciplines of, 112–13; exchange of, 249, 251; gentlemen of, 109; geography of, 249, 256; global, 3–9, 13–14, 20, 76, 87–88, 97, 110, 152, 191, 264; indigenous notions of, 277n75; institutions of, 114; liberal, 178; mass audience for, 5, 113, 151, 168, 187; material cultures of, 8–10; mental, 1, 91, 110, 116, 170; nineteenth-century, 18, 151, 154, 191, 241, 256, 260; and nonscience, 112; penal, 131; and politics, 18, 116; popular, 151, 154, 160, 164, 171, 191–92, 249; practices of, 6; problem of universality in, 328n38; racial, 13–15, 110–12, 130, 170, 260–61, 277n76, 284n143; reform, 117; and society, 6; travels of, 257, 275n29; universal, 257. *See also* atlas; history of science; industry; natural history; pseudo-science; scientific racism
Scientific American, 196. *See also* periodicals; science
scientific racism, 13–15, 110–12. *See also* race; science
Scotland, 9, 20, 33, 64, 178; a phrenological appendix from, 90–96. *See also* Britain
Secord, James, 79, 288n26
Senegal, 58. *See also* Africa
Shanghai, 2, 248. *See also* China
Shapin, Steven, 2–3, 5–6, 251–52; *Leviathan and the Air Pump* (1985) of, 251
Silliman, Benjamin, 96, 100, 103, 105. *See also* periodicals
Simla, 212–13, 216, 228. *See also* India
Simpkin, Marshall, and Company, 98–99, 104, 107–8. *See also* publishing
Sinclair, Thomas, 97. *See also* publishing
Sino-Japanese War of 1894–1895, 245. *See also* China; Japan
Sizer, Nelson, 159. *See also* phrenology
skin color, 88, 289n43. *See also* race
skull, 19–50, 66, 100, 173, 233, 260, 278n7; African, 63, 66; Ainu, 233–34; anatomy of the, 6, 35, 37, 45, 59, 88, 210; Ancient Peruvian, 102–4; authenticity of the, 23, 36; as a commodity, 279n19; comparison of the, 13–15, 36–37; as material culture,

14–17; national, 91; Native North American, 83, 86, 111; Native South American, 83, 86, 111; plaster copies of, 51–77; of a slave, 119; Swiss, 91–92, 92*f*, 97, 102, 107; Toda, 233–34; treatment in museums of the, 50; writing of phrenologists on, 22–24, 22*f*, 50. *See also* brain; burial; busts; casts; craniology; death; material culture; phrenology
slavery, 15, 58, 68–71, 122–24, 145–46, 162; abolition of, 58, 62, 65, 68–70, 76, 115–31, 172; American, 127; Atlantic, 76; defense of, 120, 122; North African trade in, 35; and phrenology, 118–24; and race, 68; and slaveholders, 117–18, 123; in the West Indies, 56–58, 61–62, 68. *See also* abolitionists; antislavery movement; colonialism; imperialism; race; violence
Smith, Andrew, 170. *See also* periodicals
Smith, Horace, 64
Smith, James McCune, 15, 129–30. *See also* physicians
Smith, John Augustine: *Select Discourses on the Functions of the Nervous System in Opposition to Phrenology* (1840) of, 103. *See also* physicians
Société d'Anthropologie de Paris, 225–26. *See also* anthropology; Paris
Société de la Morale Chrétienne, 61. *See also* France; slavery
Société Ethnologique, 113. *See also* ethnology; Paris
Société Française pour l'Abolition de l'Esclavage, 58, 61, 68. *See also* France; antislavery movement
society: American, 127; British, 149; colonial, 151, 169, 181; northern American, 120; and science, 6; southern American, 120. *See also* communications; culture; politics; reform
Society for the Acquisition of General Knowledge, 179. *See also* knowledge; Young Bengal
Society for the Civilisation of Africa, 127. *See also* Africa; antislavery movement
Society for the Diffusion of Useful Knowledge, 3–4. *See also* knowledge; phrenology
Sohier, Philemon, 51–54. *See also* busts
Somerset, Lord Charles, 169. *See also* Cape Colony
Sonthonax, Léger-Félicité, 57. *See also* Haiti
South Africa, 10, 12, 16, 197. *See also* Africa
South African Commercial Advertiser, 10, 170–72, 174. *See also* periodicals
South African Institution, 4. *See also* Cape Town
South African Museum, 170. *See also* museum
South African Public Library, 78–79, 83, 170. *See also* books
South African Quarterly Journal, 168–70. *See also* periodicals
South America, 24, 102, 109–10; independence movements of, 103; science in, 112. *See also* Americas
South Asia, 9–10, 13, 28–29, 165. *See also* Asia
South Carolina, 71. *See also* United States
Southeast Asia, 29. *See also* Asia
Southern Literary Messenger, 68. *See also* periodicals
Spencer, Baldwin: *The Native Tribes of Central Australia* (1899) of, 221–22
Spencer, Herbert: *Study of Sociology* (1901) of, 244, 246
Spurzheim, Johann, 4–6, 25, 83, 119, 123–24, 154, 303n54; *Outlines of Phrenology* (1834) of, 178; *Phrenology, or, The Doctrine of the Mind* (1832) of, 79, 153; *The Physiognomical System of Drs. Gall and Spurzheim* (1815) of, 6, 7*f*, 303n55
Sri Lanka. *See* Ceylon
St. Bartholomew's Hospital, 54. *See also* hospital
steam press, 80. *See also* printing; publishing
Stepan, Nancy, 300n214
stereotype, 81–82. *See also* casts; material culture
Stewart, Dugald, 12, 79. *See also* philosophy
St. Lawrence Island, 21. *See also* Arctic
Stone, Thomas, 12. *See also* phrenology
Stowe, Harriet Beecher, 64
St. Petersburg, 9. *See also* Russia
Straton, James, 63–64; *Contributions to the Mathematics of Phrenology* (1845) of, 64. *See also* phrenology
Struve, Gustav, 147–49; letter to George Combe of, 148*f*. *See also* revolution
Sudan, 197. *See also* British Empire; North Africa

surgery, 4, 12, 19, 21, 32, 35, 40, 43, 79, 94, 113, 137–38, 169, 178, 284n149. *See also* physicians
Switzerland, 147. *See also* Europe
Sydney, 1, 4, 10, 51, 133. *See also* Australia
Sydney Mechanics' School of Arts, 79, 81. *See also* Sydney

Tagore, Debendranath, 179, 184–85. *See also* India; religion; science
Tagore, Jyotirindranath, 262; *Twenty-Five Collotypes* (1914) of, 262*f*. *See also* phrenology
Talbot, William, 202, 209. *See also* photography
Tattvabodhini Patrika, 184. *See also* India; periodicals
Tattvabodhini Sabha, 187. *See also* India
Thomason College of Civil Engineering, 203
Thompson, George, 90, 130. *See also* abolitionists; craniology
Tidsskrift for Phrenologien, 153. *See also* Copenhagen; periodicals
The Times, 4, 147. *See also* newspapers
Times of India, 234. *See also* newspapers
Tiwanaku, 102. *See also* South America
Todas, 198–222; craniology of the, 204; female, 205*f*; lands of the, 201; phrenological account of the, 207; primitiveness of the, 206. *See also* India
Topham, Jonathan, 151
Topinard, Paul, 198, 225. *See also* anthropology
Transactions of the Phrenological Society, 140, 178. *See also* periodicials; phrenology
transformism, 226. *See also* science
transportation, 1, 4–5, 117. *See also* communications; letters; society; travel
travel: in Egypt, 37–38; of science, 275n29, 327n32. *See also* science; transportation
Tylor, Edward Burnett, 198, 219–21, 226, 234, 237, 239; *Primitive Culture* (1871) of, 200–1, 218–19. *See also* anthropology

United States, 1–13, 16, 50–53, 61, 65–76, 97–101, 111, 115, 130, 133–34, 137, 141–43, 151, 154–56, 159–61, 165, 190–96; asylum in the, 70; Civil War of the, 76, 163, 191, 241; dependency on European publishing of the, 96–97; as a global imperial power, 196; government of the, 103; industrialization of the, 126; Jacksonian democracy in the, 116; postal service in the, 130, 143; public in the, 65; reception of the *Crania Americana* in the, 90, 96, 101; scientific establishment, 100; slave plantations in the, 146; southern, 69; Spanish-American War of the, 196. *See also* Americas; North America; United States Army
United States Army, 163–64. *See also* United States
University of Cambridge, 79, 86
University of Edinburgh, 6, 12, 90, 107
University of Glasgow, 129
University of Heidelberg, 79
Upper Savage Islands, 41. *See also* Arctic

Van Diemen's Land, 132–33. *See also* prisons
van Wyhe, John, 116–17
Victorian age: Britain in the, 250; death in the, 278n5; public of the, 206. *See also* Britain; history
Vienna, 1. *See also* Europe
Vimont, Joseph: *Traité de Phrénologie Humaine et Comparée* (1832) of, 8, 72, 83, 108, 152. *See also* atlas
violence: and collecting, 38–39; colonial, 20, 25, 35, 49, 56; phrenology and, 119; of rebellion, 24–25, 27, 31–32, 56–57; of slavery, 56, 70, 119; and warfare, 26, 57. *See also* colonialism; slavery
Virey, Julien-Joseph: *Histoire Naturelle du Genre Humain* (1824) of, 58. *See also* natural history; race
Virginia, 71. *See also* United States

Wade, John: *The Extraordinary Black Book* (1831) of, 123. *See also* reform
Wajid Ali Shah, 165, 167–68. *See also* India
Waldie, Adam, 105. *See also* periodicals; publishing
Waldie's Journal of Polite Literature, 97. *See also* periodicals
Walker, David: *Appeal to the Colored Citizens of the World* (1829) of, 71
Wallace, Alfred Russel, 194–95, 221–22, 231, 233, 240, 262–64; *Studies Scientific and Social* (1900) of, 221, 223*f*; *The Wonderful Century* (1898) of, 264. *See also* photography

War of 1812, 84. *See also* United States
Warrington, 54, 64. *See also* Britain
Watson, Hewett, 10–11, 52, 88–90, 93, 96, 98, 115; *Statistics of Phrenology* (1836) of, 11, 52. *See also* phrenology
Wedgwood, Josiah, 124. *See also* antislavery movement
Weld, Theodore Dwight, 74. *See also* antislavery movement
Wells, Samuel, 155. *See also* phrenology; publishing
West Africa, 72, 126. *See also* Africa
Western Farmer, 157. *See also* newspapers
Western Reserve College (Ohio), 129
West Indies, 4, 56–62, 68; abolition of slavery in the, 122, 127; economy of the, 70. *See also* Americas; colonialism
Whewell, William, 86
Wilde, William, 35. *See also* surgery
William and Robert Chambers, 80–81, 141. *See also* Edinburgh; publishing
Williams, F. Fisk: *Guide to the Indian Photographer* (1860) of, 208–9, 213–14
Wilson, Horace, 138. *See also* science
Wintle, Reverend Henry, 116. *See also* religion
Wood, William, 194. *See also* photography; Philippines
woodcuts, 101–2, 101*f*, 105, 107–8, 113, 226, 228. *See also* illustration; images; lithography; photography
Wundt, Wilhelm, 244, 246. *See also* psychology

Xhosa War, 172, 174, 175*f*. *See also* Cape Colony

Yale University, 100, 106
Young Bengal, 179, 187. *See also* India

Zeitschrift für Phrenologie, 147, 153. *See also* Heidelberg; periodicals
zoology, 88. *See also* science
Zulus, 197–98. *See also* South Africa